Introduction to Desalination

Introduction to Desalination

Principles, Processes, and Calculations

Dr. Louis Theodore
Theodore Tutorials
East Williston, NY
USA

Dr. R. Ryan Dupont
Utah Water Research Laboratory
Utah State University
Logan, UT
USA

WILEY

This edition first published 2022

Registered Offices
John Wiley & Sons, Inc., 111 River Street, Hoboken, NJ 07030, USA
John Wiley & Sons Ltd, The Atrium, Southern Gate, Chichester, West Sussex, PO19 8SQ, UK

Editorial Office
The Atrium, Southern Gate, Chichester, West Sussex, PO19 8SQ, UK

For details of our global editorial offices, customer services, and more information about Wiley products visit us at www.wiley.com.

Wiley also publishes its books in a variety of electronic formats and by print-on-demand. Some content that appears in standard print versions of this book may not be available in other formats.

For general information on our other products and services or for technical support, please contact our Customer Care Department within the United States at (800) 762-2974, outside the United States at (317) 572-3993 or fax (317) 572-4002.

Wiley also publishes its books in a variety of electronic formats. Some content that appears in print may not be available in electronic formats. For more information about Wiley products, visit our web site at www.wiley.com.

Library of Congress Cataloging-in-Publication Data
Names: Theodore, Louis, author. | Dupont, R. Ryan, author.
Title: Introduction to desalination : principles, processes, and calculations / Dr. Louis Theodore, Manhattan College, R. Ryan Dupont.
Description: First edition. | Hoboken, NJ, USA : John Wiley & Sons, Inc., 2022. | Includes bibliographical references and index.
Identifiers: LCCN 2021060406 (print) | LCCN 2021060407 (ebook) | ISBN 9781119691679 (hardback) | ISBN 9781119691716 (pdf) | ISBN 9781119691747 (epub) | ISBN 9781119691754 (ebook)
Subjects: LCSH: Saline water conversion.
Classification: LCC TD479 .T54 2022 (print) | LCC TD479 (ebook) | DDC 628.1/67--dc23/eng/20220124
LC record available at https://lccn.loc.gov/2021060406
LC ebook record available at https://lccn.loc.gov/2021060407

Cover image: © MARK GARLICK/Getty Images
Cover design by Wiley

Set in 9.5/12.5pt STIX TwoText by Integra Software Services Pvt. Ltd, Pondicherry, India

10 9 8 7 6 5 4 3 2 1

MIX
Paper from responsible sources
FSC® C013604

Brief Contents

Preface

Since the late 1960s there has been an increased awareness of a wide range of environmental issues covering all media: air, land, and water. More and more people are becoming aware of these environmental concerns and it is important that all individuals, many of whom do not possess an understanding of environmental problems or have the proper information available when involved with environmental issues, develop capabilities in this area. All professionals should have a basic understanding of the technical and scientific terms related to these issues. One of the key issues is water scarcity and the role desalination can play in improving global water security. Hopefully, this text will serve the needs of those interested in increasing their knowledge of, and help solve, the potable water problems facing society.

The current human population on Earth is approximately eight billion, and almost certainly will increase in the future. The influence and effects of human activities on the environment have become increasingly evident at the local, state, national, and international levels, while particular issues such as potable water availability and desalination have become more pervasive and pressing. The net result is that there has been a more significant increase in both awareness and interest about this topic since the turn of this century.

The authors considered and then decided to write an introductory text on desalination that highlighted pragmatic rather than theoretical materials. It contains three parts and provides material that can be used in teaching an introductory course to those individuals interested in becoming part of this profession. As such, this book is intended to serve as a training tool for those individuals involved with, and those interested in entering this dynamic field. Although the literature is inundated with desalination texts emphasizing theory and theoretical derivations, the goal of this book is to present this subject of desalination from a strictly introductory and pragmatic perspective.

The book is the result of two years of effort by the authors. The first rough draft was prepared during the years 2017 through 2019. The manuscript underwent significant revisions during these past two years, some of it based on the experiences gained from earlier written material.

It should be noted that the authors cannot claim sole authorship to all of the essay material in this text. The present book has evolved from a host of sources, including notes, homework problems, and exam problems prepared by several faculty for undergraduate courses offered at Utah State University and Manhattan College, several other books by L. Theodore and R. Dupont, and a host of Theodore Tutorials published by Theodore Tutorials of East Williston, New York. Although the bulk of the material is original or taken from sources that the authors have been directly involved with, every effort has been made to acknowledge material drawn from other sources.

As noted in the Table of Contents, this desalination book is divided into three parts:

- Part I: Introduction
- Part II: Desalination and Water Treatment Processes
- Part III: Select Related Topics

Part I includes the following chapters: Global Water Scarcity and the Need for "New Water;" a Technical Glossary; Engineering Principles; Physical, Chemical, and Biological Properties of Water; Water Chemistry; Conservation Laws; Unit Operations; and Heat Transfer and Thermodynamic Principles. Part II then introduces the desalination process. Part II highlights specific processes including: Evaporation, Reverse Osmosis, Crystallization, Other Desalination Processes, New Desalination Processes, and Other Potable Water Treatment Processes. These seven chapters are generously supplemented with illustrative examples. Part III contains five chapters that are concerned with what the authors have termed, "Select Related Topics," i.e. Water Conservation, Economic Considerations, Optimization, Ethics, and The Future of Desalination.

The authors hope that this book will place in the hands of academic, industrial, and government personnel – including policy makers and water resource/management planners – a book that covers the fundamental principles and applications of desalination in a thorough and clear manner. The authors further hope that, on completion of the text, readers will have acquired not only a working knowledge of the principles of desalination, but also experience in their application; and, that they will find themselves approaching advanced texts, engineering and science literature, and industrial applications (even unique ones) with more confidence.

Last, but not least, the authors believe that this modest work will help the majority of individuals working and/or studying in the field of desalination to obtain a more complete understanding of this subject matter. If you have come this far and read through this Preface, you have more than just a passing interest in this subject. We strongly recommend that you take advantage of the material available in this book. We think that it will be a worthwhile experience.

East Williston, NY, January 2022 *Louis Theodore*
Smithfield, UT, January 2022 *R. Ryan Dupont*

Whisky is for drinkin', water is for fightin'
Anonymous

To all the members of East Williston Fire Department, but in particular: Billy Behan, Angelo Carbone Jr., Domenic Conte, Steve Jones, Billy McLean Jr., and a host of others.
and
In loving memory of Marie K. and Robert H. Dupont

Part I

Introduction

Chapters in this Part include the following:

1

Global Water Scarcity and the Need for "New Water"

1.1 Introduction

The title of this book includes the word *desalination*. Dictionary.com (2012) defines desalination as "the process of removing salt, esp from seawater so that it can be used for drinking or irrigation." Thus, desalination is a process that produces potable water. The reader should keep in mind, and this will be discussed later in the book, that there are other processes available for producing potable water, many of which are *currently* uneconomical. For this reason, the book primarily deals with desalination. The general topic of potable water is also considered and is addressed in several of the paragraphs to follow.

The world's total water is enormous compared with the presently conceivable needs of man, yet there is a growing water problem. Of the 1.39 billion km^3 (332 million mi^3) of water in the Earth's crust 99% is salty and useful neither for irrigation by present techniques nor for the majority of man's other needs.

Rain provides many times the world's annual water needs, but fresh water supplies vary widely over the Earth and from time to time in a given region. This accounts for the "water problem" that exists as the Earth's population continues to grow. In the United States, the actual "consumptive use" is approximately 12%–18% of the potential supply. Water prices are low, and on a statistical basis there should be no water shortages for many years. These average figures are of little comfort to the people of arid regions, however, and the United States, though relatively well supplied with water, is now beginning to turn to sources other than rainfall and runoff to meet growing water demands. The Southwest has too little water to support rapid population expansion. California has embarked on multi-billion dollar projects to bring water nearly 800 miles to Los Angeles and southern California from the wetter northern parts of the state. Outside of the United States, expensive desalination plants have been in operation for many years in the Virgin Islands, many Persian Gulf states, the West Indies, Libya, South Africa, and Israel.

Introduction to Desalination: Principles, Processes, and Calculations, First Edition. Louis Theodore and R. Ryan Dupont.

An obvious way to increase water availability is to recover fresh water from seawater or large underground stores of brackish water that are available in many arid regions, but that are generally not usable without significant treatment. Other options for increasing potable water supplies is to find ways to use the saline or brackish water for agriculture or other uses that can tolerate higher salinity levels, or to actively pursue water conservation measures to protect limited fresh water supplies that may be available.

This chapter introduces the reader to the general subject of global water supplies and water scarcity issues through a review of water availability and water sources, global water resources, global water resource issues, and the history of desalination. A series of illustrative examples are provided to highlight concepts presented throughout the chapter.

1.2 Water Availability and Water Sources

The water on the surface of the Earth can be found in eight distinct compartments which are shown in Table 1.1 along with how much of the Earth's total water budget each makes up. These eight compartments are more commonly organized into three distinct groupings. First, there is *surface water*, which includes oceans, lakes, rivers, seas, etc. In short, surface water is all forms of liquid water that lie on the surface of the Earth. The second grouping is *glacial and frozen water*. These are resources that are located on the surface; however, these resources are in frozen, long-term storage. Long-term storage is defined as water that has historically taken thousands of years to move through the water cycle. The third and final grouping is *groundwater*. This is the water that is found below the surface of the Earth in conditions where the percent saturation in soil

Table 1.1 Breakdown of the total water budget on the Earth.

Water compartment	% of Total Earth's water budget
Oceans	97.2
Frozen glacier water & other ice	2.15
Ground or subsurface water	0.61
Freshwater lakes	0.009
Inland seas	0.008
Soil moisture	0.005
Atmospheric water	0.001
Rivers and streams	0.001

pore water is 100% (Mullen 2021). The process known as the hydrologic, or *water cycle*, has played a crucial role in the formation and distribution of all water resources on Earth today. The water cycle is used to describe the constant circulation of water through its three phases of solid, liquid, and gas as it moves within the eight compartments making up the water resource pool on the Earth.

1.3 Global Water Resources

Usable water is the key to sustainable life. However, out of all the water on Earth today, it is estimated that only 2.5 to 3.0% is usable. That still is quite a large amount of water, considering the Earth's total global water supply is over 1.39 billion km^3. However, when one delves deeper into this estimated usable supply, it is found that a little over 0.3% of the usable water on Earth is easily accessible (Gleick 1996). This usable water is found in rivers, streams, lakes, and swamps. With the ease of access to these forms of usable water being so high, it is estimated that over 50% of the world's population lives within 3.0 km of these surface water sources (Kummu et al. 2011). The majority of the available water that is not considered easily accessible is found in ground or subsurface water in natural aquifers.

Natural aquifers are the most common forms of groundwater storage. Today, these natural aquifers are thought to hold roughly 34.2 million km^3 of total water. Out of these 34.2 million km^3 of water, it is estimated that close to 10.4 million km^3 of this water is usable (Gleick 1996). Since these aquifers are below the surface of the Earth, they do require the construction of wells for water recovery. These wells can range from tens to hundreds of feet deep, depending on where the water table is located. It is also important to note that these wells will generally require the installation of a pumping system to bring the water to surface. Table 1.2 provides a comparison between the total water on Earth to the amount that is both groundwater and fresh-groundwater.

Table 1.2 Fresh groundwater, total groundwater, and total global water resources.

Water source	Water volume, mi^3	Water volume, km^3	% of Total water resource	% of Total fresh water resource
Fresh groundwater	2,526,000	10,530,000	0.80	30.1
Total groundwater	5,614,000	23,400,000	1.7	
Total global water resource	332,500,000	1,386,000,000		

As small as the ratio of total volume of usable water to unusable water is in the world, there are still considerable amounts of usable water resources present among the seven continents. The amount of usable water on each continent can be divided among the three water categories as just described earlier, i.e. groundwater; glacier/permanent ice caps; and surface water (i.e. wetlands, lakes, reservoirs, and rivers). These categories, along with the volumes of water in each for each continent across the globe are summarized in Table 1.3.

On examining Table 1.3 it becomes immediately obvious that Antarctica has by far the largest amount of fresh water of any continent. However, this fresh water is in the form of glaciers and permeant ice caps, thus making it unusable. When comparing usable water resources, the top three continents are Asia, with roughly 7,830,622 km^3 of water, Africa, with approximately 5,531,776 $km^3$of water, and North America, with 4,327,003 km^3 of total usable water resources. Furthermore, these three continents contain over 75% of the usable water on Earth. With this large amount of water being contained in three of the seven continents, mankind has adapted through the development of artificial reservoirs. These artificial reservoirs are most commonly constructed by the creation of a dam around an existing natural water source. Today, it is estimated that there are more than 515,000 of these man-made reservoirs of greater than 2.5 ac in size that cover roughly 258,000 km^2 of the Earth's surface (Osborn 2019).

Table 1.4 summarizes the annual renewable fresh water resources for the top 25 countries in the world. Renewable fresh water resources are an estimate of the average precipitation, recharged groundwater, and fresh surface inflows

Table 1.3 Global water resources by continent (Modified from Rekacewicz 2006).

Continent	Glacier & permanent ice caps, km^3	Groundwater, km^3	Wetlands, large lakes, reservoirs & rivers, km^3	Totals, km^3	Total usable water resource, km^3
North America	90,000	4,300,000	27,003	4,427,003	4,327,003
South America	900	3,000,000	N/A	3,00,900	3,000,000
Europe	18,216	1,60,000	2,239	1.620,745	1,602,529
Africa	0	5,500,000	31,776	5,531,776	5,531,776
Australia	180	1,200,000	221	1,200,401	1,200,221
Asia	60,984	7,800,000	30,622	7,891,606	7,830,622
Antarctica	30,109,800	N/A	N/A	30,109,800	0

Table 1.4 Renewable annual fresh water resources for top 25 countries (Table 1.4 Advameg, Inc. 2010).

Rank	Country	Total fresh water renewable resource, km^3
1	Brazil	8,233
2	Russia	4,498
3	Canada	3,300
4	United States	3,069
5	Indonesia	2,838
6	China	2,830
7	Colombia	2,132
8	Peru	1,913
9	India	1,908
10	Democratic Republic of the Congo	1,283
11	Venezuela	1,233
12	Bangladesh	1,211
13	Burma	1,046
14	Chile	922
15	Vietnam	891
16	Republic of the Congo	832
17	Argentina	814
18	Papua New Guinea	801
19	Bolivia	622
20	Malaysia	580
21	Philippines	479
22	Cambodia	476
23	Mexico	457
24	Ecuador	432
25	Japan	430

from surrounding countries that each of these countries receives in any given year. What is interesting to note from these data is that despite a country having a great deal of renewable fresh water resources, water shortages and unequal distribution of resources can still take place. Consider Brazil, which tops the list with over 8,233 km^3 of renewable fresh water each year. Despite what would appear to be an abundance of fresh water, Brazil is experiencing a long-term water crisis caused by poor water infrastructure and water resource

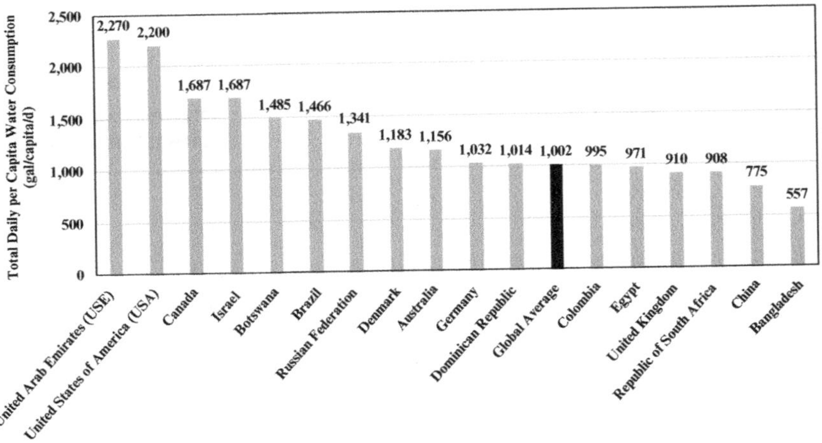

Figure 1.1 Total daily per capita water consumption by selected country. Data from Grace Communication Foundation (2019).

management which has left over 5 million, or more than 2% of its citizens without access to safe potable water (Water.org 2019).

Even in what is considered a highly advanced country like the United States, which receives 3,069 km^3 a year of renewable fresh water and has historically invested in water infrastructure as well as the management of its water resources, it is estimated that as of 2014, 1.6 million people (0.5% of its population) were without a toilet, a tub or shower, or running water (UNC Environmental Finance Center 2017). Despite availability of renewable fresh water resources, equitable and affordable supply and access to these resources remain a significant challenge, particularly to minority and low-income communities, even in the most developed countries in the world.

Figure 1.1 summarizes the total per capita water consumption (consumption of surface or groundwater for domestic, commercial and industrial purposes; water for crop production; and water required for the assimilation of pollutants produced from domestic, commercial, and industrial activities) for the select countries in the world whose daily consumption is above and below the global average (Mekonnen and Hoekstra 2011). The global average for total per capita daily water usage is roughly 1,000 gal/capita/d (Grace Communication Foundation 2019). This is water that is used for domestic purposes such as laundry, cooking, drinking, washing, etc., as well as water used for food production, and the production of consumer goods, like cars, plastic bags, clothing, etc. At the top of this list with 2,270 gal/capita/day is the United Arab Emirates, 76% of which is associated with products or ingredients of consumer goods imported annually into the UAE (virtual water). In the United States, of the 2,200 gal/capita/day of total water use, only 20% is associated with imported external water consumption. The significant differences seen in the daily internal versus external per capita total water consumption are directly attributed to the consumption habits of each country's citizenry.

1.4 Global Water Resource Issues

The root of all major issues pertaining to global water resources today stems from a lack of water accessibility. It is estimated that over 1 billion people live without access to safe water (WHO 2017). This number more than doubles to over 2.3 billion people when basic sanitation services are considered. It is this lack of access to safe drinking water and adequately treated wastewater that drives other major water-related issues in the world today. As indicated earlier, this access issue stems directly from the lack of necessary water infrastructure as well as the poor maintenance and management of the limited infrastructure that currently exists.

Improved water resource management and water storage capacity makes an economy more resilient to external shocks, such as rainfall variability, and thus provides a stable and sustainable base for increased food and industrial productivity and production to maintain economic growth and development. Consider a heavily water-dependent country like Kenya where frequent floods and droughts take a heavy toll on the economy (SIWI 2005). These floods and droughts translate to a direct long-term fiscal liability and decline in economic growth of about 2.4% of Kenya's annual GDP (SIWI 2005). Much of this loss in economic output comes from lack of adequate water storage capacity, a deficit of more than $10 billion in 2004 (SIWI 2005). If Kenya, as for many other African countries, had invested in storage facilities for water, they would be able to store the excess water received in heavy flood years for use in heavy drought years, thus, avoiding flood liabilities and water scarcity cost, allowing the country to avoid the drops in economic growth that have occurred in the past.

The SIWI (2005) indicated that national economies are more resilient to rainfall variability, and economic growth is increased when water storage capacity is improved. Investing in adequate water resource management in developing countries would be expected to produce significant economic gains of between two to four times the infrastructure investment. The SIWI (2005) concluded that an even greater return on investment of $8.00–$200 would be saved in water treatment costs for every $1.00 invested in watershed protection. Investments in improved water and sanitation infrastructure in developing countries also has significant economic benefits when considering increased productivity from the improved health of workers in households and the commercial and industrial sectors, ranging from hundreds of millions to billions of dollars annually (SIWI 2005).

In today's world, the term *hydro-politics* is becoming more and more known. Hydro-politics can be defined as debate or conflict amongst individuals, parties, or organizations within governments about water-related issues. These debates/conflicts may even take place between multiple countries, not just the individuals within those countries. Commonly, these types of conflicts are solved by establishing a system known as *Riparian Water Rights*. This is a system used to allocate water among different countries or jurisdictions within countries that are in control of the land through which the water travels. These types of rights are considered common law amongst many countries in the world such as the United States, Canada, Australia, and Great Britain.

Even though many of the largest countries in the world have used this system to manage water conflicts for years, there are many other countries where these systems are relatively new. A prime example of this is conflict between Egypt and Ethiopia. For centuries, these two countries have been in constant conflict over who owns the developmental rights for the water in the River Nile. It wasn't until 2015 that these two counties set aside their differences and signed an agreement ensuring shared and equitable access to the river (Lufkin 2017).

Even with compromises such as the one between Egypt and Ethiopia, some individuals, such as Ismail Serageldin (former Vice President of the World Bank), see water becoming the next so called "Oil War." Ismail Serageldin famously said, "The wars of this century were fought over oil, the wars of the next century will be fought over water" (Barnaby 2009). However, when one sees countries coming together to solve their water resources conflicts, forming compromises like Egypt and Ethiopia, the potential for water wars seem less likely, and the arguments by Barnaby (2009), Fröhlich (2012), and McCrary (2018) appear to have more validity. These latter authors believe that physical conflict over water will not happen because established systems for international trade and international agreements leading to mutual benefits ease these conflicts, thus preventing the scale of these problems from becoming large enough to cause military style confrontations. Certainly, the expansion of available water supplies through the expansion of desalination technologies will contribute to the reduction in multinational conflicts over water and ensure more adequate water supplies for all peoples.

1.5 History of Desalination

Three prime candidate desalination processes have emerged and are briefly reviewed in the sections to follow.

1.5.1 Evaporation Processes

The oldest and best developed process for saline water conversion is the evaporation (or what some define as distillation) method. While there are many desalination technologies in use or being developed today, desalination began using evaporative processes. These evaporative desalination techniques were recognized over 2,000 years ago when Aristotle wrote in 320 B.C., "saltwater, when it turns into vapor, becomes sweet and the vapor does not form saltwater again when it condenses." Evaporation remains the major method today for commercial production of fresh water from seawater. In principle, evaporation is the simplest method. Seawater is boiled in an evaporator (Flynn et al. 2019; Theodore 2014; Theodore et al. 2017) by passing hot steam through a steam chest where the steam condenses on the inside of the tubes of the chest and is usually returned to the boiler. The vapors rising from the seawater feed are cooled in a condenser and thus converted into pure liquid water which is collected in a storage vessel. The resulting concentrated brine solution is continuously or intermittently withdrawn from the evaporator. The most advanced commercial evaporation processes are the following:

1. Multiple-Effect Evaporation.
2. Flash Evaporation.
3. Vapor Compression Evaporation.

Details of these processes are provided in Chapter 10.

1.5.2 Membrane Processes

Electrodialysis was the membrane separation process employed for desalination a century ago. Here, the ions forming the salt are removed from the saltwater by electric forces and concentrated in separate compartments. However, the higher the salinity of the raw water, the more electric power is needed for this process. Hence, this process was applied primarily in the treatment of moderately brackish waters containing 1,000 to 2,000 mg/L of dissolved salts. In recent years, reverse osmosis (RO) (Theodore and Ricci 2010) has displaced electrodialysis as the primary membrane separation desalination process, leaving the latter as the choice for medical kidney applications.

An RO system consists of an intake, a pre-treatment component, a high-pressure pump, a membrane apparatus, remineralization, and pH adjustment components, as well as a disinfection step. Generally, a pressure of about 1.7 to 6.9 MPa is required to overcome the osmotic pressure of saltwater.

This advanced separation technique may be used whenever low molecular weight solutes such as inorganic salts or small organic molecules (e.g. glucose) are to be separated from a solvent (usually water). In normal (as opposed to reverse) osmosis, water flows from a less concentrated salt solution to a more concentrated salt solution as a result of an innate concentration driving force (thermodynamically referred to as the chemical potential). As a result of the migration of water through the membrane, an "osmotic pressure" is created on the side of the membrane to which the water flows. In RO, the membrane is permeable to the solvent or water and relatively impermeable to the solute or salt. In order to make water pass through an RO membrane in the desired direction (i.e. away from a concentrated salt solution), a pressure must be applied that is *higher* than the osmotic pressure. The membranes used for RO processes are characterized by a high degree of semi-permeability, high water fluxes, good mechanical strength, chemical stability, and relatively low operating and capital costs. Early RO membranes were composed of cellulose acetate, but restrictions on process stream pressure, temperature, and organic solute rejection spurred the development of noncellulosic and composite material membranes ("sandwiches"). These membranes may be configured into a variety of geometries for system operation, including: plate and frame, tubular, spiral wound (composite), and hollow fiber.

1.5.3 Crystallization Processes

Crystallization processes were also employed over 2,000 years ago. Today, these processes are important mass transfer operations that are often employed in the preparation of a pure product. In the process, a crystal usually separates out as a substance of specific composition from a solution of varying composition. Any impurities in the liquid (often referred to as the mother liquor) are carried in the crystalline product only to the extent that they adhere to the surface or are occluded (retained) within the crystals that may have grown together during or after the crystallization operation. The separation of a solid from a solution onto a crystal occurs only if there is a state of imbalance involving a mass driving force; namely, a decrease in chemical potential (or concentration) between the bulk of the liquid solution and the crystal interface. This effectively means that the solution must be supersaturated. There are several different ways that crystallization can occur. The four most often encountered in practice are (Theodore 2014):

1. Cooling.
2. Evaporation.
3. Cooling and evaporation (also referred to as adiabatic evaporation).
4. A salting out process.

Process 1 is the most commonly employed, provided the solubility of the component being crystallized decreases with decreasing temperature.

World-wide development of desalination techniques in the last half century has been driven out of necessity due both to water scarcity and population growth. The private sector has primarily led the investment in research and development since water has begun to be seen not as a commodity, but as a product to be sold at a profit. This development by the private sector has led to a significant drop in the cost of water generated using these desalination techniques.

1.6 Summary Observations

The following summary observations can be made related to water, seawater desalination, and water treatment in general.

1. Water demand at both the national and international levels is increasing at a faster rate than originally realized.
2. Water demand at both the national and international levels is increasing at a faster rate than population growth.
3. There is indeed growing water scarcity.
4. One approach to solving this water scarcity problem is to develop and/or find new sources of water.
5. Nonconventional water sources will become more attractive in the future.
6. Water conservation measures should be expanded in the future.
7. Seawater and brackish groundwater are nonconventional water sources serving as feedstock for desalination processes.
8. Poor seawater quality can limit RO applications but generally does not have as great an effect on evaporation and crystallization processes.
9. Desalination processes featured evaporation in earlier times, while RO is generally the process of choice today.
10. The increasing demand for potable water will ensure that improvements in current technologies and development of new approaches for potable water production will continue to develop in the future.

1.7 Illustrative Examples

Five illustrative examples complement the material presented in this chapter.

1.7.1 Illustrative Example 1

Describe the role process water plays in industrial processes/applications.

Solution. Process industries rank above all others as users of water. No process plant could operate without water as a cooling medium and as a direct raw material in certain phases of a process. The local water supply, therefore, must be studied before an area can even be considered as a possible site for a processing facility. A detailed estimate of present and future water requirements must precede a plant siting study. Then the availability of water in the region being considered should be carefully investigated. If well water is to be used, a complete study of the history of the groundwater table is necessary. If groundwater supplies are adequate, they are preferred because of their lower temperature and generally higher water quality as compared to surface water sources.

Surface waters from streams or lakes also require careful consideration since both their quantity and quality can vary greatly on a seasonal basis. Streams discharging into the ocean during times of low runoff can become saline as a result of seawater intrusion. Under such conditions, plant design may have to include large storage facilities for fresh water to be collected during periods of high runoff and used during dry seasons when the stream becomes saline.

Companies moving into relatively nonindustrialized areas often fail to consider the possibility of other plants following suit. The availability of the water supply should be adequate not only for the future needs of the proposed plant but also for the anticipated needs of other industries that might move into the area. It is also desirable to consider alternate sources of supply that may be required if the preferred water source becomes depleted. Water quality must be studied as well as water quantity. Chemical and bacteriological examination will indicate the extent of treatment required and can aid in the development of water cost figures for comparison with other locations. The possible contamination of the water source by other industries in the area should be anticipated. Note that "contamination" may consist of raising the temperature of the water to a level that renders its use as a cooling medium impossible.

1.7.2 Illustrative Example 2

A range of water needs typically exist within a facility, all of which may have different water quality requirements for specific uses. Describe five of these common needs.

Solution. Five common industrial facility water uses include:

1. Cooling Water. Natural and forced-draft cooling towers are generally used to provide the cooling water required at a site unless water can be drawn from a convenient river or lake in sufficient quantity to meet a facility's cooling demands. Seawater or brackish water can be used at coastal sites but, if used directly, necessitates more expensive materials of construction

for heat exchangers because of potential corrosion problems resulting from the high dissolved solids content of this water. Often, this cooling water does not have to be of high purity and can be taken from blowdown streams from boilers or process lines that require much more stringent water quality conditions.

2. Potable and General Use Water. The water required for general purposes on a site is usually taken from a local municipal or private supplier, unless a cheaper source of suitable quality water (e.g. a river, lake, or well) is available. If the cost of this supply is low, incentives for water use reduction will also be low. It should be remembered, however, that the cost of the supply may be a small portion of the overall cost of managing this water, particularly if it is used in process rinsing, general facility cleaning and wash-down, etc. More water use requires more energy to convey it from the source and within a plant. When the water becomes contaminated, costs for its handling increase exponentially. In addition, waste treatment becomes progressively less efficient and more costly as a waste stream is diluted, and load penalties may be incurred based on the volume of waste discharged to publicly owned treatment works (POTWs). It should be evident, then, that numerous benefits can be associated with increased water use efficiency within a plant.
3. Demineralized Water. Water from which all the minerals have been removed by ion exchange must be used where ultrapure water is needed to meet process demands and strict boiler feedwater requirements. Mixed and multiple-bed ion exchange units are used for this purpose, with resins exchanging multivalent cations for hydrogen. Boiler water as condensed steam and process water must be removed on a routine basis (blow down) to prevent the build-up of unwanted constituents within these systems, and this ultrapure water may be effectively reused within a plant for other unit operations that demand much less stringent water quality characteristics, i.e., cooling water, process rinse water, and so on.
4. Refrigeration. Refrigeration is needed for processes that require temperatures below those that can be economically obtained with cooling water. Chilled water can be used to lower process temperatures down to approximately 10°C. For lower temperatures, to -30°C, salt brines (NaCl and $CaCl_2$) are used to distribute the "refrigeration" around the site from a central refrigeration unit. Vapor compression machines are normally used for this purpose. As with boilers, evaluation of the operating conditions of a chiller can lead to significant pollution prevention opportunities. Consider, for example, that a 1% improvement in chiller efficiency can be expected for each 1°F increase in the chiller set point.
5. Steam. The steam for process heating is generated in either fire - or water-tube boilers, using the most economical fuel available. The process temperatures required can usually be obtained with low-pressure steam

(typically 25 psig), with higher steam pressures needed only for high process temperature requirements. A significant pollution prevention, energy conservation, and cost reduction opportunity may exist in a facility's steam generation system, as many facilities operate at higher than necessary steam pressures, increasing their energy demands for steam production and the cost of producing this steam. In addition, repair and replacement of leaking steam and condenser lines will prevent the wasting of steam and the associated energy and cost needed to produce this wasted steam.

1.7.3 Illustrative Example 3

Briefly describe the wastewater streams typically generated in the petroleum industry.

Solution. Throughout the life cycle of crude oil, through extraction, transportation, and refining, roughly 5 to 7 gal of water are used for every 1 gal of crude oil processed. Of this total, roughly 1 to 2.5 gal of water are used for every 1 gal of crude oil stock. In the United States, 1 to 2 billion gal of water are used per day in the refining industry. Water consumption varies for different regions and different qualities of crude oil. These water estimates do not include unconventional shale formations, where water usage is generally much higher. Water consumption in refining is a result of evaporation and drilling, blowdown, process discharge, and some incorporation into finished products. All water that is incorporated into the refining process is treated or filtered onsite before it is either reused on site or discharged into the environment.

Wastewater streams include separation process wastewater, tank bottoms, cooling tower condensate blowdown, source water treatment, and storm water. American Petroleum Institute (API) separators are commonplace in most refineries and separate sludge and hydrocarbon components out of process wastewater. Refineries today have installed dike systems to keep process wastewater from draining into storm sewers.

1.7.4 Illustrative Example 4

Refer to Table 1.5 and calculate the desalination mass capture efficiency of salt in seawater to convert typical seawater to potable water. Assume a basis of 100 g of seawater.

Solution. For 100 g of seawater,

Salt in seawater = 3.5 g

Salt in potable water = 0.05 g

The required salt capture efficiency, *SCE*, is therefore:

$$SCE = \frac{(3.5 - 0.05)}{3.5}(100) = 98.6\%$$

1.7.5 Illustrative Example 5

Mekonnen and Hoekstra (2011) provide a summary of national water footprints of production and consumption for countries around the world. They provide a guideline for the determination of the representative water footprint of various commodities that may be both produced within a country as well as imported for consumption from elsewhere around the globe. An example that is provided determines the average water footprint for tomatoes consumed in Germany. Using the data presented in Table 1.6, determine what the water footprint for tomato consumption in Germany is assuming the total quantity imported is consumed, using a weighted average of each water footprint for tomatoes imported from other countries.

Solution. Using a weighted average of the water footprint and Germany's import levels from the countries listed in Table 1.6 yields the following results using Equation 1.1 for the calculations.

$$\text{Water Footprint} = \frac{\int_1^n (\text{Tomato Consumption, T/yr})_i (\text{Water Footprint, m}^3/\text{T})_i}{\text{Total Tomato Consumption, T/yr}} \quad (1.1)$$

where Tomato Consumption = annual T of tomatoes produced or imported and consumed from each country, *i*; Water Footprint = m^3 water/T tomatoes produced from each country, *i*; and *n* = total number of countries, including itself, from which Germany imports tomatoes. Substituting values from Table 1.6 into Equation 1.1 yields:

$$\text{Water Footprint} = \frac{(47{,}000)}{(36) + (252{,}000)(10) + (244{,}000)(83) + (72{,}000)(109) + (52{,}000)(110)(\text{T/yr})(\text{m}^3/\text{T})(667{,}000\,\text{T/yr})}$$

$$\text{Water Footprint} = \frac{38{,}032{,}000}{667{,}000} = 57\text{m}^3/\text{T}$$

Table 1.5 Seawater and potable water major components.

	H_2O, wt%	Salt, wt%
Seawater	96.5	3.5
Potable water	≥99.95	≤0.05

Table 1.6 Water footprint and quantities of German tomato consumption as a function of place of origin.

Place of origin	Quantity imported or consumed, T/yr	Water footprint, m^3/T
Germany	47,000	36
The Netherlands	252,000	10
Spain	244,000	83
Italy	72,000	109
France	52,000	110
Total	667,000	

1.8 Problems

1. In terms of the question "how much water is there?" explain why only approximately 0.3% of the entire water on the Earth is considered usable fresh water.
2. With so much water on the planet explain why there is a water availability and water supply crisis.
3. List and briefly describe the four major aquatic habitats in the United States and indicate concerns regarding contamination of these resources.
4. SIWI (2005) and Hares (2017) have made strong cases for the economic benefit of water infrastructure investments around the world. In the United States, the US Water Alliance (2017) outlined the way forward for ensuring all people have access to clean, affordable water, and all benefits from water infrastructure investments to make water supplies more reliable and resilient in the face of climate change pressures. Review the US Water Alliance (2017) national briefing paper on water equity and summarize the "Three Pillars" of water equity put forth in this document.
5. A comparison of global drinking water supply and sanitation services provided by WHO (2017) suggests that in 2015, North America had some of the highest rates of improved water supplies (96%) and basic sanitation (97%) and safely managed improved sanitation facilities (78%) in the world. The statistics for the United States are even higher at 99% of the population using improved water supplies, while 89% were using improved sanitation facilities. For those in the United States either not having access to improved water supplies or improved sanitation facilities, UNC Environmental Finance Center (2017) has indicated that conditions for them are similar to individuals in developing countries. Review the UNC

Environmental Finance Center (2017) document on *Clean Water Access Challenges in the United States* and summarize the six major clean water access challenges they identified in twenty-first-century America.

References

Advameg, I. (2010). Encyclopedia of the Nations. Total renewable water resources – The World Factbook – CIA. http://www.nationsencyclopedia.com/WorldStats/CIA-Total-renewable-water-resources.html (accessed July 9, 2021).

Barnaby, W. (2009). Do nations go to war over water? *Nature* 458: 282–283.

Dictionary.com (2012). *Desalination.* Collins English Dictionary – Complete and Unabridged. New York, NY: Harper Collins Publishers. https://www.dictionary.com/browse/desalination. (accessed July 8, 2021).

Flynn, A.M., Akashige, T., and Theodore, L. (2019). *Kern, Process Heat Transfer*, 2nd Edition. Salem, MA: Schrivener-Wiley.

Fröhlich, C.J. (2012). Water: Reason for conflict or catalyst for peace? The case of the Middle East. *L'Europe en Formation* 365 (3): 139–161. https://www.cairn.info/revue-l-europe-en-formation-2012-3-page-139.htm.

Gleick, P.H. (1996). Water resources. In: *Encyclopedia of Climate and Weather* (ed. S.H. Schneider), 2. 817–823. New York: Oxford University Press.

Grace Communication Foundation (2019). Water footprint comparisons by country. https://www.watercalculator.org/footprints/water-footprints-by-country. (accessed July 9, 2021).

Hares, S. (2017). The cost of clean water: $150 billion a year, says world bank. In: *Reuters* London, UK: Thomson Reuters Corporation. https://www.reuters.com/article/us-global-water-health/the-cost-of-clean-water-150-billion-a-year-says-world-bank-idUSKCN1B812E.

Kummu, M., de Moel, H., Ward, P.J., and Varis, O. (2011). How close do we live to water? A global analysis of population distance to freshwater bodies. *PLoS One* 6 (6): e20578. https://www.ncbi.nlm.nih.gov/pmc/articles/PMC3110782.

Lufkin, B. (2017). Future – Why 'hydro-politics' will shape the 21st century. BBC Future Now. http://www.bbc.com/future/story/20170615-why-hydro-politics-will-shape-the-21st-century (accessed July 9, 2021).

McCrary, K. (2018). The politics of water development: A source of conflict? International association for political science students. http://www.iapss.org/wp/2018/02/28/the-politics-of-water-development-a-source-of-conflict (accessed July 9, 2021).

Mekonnen, M.M. and Hoekstra, A.Y. (2011). *National Water Footprint Accounts: The Green, Blue and Grey Water Footprint of Production and Consumption. Volume 1: Main Report.* Value of Water Research Report Series No. 50. Delft,

The Netherlands: UNESCO-IHE Institute for Water Education. https://waterfootprint.org/media/downloads/Report50-NationalWaterFootprints-Vol1.pdf.

Mullen, K. (2021). *Information on Earth's Water*. Westerville, OH: National Ground Water Association. https://www.ngwa.org/what-is-groundwater/About-groundwater/information-on-earths-water (accessed July 8, 2021).

Osborn, L. (2019). Number of reservoirs in the world. Current results. Weather and Science Facts. https://www.currentresults.com/Environment-Facts/number-reservoirs-in-world.php (accessed July 8, 2021).

Rekacewicz, P. (2006). *Freshwater Resources: Volume by Continent*. Arendal, Norway: GRID-Arendal. http://www.grida.no/resources/5608.

Stockholm International Water Institute (2005). *Making Water a Part of Economic Development. The Economic Benefits of Improved Water Management and Services*. Stockholm, Sweden: SIWI and the World Health Organization. https://www.siwi.org/wp-content/uploads/2015/09/waterandmacroecon.pdf.

Theodore, L. (2014). *Chemical Engineering: The Essential Reference*. New York, NY: McGraw-Hill.

Theodore, L., Dupont, R.R., and Ganesan, K. (2017). *Unit Operations in Environmental Engineering*. Salem, MA: Schrivener-Wiley.

Theodore, L. and Ricci, F. (2010). *Mass Transfer Operations for the Practicing Engineer*. Hoboken, NJ: John Wiley & Sons.

University of North Carolina Environmental Finance Center (2017). *An Overview of Clean Water Access Challenges in the United States*. Chapel Hill, NC: University of North Carolina at Chapel Hill Environmental Finance Center. https://efc.sog.unc.edu/wp-content/uploads/sites/1172/2021/07/An-Overview-of-Clean-Water-Access-Challenges-in-the-United-States.Final_.pdf.

US Water Alliance (2017). *An Equitable Water Future. A National Briefing Paper*. Washington, D.C.: US Water Alliance. http://uswateralliance.org/sites/uswateralliance.org/files/publications/uswa_waterequity_FINAL.pdf.

Water.org (2019). *Brazil's Water and Sanitation Crisis*. Kansas City MO: Water.org. https://water.org/our-impact/brazil.

World Health Organization (2017). *Progress on Drinking Water, Sanitation and Hygiene 2017. Update and SDG Baselines. Geneva*. Switzerland: World Health Organization (WHO) and the United Nations Children's Fund (UNICEF). https://apps.who.int/iris/bitstream/handle/10665/258617/9789241512893-eng.pdf;jsessionid=BFB5D93E1AC22CFE3A7EAB3C0F603BF9?sequence=1.

2

Technical Glossary

2.1 Introduction

This chapter provides definitions for many, but not all, of the terms that the reader will encounter in this book. It should be noted that some of the definitions specifically refer to the United States and that a good number of definitions pertain to wastewater and wastewater-related terms. It should also be noted that many of the terms have come to mean different things to different people, and this will become evident as one delves deeper into the applicable literature. Finally, the bulk of the material in this chapter was drawn from the earlier work of Theodore et al. (1997).

2.2 Glossary

- **Absolute humidity** – The amount of water vapor present in a unit mass of air, which is usually expressed as kilograms of water vapor per kilogram of dry air or pounds of water vapor per pound of dry air.
- **Absolute pressure** – The actual pressure exerted on a surface that is measured relative to zero pressure; it equals the gauge pressure plus the atmospheric pressure.
- **Absolute pressure gauge** – A device that measures the pressure exerted by a fluid relative to a perfect vacuum.
- **Absolute scale** – A temperature scale that is based on absolute zero and that uses units of measurement equivalent to Centigrade degrees on the Kelvin scale or to Fahrenheit degrees on the Rankine scale.
- **Absolute temperature** – The temperature expressed in degrees Kelvin (K) or degrees Rankine (°R).
- **Absolute temperature scale** – A scale (e.g. Kelvin, Rankine) in which temperatures are measured relative to absolute zero.

Introduction to Desalination: Principles, Processes, and Calculations, First Edition. Louis Theodore and R. Ryan Dupont.

- **Absolute vacuum** – A void that is completely empty of matter.
- **Absolute zero temperature** – The temperature of zero degrees on either the Kelvin or Rankine scale at which molecular motion is thought to cease.
- **Absorbate** – A substance that is taken up and retained by an absorbent.
- **Absorbent** – Any substance that takes in or absorbs other substances.
- **Absorber** – A device in which a gas is absorbed by contact with a liquid.
- **Absorption** – The process in which one material (the absorbent) takes up and retains another (the absorbate) to form a homogenous solution; it often involves the use of a liquid to remove certain gas components from a gaseous mixture.
- **Absorption tower** – A vertical tube or pipe in which a rising gas is partially absorbed by a liquid, usually in the form of falling droplets.
- **Acid** – A material containing hydrogen that produces at least one hydrogen ion when dissolved in a water solution; it can react with and neutralize a base to form a salt.
- **Acre-foot** – A measure of volume; 43,560 cubic feet, 325,900 gallons, or the volume of water covering 1 acre, 1 foot deep.
- **Action level** – The level of a contaminant that, if exceeded, triggers treatment or other requirements that a water system must follow.
- **Acute health effect** – An immediate effect (i.e. within seconds, minutes, hours, or days) that may result from exposure to certain drinking water contaminants (e.g. pathogens).
- **Adiabatic** – A term used to describe a system in which no gain or loss of heat occurs.
- **Adjudication** – A process performed by the courts that determines ownership of groundwater and assigns a Watermaster to manage or enforce pumping rights and sometimes water quality; this is often a multiyear process.
- **Advanced wastewater treatment** – Any process that is employed for the treatment of wastewater that follows secondary treatment and serves to improve the quality of effluent prior to reuse or discharge; this may include the removal of phosphorus, nitrogen, suspended solids, dissolved organic compounds, dissolved solids, etc.
- **Aerobic** – A term used to describe a system that requires oxygen to sustain itself.
- **Agricultural irrigation** – Water distribution systems and practices in agriculture.
- **Air stripping tower** – A tower used to remove volatile organic chemicals (such as solvents) from contaminated water by causing them to evaporate; polluted water is sprayed downward through a tower filled with packing materials while air is blown upward through the tower.
- **Algae** – A large group of aquatic plants that contain chlorophyll; they vary from a single cell to multicellular organisms; algae exist in saltwater and

freshwater, and can adversely affect water quality when excessive growth and death occurs (eutrophication), resulting in the lowering of the oxygen content in water.

- **Algae bloom** – A sudden proliferation of algae in water bodies, usually stimulated by the excessive input of nutrients (e.g. nitrogen and phosphorous); this can adversely affect water quality by causing a lowering of the oxygen content in the water.
- **Ambient** – A term used to describe the surrounding area or environment.
- **Anaerobic** – A chemical/biochemical reaction, process, or microorganism that occurs in the absence of oxygen.
- **Appropriative water rights** – Holders can use available water that is not taken by anyone else.
- **Aquatic** – A term used to describe any organism growing in, living in, or frequenting water.
- **Aquatic growth** – Any floating, drifting, or attached organism in a body of water.
- **Aqueduct** – A conduit or channel employed to convey water from one location to another.
- **Aqueous solution** – A solution with water as the liquid phase.
- **Aquifer** – A geologic formation, group of formations, or part of a formation that is capable of yielding a significant amount of groundwater to wells or springs.
- **Artesian well** – A well tapping a confined aquifer in which the static water level is above the bottom of the upper confining unit; a flowing artesian well is a well in which the water level is above the land surface.
- **Audit** – The examination of something with the intent to check, verify, or inspect.
- **Available oxygen** – The quantity of dissolved oxygen available for oxidation of organic matter in a water body.
- **Backwash** – Water used in a filtration system to separate clogging material from a filter media to clean the filter so it may be put back into service.
- **Backflow** – A flow condition induced by a differential in pressure that causes the flow of water or other fluid into the distribution pipes of a reservoir from any source other than its intended primary source.
- **Base** – Any compound that dissociates in aqueous solution to yield hydroxyl ions; it is employed to neutralize acids.
- **Batch process** – An unsteady-state process that is not continuous; its operations are carried out with discrete quantities of material.
- **Bedrock** – The solid rock mass located below the loose material on the Earth's crust such as soil, glacial drift, or alluvium.
- **Biota** – All of the living organisms that exist in an area.
- **Blackwater** – A water that contains human wastes.

- **Blowdown** – The cyclic or constant removal of a portion of any process flow to maintain the constituents of the flow at a desired level.
- **Boiler feedwater** – The water fed into a boiler to replace that evaporated in the generation of steam.
- **Boiling point** – The temperature of a liquid at which its vapor pressure is equal to that of the atmospheric pressure of the environment; it is 100°C (212°F) for water at sea level.
- **Bound water** – The water molecules that are tightly held by various chemical groups in a larger molecule.
- **Brackish water** – A water with a salt content in the range between that of freshwater and seawater.
- **Brine** – A concentrated solution of salt and water that remains after the removal of a distilled product.
- **By-pass** – The avoiding of a particular portion of a process or system.
- **By-product** – A material that is not one of the primary products and is not solely or separately produced by the production process.
- **Calibration** – The determination, checking, or adjustment of the accuracy of any instrument that gives quantitative measurements.
- **Carcinogen** – Any substance that can cause cancer.
- **Carryover** – The entrainment of liquid or solid particles in the vapor evolved by a boiling liquid or from a process unit.
- **Catch basin** – A chamber or well, usually built at the curbline of a street, that admits surface water for discharge into a storm water drain.
- **Chemical analysis** – The analysis by chemical methods that provides the composition and concentration of a substance.
- **Chemical sludge** – The sludge obtained by treatment of wastewater with chemicals.
- **Chemical treatment** – Any one of a variety of technologies that use chemicals or a variety of chemical processes to treat a system or waste.
- **Chlorine demand** – The amount of chlorine necessary to produce a free chlorine residual in a water sample.
- **Chlorine water** – A clear, yellowish liquid that degrades on exposure to air and light; it is employed as a deodorizer, disinfectant, and antiseptic.
- **Clarification** – The removal of suspended solids from wastewater by gravity settling; this process is often accelerated through coagulation of and flocculation of small solids with chemicals.
- **Closed loop** – A term used to describe an enclosed process.
- **Closed loop cooling tower** – Water-conserving cooling tower system in which water used for cooling is recycled through a piping system that cools the water.

- **Coastal waters** – The waters of the coastal zone, except for the Great Lakes and specified ports and harbors on inland rivers.
- **Coastal zone** – The lands and water adjacent to the coast that exert an influence on the uses of the sea and its ecology, or whose uses and ecology are affected by the sea.
- **Coastline** – The line separating the land surface and the water surface of the sea.
- **Coliform** – A group of related bacteria whose presence in water may indicate contamination by disease-causing microorganisms; indicators organisms of recent fecal contamination.
- **Communicable disease** – An illness that is caused by a specific infectious agent or its toxic products, and that arises through transmission from a reservoir to a susceptible host.
- **Community water system** – A public water system that serves at least 15 service connections employed by year-round residents or regularly serves at least 25 year-round residents.
- **Condenser** – Any device that cools gases or vapors to liquid.
- **Conduit** – Any artificial or natural duct, either opened or closed, employed for conveying fluids.
- **Confined aquifer** – An aquifer in which groundwater is confined under pressure.
- **Confined groundwater** – Water in an aquifer that is bounded by confining beds and is under pressure significantly greater than atmospheric pressure.
- **Confining bed** – A layer or mass of rock having very low hydraulic conductivity that hampers the movement of water into and out of an adjoining aquifer.
- **Connate water** – Water entrapped in the interstices of sedimentary rock at the time of its deposition.
- **Conservative pollutant** – A pollutant that does not decay, does not react, is persistent, and is not biodegradable.
- **Contaminant** – Any physical, chemical, biological, or radiological substance that has a harmful effect on human health or the environment when contained in air, water, or soil.
- **Continuous sampling** – The continuous withdrawal for analysis of a sample from some larger quantity of liquid, air, or solid.
- **Cooling tower** – A hollow, vertical structure, perhaps with internal baffles, to disperse water so it is cooled by flowing air and by evaporation at ambient temperature.
- **Cooling tower makeup** – Water added to a recirculating cooling tower water stream to compensate for water evaporation losses.
- **Cooling water** – Water typically used to cool heat-generating equipment or to condense gases in a thermodynamic cycle.

- **Cooling water blowdown** – The procedure used to reduce total dissolved solids by removing a portion of poor-quality recirculating water.
- **Cooling water drift** – Unevaporated water carried out of a cooling tower by the airflow; it has the same composition as the recirculating water.
- **Cooling water evaporation** – Cooling water recycling approach in which water loses heat when a portion of it is evaporated.
- **Creep** – The movement of water under or around a structure built on permeable foundations.
- **Cryogenics** – The production and utilization of extremely low temperatures.
- **Crystallization** – The change of state of a substance from a liquid to a solid by the phenomenon of crystal formation by nucleation and accretion (e.g. the freezing of water into ice).
- **Deaeration** – A process by which dissolved air and oxygen are removed from water.
- **Decantation** – The separation of a liquid from a solid or a higher density liquid with which it is immiscible by drawing off the fluid.
- **Dechlorination** – The removal of chlorine from a substance by chemically replacing it with hydrogen or hydroxide ions in order to detoxify the substances.
- **Decontamination/detoxification** – A process that converts toxic wastes into nontoxic compounds.
- **Deep-well injection** – A method of ultimate disposal that involves depositing liquid waste into a deep well beneath the surface of the Earth for permanent storage.
- **Dehumidifier** – A device incorporated into many air conditioning systems to dry incoming air by passing it across a bed of a hygroscopic substance or through a spray of very cold water.
- **Dehydration** – The chemical process where water in a chemical or material is removed.
- **Deionized water** – Common industrial water devoid of dissolved salts and organics, used to remove contaminants from products and equipment.
- **Demineralization** – The process of removing dissolved minerals from water by physical, chemical, or biological means.
- **Demister** – A device composed of plastic threads, wire mesh, or glass fibers employed to remove liquid droplets entrained in a gas stream.
- **Desalination** – The extraction of freshwater from sea or other salt water by the removal of salts, usually by evaporation, reverse osmosis, or crystallization.
- **Desert** – A terrestrial environment where evaporation exceeds precipitation, with consequent lack of vegetation.

- **Detoxification** – The destruction of the toxic aspects of a substance.
- **Dew point** – The temperature at which the first droplet of water forms on the progressive cooling of a mixture of air and water vapor; at the dew point, the air becomes saturated with water.
- **Dialysis** – The separation of smaller molecules from larger ones in a solution by means of the diffusion from a concentrated solution to a dilute solution across a semipermeable membrane.
- **Discharge** – The volume of water that flows past a given area in a given time.
- **Dissolved solids (TDS)** – Total dissolved solids, minerals, or salts in water.
- **Disinfectant** – Any substance that destroys harmful microorganisms or inhibits their activity.
- **Disposal well** – A well employed for the disposal of waste into a subsurface stratum.
- **Dissolved oxygen (DO)** – The oxygen freely available in water; it is one of the most important indicators of the quality of a water supply since oxygen is necessary for the life of aquatic organisms.
- **Distillation** – A process of separating the constituents of a liquid mixture by means of partial vaporization of the mixture and separate recovery of vapor and residue due to a difference of vapor pressure.
- **Distilled water** – A water of high purity, prepared by repeated distillation.
- **Domestic wastewater** – Wastewater generated from household activities that includes sanitary waste, liquid waste from food preparation and laundering, bathing and showering waste, and other liquid cleaning wastes generated from households.
- **Downstream** – The regions of a river system located in a hydraulically lower location than a given position along a stream or river. A section of the river system is hydraulically lower if gravity transports the water in the direction of or nearer to the mouth of the stream or river.
- **Downtime** – A term used to describe the periods when a system is unavailable or not operating.
- **Drain** – Any channel that carries off surface water.
- **Drawdown** – The difference between the water level in a well before pumping and the water level in the well during pumping.
- **Drilled well** – A well constructed by either percussion or rotary hydraulic drilling.
- **Drinking water supply** – Any raw water source that is or may be employed by a public water system or as drinking water by one or more individuals.
- **Dry cooling** – Cooling system using air instead of water as the cooling fluid to eliminate the use and evaporative loss of water.

- **Dry well** – A shallow well used in storm water systems for the infiltration of collected storm water for shallow groundwater recharge; the well is dry during dry weather conditions.
- **Duct** – A round or rectangular conduit, usually metal or fiberglass, employed to transport fluids, usually air.
- **Dystrophic lake** – An acidic, shallow body of water that contains high concentrations of humic substances and organic acids that are often brown in color; it contains many plants but few fish.
- **Ebb tide** – The tide occurring at the ebb period of tidal flow; it is sometimes referred to as falling tide to describe the direction of the current.
- **Ecosphere** – The global sum of all ecosystems on Earth.
- **Ecosystem** – The interacting system of a biological community and its non-living surroundings.
- **Eddy** – A circular movement occurring in flowing water caused by currents in the water induced by obstructions or changes and irregularities in the banks or bottom of the channel, or by differences in temperature.
- **Effluent** – Any fluid emerging from a pipe or similar outlet that enters the environment; it usually refers to treated wastewater from municipal or industrial treatment plants.
- **Electrolysis** – The use of a direct electric current to drive an otherwise non-spontaneous chemical reaction.
- **Elution** – The process of moving a substance through a bed by means of a slow-moving stream.
- **Elutriation** – A process for separating particles based on their size, shape, and density, using a stream of gas or liquid flowing in a direction usually opposite to the direction of sedimentation. This method is mainly used for particles smaller than 1 μm.
- **Empirical** – Anything derived from experimentation or observation and not from fundamentals or theory.
- **Endothermic** – A term used to describe a process or change that occurs with absorption of heat.
- **Enrichment** – The addition of constituents, generally nutrients, from wastewater treatment plant effluent or agricultural runoff to surface water.
- **Environment** – The system of external conditions affecting the existence and development of an individual or organism.
- **Environmental audit** – An independent assessment of the current status of a company's compliance with applicable environmental requirements.
- **Epidemiology** – The study of diseases as they affect population, including the distribution of disease, or other health-related states and events in human populations, the factors that influence this distribution, and the application of this study to control health problems.

- **Erosion** – The breakdown of solid rock into smaller particles and its removal by wind, water, or ice; the removal of surface soil by wind, water, or ice.
- **Estuary** – A region of interaction between rivers and near-shore ocean water where tidal action and river flow create a mixing of fresh and saltwater.
- **Euphotic zone** – The upper zone of a sea or lake into which sufficient light can penetrate for active photosynthesis to take place.
- **Eutrophic lake** – A body of water that is characterized by an abundant accumulation of nutrients that support a dense growth of algae and other organisms, the decay of which depletes the shallow waters of oxygen in summer.
- **Eutrophication** – The process by which a body of water becomes enriched in dissolved nutrients (such as phosphates) that stimulate the growth of aquatic plant life usually resulting in the depletion of dissolved oxygen.
- **Evaporation** – The conversion of a liquid into a vapor.
- **Evaporation pond** – A shallow pond or impoundment with a large surface area that is designed to contain wastewater and allow it to evaporate with no discharge to the environment.
- **Excursion** – An unintentional occurrence, such as a discharge of pollutants above the permitted amount, due to unplanned reasons.
- **Exempted aquifer** – An underground body of water defined in the underground injection control program as an aquifer that is "not reasonably expected" to be a source of drinking water and that is exempted from regulations barring underground injection activities.
- **Exothermic** – A term used to describe a reaction or process during which heat is released.
- **Faucet aerator** – Device that can be installed in a sink to reduce water use.
- **Feed** – The material supplied to a processing unit for treatment or processing.
- **Feedforward control system** – A system in which changes are detected at the process input and an anticipated correction signal is applied before process output is affected.
- **Feedstock** – The raw materials supplied to manufacturing or processing plants for use in the production of goods or materials.
- **Film-type condensation** – The process by which a saturated vapor comes into contact with a surface whose temperature is below the saturation temperature and uniformly condenses on the surface.
- **Filter membrane** – A thin film containing many fine pores that is employed to filter a liquid or solid stream.
- **Filter press** – A mechanical device that forces sludge, usually conditioned with coagulant addition, between moving filter membranes to squeeze water out of it and produce a dry sludge cake of between 15 and 25% solids.

- **Finished water** – Water that has been treated and is ready to be delivered to customers.
- **Flash drum** – A unit in which volatile components are vaporized and separated from a liquid stream for further fractionation.
- **Flash evaporator** – A distillation device in which saline water is injected in a superheated state into a vessel under vacuum and in which boiling occurs without the usual heat source.
- **Flocculation** – The process by which solids in water or sewage are made to increase in size by biological or chemical action so that they can be more easily separated from the water by sedimentation.
- **Fluid** – Any material or substance that changes shape or direction uniformly in response to an external force imposed on it; the term applies to liquids and gases.
- **Fluidization** – A technique in which a finely divided solid is caused to behave like a fluid by suspending it in a moving gas or liquid.
- **Fluoridation** – The addition of fluoride to public water supplies to prevent or delay the onset of dental decay.
- **Fractional distillation** – A distillation process in which countercurrent distillation is employed to obtain a product as nearly pure as possible; it is also any distillation process in which the product collected may be a series of separate components of similar boiling range.
- **Freeze-drying** – A method of dehydration or separating water from materials; the material is first frozen and placed in a high vacuum so that the water (ice) vaporizes (sublimes) in the vacuum without melting and the non-watery components are left behind in an undamaged state.
- **Freshwater** – The water that generally contains less than 1,000 mg/L of dissolved solids.
- **Froth** – A foamy mass of bubbles that exists on a body of water.
- **Gas permeation** – The movement of gas from the high pressure side of a membrane to the low pressure side.
- **Geohydrology** – The branch of hydrology involving the study of groundwater and its physical and chemical interactions with the physical environment.
- **Geology** – The science that deals with the origin, history, and structure of the Earth, as recorded in rocks, along with forces and processes that modify rocks.
- **Geothermal energy** – The energy derived from the superheated water and steam trapped in underground reservoirs.
- **Geothermal gradient** – The change of temperature in the Earth with depth, usually expressed in degrees per unit of depth.
- **Glacial drift** – The rock material transported and deposited directly by glaciers or indirectly by ice or water emanating from a glacier.

- **Glacier** – A large body of ice originating on land by the compaction and recrystallization of snow, and showing evidence of present or past movement.
- **Grab sample** – A sample that is collected at such a time and place so that it is ideally most representative of a total discharge.
- **Gray water** – Domestic wastewater composed of water from kitchen sinks, bathroom sinks and tubs, clothes washers, and laundry tubs.
- **Greenhouse effect** – The gradual rise in the average global temperature due to the absorption of infrared radiation reflected by the Earth's surface by increasing amounts of carbon dioxide, methane, nitrous oxide, and other greenhouse gases in the air.
- **Groundwater** – Water that exists underground in saturated zones beneath the land surface.
- **Groundwater basin** – A porous formation with sides and bottom of relatively impervious material in which groundwater is held or retained.
- **Groundwater hydrology** – The branch of hydrology dealing with groundwater sources, movement, recharge, depletion, etc.
- **Groundwater reservoir** – An area below the ground surface in which groundwater is stored, also called an aquifer.
- **Groundwater recharge** – The input of surface water to resupply groundwater reservoirs through infiltration at the soil surface or from the bottom of rivers and streams; the use of reclaimed wastewater, by surface spreading or direct injection, to replenish freshwater aquifers; to resupply freshwater aquifers to prevent saltwater intrusion, to control or prevent ground subsidence, and to augment non-potable or potable groundwater aquifers.
- **Groundwater system** – A groundwater reservoir and its contained water; also, the collective hydrodynamic and geochemical processes at work in the reservoir.
- **Habitat** – A dwelling place of a species or community, providing a particular set of environmental conditions (e.g. forest floor, seashore, etc.).
- **Hard water** – Any water containing high levels of dissolved divalent cations, primarily calcium and magnesium, that interfere with some industrial processes and prevent soap from lathering.
- **Hardness (water)** – A property of water causing formation of an insoluble residue when the water is used with soap and forming a scale in vessels in which water has been allowed to evaporate; it is due primarily to the presence of divalent ions of calcium and magnesium.
- **Heat exchanger** – A unit or vessel in which a hot fluid stream transfers part of its energy to a cooler fluid stream.
- **Heat sink** – A structure designed to absorb heat.
- **Heterogenous** – A term used to describe a mixture of different phases (e.g. liquid–vapor, liquid–vapor–solid).

- **High dam** – A dam that is taller than 165 feet; the first high dam was Hoover Dam in the United Sates; four other notable high dams are Grand Coulee Dam in the United States, Aswan High Dam in Egypt, Three Gorges Dam in China, and the Sardar Sarovar Dam in India.
- **Holding pond** – A pond or reservoir, usually made of earth, that is built to store polluted runoff.
- **Holdup** – A volume of material held or contained in a process vessel or line.
- **Homogenous** – A term used to describe a mixture or solution comprised of two or more compounds or elements that are uniformly dispersed in each other.
- **Hot brine** – A slightly salty subterranean water, the temperature of which is markedly higher than that dictated by the normal geothermal gradient; it can be employed as a source of geothermal energy.
- **Hot rock** – A subterranean rock, the temperature of which is higher than would be dictated by the normal geothermal gradient; it can be employed as a source of geothermal energy.
- **Humidifier** – A device for increasing the water content of air; it is usually incorporated into an air conditioning system.
- **Hydration** – The chemical process of combination or union of water with other substances.
- **Hydraulic fill** – An earth structure or grading operation in which the fill material is transported and deposited by means of water being pumped through a flexible or rigid pipe.
- **Hydrogeology** – The geology of groundwater, with particular emphasis on the chemistry and movement of water.
- **Hydrologic cycle** – The constant movement and cycling of water by evaporation, precipitation, and condensation in the Earth-atmosphere system.
- **Hydrology** – The science dealing with the properties, distribution, and circulation of water in relation to the land surface.
- **Hydrolysis** – The reaction of a salt with water to form an acid and a base.
- **Hydrophilic** – A term used to describe a substance with an affinity for water.
- **Hydrophobic** – A term used to describe a substance that separates from water or surfaces that repel water.
- **Hydrosphere** – The part of the Earth that is composed of water, including oceans, seas, lakes, rivers, icecaps, etc.
- **Hydrothermal** – A term used to describe any geological process involving heated or superheated water.
- **Ice** – The allotropic, crystalline form of water.
- **Imported water** – Water supply that is conveyed from one watershed to be used in another.

- **Impurity** – The presence of one substance in another, often in such low concentration that it cannot be measured quantitatively by ordinary analytical methods.
- **In situ** – A term used to describe any reaction occurring in place, and a term used to describe a fossil, mineral, or rock found in its original place of deposition, growth, or formation.
- **Induced draft** – The negative pressure created by the action of a fan, blower, or other gas-moving device.
- **Industrial wastewater** – The wastewater generated from industrial processes.
- **Industrial water** – The water that is withdrawn from a source for sole use in an industrial process.
- **Infiltration** – The penetration of water through the ground surface into sub-surface soil or the penetration of water from the soil into sewer, or other pipes through defective joints, connections, or other leaks in a system, or a land application technique where large volumes of wastewater are applied to land, allowed to penetrate the surface, and percolate through the underlying soil.
- **Inflow** – An entry of extraneous stormwater into a sewer system from sources above the ground surface, typically through manhole covers or direct pumping, i.e. via sump pumps.
- **Influent** – Any untreated wastewater stream flowing into a wastewater treatment plant or non-potable water entering a water treatment plant.
- **Injection well** – A well into which fluids are injected for purposes such as waste disposal, remediation, hydraulic control, etc.
- **Injection zone** – A geological formation, group of formations, or part of a formation receiving fluids through a well.
- **Inland waters** – The waters of the United States in the inland zone, waters of the Great Lakes, and specified ports and harbors on inland rivers.
- **Inland zone** – Land inside the coastal zone, excluding the Great Lakes and specified ports and harbors on inland rivers.
- **Insoluble** – A term used to describe a substance that is incapable of being dissolved in a liquid.
- **Instantaneous sampling** – The collecting of a sample in a very short period of time so that the sampling time is insignificant in comparison with the duration of the operation or the period being studied.
- **Instream use** – The water use taking place within a stream channel, hydroelectric power generation, navigation, water quality improvement, fish propagation, or recreation.
- **Interstitial water** – The water contained in the interstices of rocks, where the origin of the water is unknown or unspecified.

- **Ion exchange** – A mass transfer process that involves the interchange of ions between a liquid and a solid material; it can be employed to concentrate and recover desired materials or to remove undesired ions from a water supply.
- **Irrigation** – A technique for applying water or wastewater to land areas to supply the water and nutrient needs of plants.
- **Irrigation districts** – Special units of local government that control the bulk of surface water supplies, primarily in the Western states in the United States.
- **Irrigation field practices** – Techniques that keep water in the field, more efficiently distribute water across the field, or encourage the retention of soil moisture.
- **Irrigation withdrawals** – Withdrawal of water for application on land to assist in the growing of crops and pastures.
- **Isothermal** – A term used to describe a process that exhibits no change in temperature, i.e. constant temperature.
- **Lagooning** – The placement of solid or fluid material in a basin, reservoir, or artificial impoundment for purposes of treatment, storage, or disposal.
- **Lake** – An inland body of fresh or saltwater of considerable size, occupying a basin on the Earth's surface.
- **Land application** – The discharge of wastewater onto the ground surface for treatment or disposal.
- **Leaching** – A process by which something is removed by percolating, or trickling liquid; it usually refers to the removal of components in soil by water.
- **Leachate** – Any liquid, including any suspended components in the liquid, that has percolated through or drained from a solid, such as liquid generated from waste in a landfill.
- **Leakage** – An undesired and gradual escape or entry of a fluid into or out of a tank or vessel.
- **Levee** – A dike or ridge at the side of a river, intended to protect the land side from flood waters or to confine the stream flow to its regular channel.
- **Limnetic zone** – The well-lit, open-water surface region of a lake away from shore.
- **Limnology** – The scientific study of the physical, chemical, and biological components of freshwater with emphasis on plants and animal life.
- **Liquid** – An amorphous, noncrystalline state of matter; the molecules of which are much more highly concentrated than in gases and usually less concentrated than in solids.
- **Liquid permeation** – The movement of liquid components from one side of a membrane to the other.

- **Liquid-liquid extraction** – The separation of a solute based on its relative solubility in two immiscible liquids, usually water and an organic solvent.
- **Low-flow showerhead** – A showerhead that produces 2.5 gallons per minute or less, as compared to the 4.5 gallons per minute produced by most older standard showerheads.
- **Low-flush toilet** – A toilet that requires 1.6 gallons of water per flush or less, as compared to the 3.5 to 5 gallons of water required to flush most older, standard toilets.
- **Magma** – The molten material within the Earth's crust that is composed of silicates and volatiles (water and gases) in complex solution.
- **Make-up water** – The water that is employed to replenish a system that loses water through leakage, evaporation, etc.
- **Manifold** – A pipe fitting with numerous branches to convey fluids between a large pipe and several smaller pipes or to permit the choice of diverting flow from one of the several sources or to one of the many discharge points.
- **Mantle** – The part of the interior of the Earth between the crust and the core.
- **Maximum contaminant level (MCL)** – A maximum allowable level of a compound within a treated drinking water deemed safe for human consumption; concentrations above this level are deemed unsafe and unacceptable for human consumption.
- **Membrane** – A thin sheet of material through which a gaseous or liquid solution may pass.
- **Membrane hydrolysis** – A hydrolysis process that occurs when a colloidal electrolyte is separated from pure water by a membrane.
- **Membrane selectivity** – The ability of a membrane to allow passage of only anions or cations.
- **Microorganism** – An organism of microscopic size generally considered to include bacteria, algae, protozoa, fungi, and rotifers but excluding viruses; they (primarily bacteria) are employed in biological wastewater treatment processes to remove biodegradable organic matter and suspended solids. Some species are human pathogens and are removed by water and wastewater treatment to prevent disease transmission and protect public health.
- **Mine water** – The water encountered in mining operations that, if discharged into surface streams or other bodies of water, often contaminates them and makes them unfit for beneficial use.
- **Mineral spring** – A spring that contains high levels of mineral salts.
- **Miscibility** – The ability of a liquid or gas to dissolve uniformly in another liquid or gas.

- **Monitoring** – A periodic or continuous surveillance or testing to determine the level of compliance with statutory requirements and pollutant levels in various media, or in living things.
- **Monitoring well** – A well drilled to collect groundwater samples for analysis to determine the amounts, types, and distribution of contaminants in the groundwater.
- **Mother liquor** – A concentrated solution that is substantially freed from undissolved matter by filtration, centrifuging, or decantation; the product can be obtained by evaporation and/or crystallization.
- **Municipality** – A city, town, borough, county, parish, district, or other public body.
- **Natural resources** – The land, fish, wildlife, biota, air, water, groundwater, drinking water supplies, and other such resources belonging to, managed by, held in trust by, pertaining to, or otherwise controlled by the United States.
- **Navigable water** – Any stream, lake, arm of the sea, or other natural body of water that is navigable and of sufficient capacity to float watercraft for the purposes of commerce, trade, transportation, or recreation, as indicated under the Clean Water Act.
- **Nitrates** – Oxidized nitrogen that forms a monovalent anion; is the form of nitrogen that is produced from the aerobic oxidation of ammonia (nitrification); and that serves as an electron acceptor for the anaerobic oxidation of organic matter (denitrification); it is a nutrient that can result in eutrophication of lakes and ponds and is the cause of "blue baby" syndrome in humans if in high concentrations in drinking water: its MCL is 10 mg/L in drinking water.
- **Nonpoint source** – A source of pollutant discharge that is not traceable to a discrete location.
- **Non-potable water** – Water considered unsafe and/or unpalatable for drinking (see also potable water).
- **Nutrient** – Any element or compound that is essential to the life and growth of plants or animals, either as such or as transformed by chemical or enzymatic reactions; nitrogen and phosphorous are the major nutrients of concern in polluted waters.
- **Ocean water (seawater)** – A uniform solution contained approximately 96.5% water and 3.5% ionized salts; ingestion of substantial amounts will create a bodily chloride imbalance with harmful effects.
- **Oceanic** – A term used to describe the parts of the oceans deeper than 200 meters.
- **Oceanography** – The science dealing with oceans, including their form, physical and chemical features, and related phenomena.

- **Open channel** – Any natural or artificial waterway or conduit in which a fluid flows with a free surface exposed to atmospheric pressure.
- **Osmosis** – The passage of a pure liquid (usually water) through a semipermeable membrane from a solution of low concentration into a solution of a higher concentration (e.g. the flow of pure water into a solution of salt and water); see also reverse osmosis.
- **Osmotic pressure** – The pressure that results from osmosis.
- **Outfall** – The place where an effluent is discharged into a receiving water.
- **Outfall sewer** – A pipe or conduit that transports wastewater effluent to a final point of discharge.
- **Oxygenation** – The increase of the dissolved oxygen content within water through aeration.
- **Ozonation** – The addition of ozone to a water supply to reduce taste and odor problems and for disinfection.
- **Ozonator** – A device that creates ozone from oxygen and adds it to water.
- **Parts per billion (ppb)** – The fraction (ppb_m for mass fraction and ppb_v for volume fraction) multiplied by 10^9; it is a unit used to measure extremely small concentrations of a substance; equivalent to units of μg/L in water.
- **Parts per million (ppm)** –The fraction (ppm_m for mass fraction and ppm_v for volume fraction) multiplied by 10^6; it is a unit used to measure small concentrations of a substance; equivalent to units of mg/L in water.
- **Pathogenic waste** – A discarded waste that contains organisms capable of causing disease.
- **Percolation** – The flow of a liquid downward through a filtering medium or soil layer.
- **Permafrost** – The portion of the Earth which is permanently frozen, such as the Artic regions and portions of Alaska.
- **Permeability** – The degree to which a liquid can move freely through soils.
- **Physical quality** – The physical characteristics possessed by a material; it includes temperature, color, odor, and turbidity.
- **Physical treatment** – A water or wastewater treatment process that utilizes physical means for pollutant removal; processes include screening, grinding, settling, filtration, and centrifugation.
- **Physicochemical** – A term used to describe processes that involve both physical and chemical characteristics; adsorption is one such process that involves both physical and chemical attraction of an adsorbate to an adsorption site.
- **Point source of pollution** – Pollution originating from a discrete source, such as the outflow from a pipe, ditch, tunnel, concentrated animal-feeding operation, or floating craft.

- **Pollutant** – Any harmful substance present in air, water, or soil.
- **Pollution** – The direct or indirect alteration of the physical, thermal, biological, or radioactive properties of any part of the environment in such a way as to create a hazard or potential hazard to the health, safety, or welfare of any living species.
- **Potable water** – Water that is safe for human consumption.
- **POTW** – Publicly owned treatment works, or municipal wastewater treatment plants.
- **Pre-aeration** – A water treatment process in which the removal of dissolved gases, and the addition of oxygen are performed; typically, in aeration towers or aeration tanks.
- **Precipitate** – A solid that separates out from a liquid due to some physical or chemical change in the liquid.
- **Pretreatment** – Any process employed to partially remove pollutants from water prior to any subsequent treatment process.
- **Psychrometric chart** – A chart employed to determine the properties of moist air as a function of temperature.
- **Public water system (PWS)** – Any system that provides piped water for human consumption to at least 15 service connections or regularly serves 25 individuals.
- **Purging** – A cleansing or removal of impurities, foreign matter, or undesirable contaminants from a process through periodic withdrawals of liquids or solids.
- **Purification** – The removal of undesirable constituents from a substance by one or more separation techniques.
- **Rank** – The stage reached by coal in the course of its carbonation; the chief ranks, in order of increasing carbon content, are lignite, subbituminous coal, bituminous coal, and anthracite.
- **Raw water** – The untreated water that enters the first treatment unit of a water treatment plant.
- **Receiving water** – Any body of water (e.g. river, lake, ocean, stream, etc.) into which treated wastewater is discharged to.
- **Recirculating cooling water** – The recycling of cooling water to greatly reduce water use by reusing the water to perform several cooling operations.
- **Reclaimed water** – Treated wastewater that is reused for generally non-potable uses to supplement or replace other raw water supplies.
- **Recycled water** – Wastewater that has been treated for reuse and is recycled, generally for non-potable uses within a home or industrial facility.
- **Red tide** – A proliferation and accumulation of certain microscopic algae, predominantly dinoflagellates, in coastal waters; some species produce

toxins that are labeled harmful algae blooms, or Harmful Algal Blooms (HABs), that pose a serious and recurring threat to human health, wildlife, marine ecosystems, fisheries, and coastal aesthetics.

- **Red water** – A rust-colored water, usually resulting from the presence of precipitated ferric iron salts.
- **Release** – Any spilling, leaking, pumping, pouring, emitting, emptying, discharging, injecting, escaping, leaching, dumping, or disposing into the environment.
- **Renewable water supply** – The rate of supply of water (volume per unit time) potentially or theoretically available for use in a region on an essentially permanent basis.
- **Reproducibility** – The ability to repeat an experiment, reaction, measurement, or process and produce the same results.
- **Reservoir** – Any body of water employed for the storage, control, or regulation of water.
- **Reverse osmosis** – A water treatment process employed to separate water from pollutants by the application of pressure to force the water through a semipermeable membrane.
- **Rinse** – The removal of foreign materials from a surface by using a flow of liquid.
- **Rinse water** – Water used to remove debris and contaminants from products and equipment.
- **Riptide** – A strong surface current of short duration flowing outward from the shore.
- **River basin** – The land area drained by a river and its tributaries.
- **River bed** – The bottom of a river.
- **Runoff** – The water from precipitation that exceeds an areas infiltration and storage that flows over the ground into a surface water body.
- **Rural area** – The area outside the limits of any city, town, village, or other designated residential or commercial area.
- **Saline water** – Water that generally is considered unsuitable for human consumption or for irrigation because of its high content of dissolved solids, generally greater than 10,000 mg/L of dissolved solids; 35,000 mg/L dissolved solids is normally assigned to seawater.
- **Salinity** – The amount of salts or minerals dissolved in water.
- **Salinization** – A process in which a soluble salt accumulates in soils.
- **Salt** – A chemical compound formed when the hydrogen ion of an acid is replaced by a metal, or when an acid reacts with a base in an aqueous solution.
- **Salting out** – A reduction in the water solubility of certain molecules in a solution of very high ionic strength.

- **Saltwater intrusion** – The displacement of fresh groundwater by higher density salt water near coastal regions.
- **Sample** – A representative specimen of a liquid, solid, or gas collected for the purpose of determining its composition.
- **Sampling** – A method employed to obtain representative test samples; it consists of the collection, isolation, and the possible concentration of a small fractional part of a larger volume of a media.
- **Sand bar** – A ridge of sand built up by deposition to the surface or near the surface of a river or along a beach.
- **Sand dune** – A mound or ridge of loose sand blown by prevailing winds.
- **Sanitary survey** – An on-site review of water sources, facilities, equipment, operation, and maintenance of a public drinking water system to evaluate the adequacy of those components for producing and distributing safe drinking water.
- **Saturated rock** – A rock that has all of its void spaces filled with fluid.
- **Saturated soil** – A soil that has all of its void spaces filled with fluid.
- **Saturated zone** – A subsurface soil or rock zone in which all the interstices or voids are filled with water.
- **Saturation temperature** – The minimum temperature at which air is saturated with water vapor; the boiling point of water.
- **Scour** – The action of a flowing liquid as it erodes and carries away material on the sides or bottom of a channel.
- **Screening** – The use of screens to separate and remove coarse floating and suspended solids from water.
- **Sea** – A large body of saltwater, second in rank to an ocean, that is generally part of, or connected to an ocean at some point.
- **Sea level** – The surface of the sea that is employed as a reference for elevation.
- **Seawall** – A coastal wall built to provide protection against erosion and/or flooding from the ocean.
- **Seawater intrusion** – The movement of seawater into freshwater aquifers near the coast when these freshwater aquifers are over pumped.
- **Secondary drinking water regulation** – A regulation that sets a maximum acceptable level for contaminants that adversely affect the taste, odor or appearance of water or otherwise adversely affect the public welfare.
- **Sediment** – The solid material or deposits that have settled from a fluid.
- **Sedimentation tank** – A tank in which water or wastewater containing settleable solids is retained for a period of hours to allow these solids to move to the bottom of the tank by gravity; the settled solids are removed from the bottom and the floating solids are skimmed off the top for further treatment and disposal; also called clarifiers. See also Settling tank.

- **Semiconfined aquifer** – An aquifer that is partially confined by a layer (or layers) of low permeability soil or rock through which recharge and discharge nevertheless may occur.
- **Semipermeable membrane** – A membrane that allows substances of a certain size to pass through it while preventing the passage of larger ones.
- **Settleable solids** – The materials that are of sufficient size and density to sink to the bottom of a sedimentation tank.
- **Settling tank** – A tank used in water and wastewater treatment to hold water for a period of hours, during which heavier particles sink to the bottom for removal, treatment, and disposal; also called a clarifier. See also Sedimentation tank.
- **Sewage** – The wastewater produced by residential, commercial, institutional, and industrial facilities.
- **Sewer** – The system of pipes or conduits employed to collect and deliver wastewater to treatment plants or storm water surface water bodies.
- **Shore** – The land bordering any body of water.
- **Slow sand filtration** – A treatment process that involves the passage of raw water through a bed of sand at low velocity which results in the substantial removal of chemical and biological contaminants through the development of a biolayer (schmutzdecke) at the sand surface.
- **Sludge** – The thick, semisolid waste that accumulates as a result of the chemical coagulation, flocculation, and settling which occurs during drinking water treatment; the thick, semisolid biomass that is produced in the biological treatment of wastewater.
- **Sludge cake** – The dewatered sludge from a treatment plant that has a solids content of 18 to 30% solids.
- **Sludge dewatering** – The process of removing water from sludge using methods such as air drying, pressure filtration, vacuum filtration, centrifugation, or belt presses.
- **Sludge dryer** – A mechanical device for the removal of a large percentage of moisture from sludge by heat.
- **Sludge filter** – Mechanical devices in which wet sludge, usually conditioned by a coagulant, is dewatered by means of vacuum (vacuum filter) or elevated pressure (pressure filter).
- **Slurry** –A high solids content mixture of particulate matter and liquid.
- **Sodium chloride (NaCl)** – A colorless, transparent, crystalline solid or white, crystalline powder; it is noncombustible.
- **Soft water** – A water with a low concentration of divalent, primarily calcium and magnesium, cations.
- **Softening** – The chemical precipitation of divalent cations which cause the hardness of water; chemical coagulants, alum or ferric chloride, are

normally used as coagulants, with flocculation and sedimentation following coagulant addition.

- **Soil drainage** – The removal of excess water from a soil by gravity.
- **Soil moisture** – Water content in the soil, generally given as volume or weight percent.
- **Solubility** – The ability of one substance to be dissolved by another.
- **Spray irrigation** – The application of water to a land surface via spray droplet application.
- **Steam drum** – A vessel in a boiler in which the saturated steam is separated from the steam–water mixture and into which the feedwater is introduced.
- **Still** – An apparatus to purify liquids through heating to selectively boil and then cool to condense the vapor (e.g. to prepare alcoholic beverages, distilled water, etc.).
- **Storm drain** – A drain employed for conveying storm water runoff to a sewer.
- **Storm sewer** – A piping system employed exclusively for the transport of stormwater from streets, buildings, and surface runoff.
- **Stormwater runoff** – The portion of the volume of a rainfall event that exceeds the infiltration and storage capacity of a watershed.
- **Subterranean water** – The water that occurs in open spaces within rock materials of the Earth's crust.
- **Superheated steam** – Steam at a temperature above its boiling point at a given pressure.
- **Supersaturation** – An unstable condition in which a solvent contains more dissolved matter or gas than is present in a saturated solution of the same components at an equivalent temperature.
- **Surface water** – All water that is above the surface of the ground and is naturally open to the atmosphere.
- **Surge irrigation** – The intermittent application of water to irrigation pathways; this method pulses water down the furrow and creates more uniform irrigation water distribution.
- **Suspension** – A system in which very small particles are uniformly dispersed in a liquid or gaseous medium.
- **Tank** – A stationary device that is essentially a container, e.g. designed to contain an accumulation of waste or liquid, that is constructed primarily of non-earthen materials which provide structural support.
- **Temperature gradient** – The change in temperature with distance or position.
- **Tertiary treatment** – The advanced treatment of wastewater beyond secondary treatment; it may involve combinations of physical, chemical, and biological treatment processes to remove solids, nutrients, metals, salts,

nonbiodegradable organics, etc. to prepare the waste for disposal into highly sensitive environments or for reuse.

- **Thickener** – A small circular or rectangular sedimentation tank, designed to increase the concentration of solids in a suspension.
- **Thickening agent** – Any of a variety of substances employed to increase the viscosity of liquid mixtures and solutions without changing its other properties.
- **Tidal wave** – An exceptionally large wave, tsunami, or increase in the water level along a shore due to strong winds, volcanic eruption, or earthquake.
- **Tide** – The periodic rising and falling of water that results from the gravitational attraction of the moon and sun acting on the rotating Earth.
- **Total organic carbon (TOC)** – The total amount of organic carbon present in water as organic compounds (e.g. amino acids, hydrocarbons, proteins, etc.).
- **Toilet displacement device** – Object placed in a toilet tank to reduce the amount of water used per flush; for example, weighted plastic jugs filled with water or toilet dams that hold back a reservoir of water when the toilet is flushing.
- **Toxic** – A term used to describe a poisonous substance that has a harmful effect on an organism by ingestion, inhalation, or skin absorption.
- **Trace** – A very small quantity of a constituent, the amount of which cannot be determined precisely because of its low concentration.
- **Transpiration** – The process by which water passes through living organisms, primarily plants, and into the atmosphere.
- **Treatment** – Any method, technique, or process that is designed to change the physical, chemical, and/or biological composition of a waste so as to neutralize it, recover energy or material resources from it, render it nonhazardous or less hazardous, or make it safer to transport, store, reuse, or dispose of.
- **Tributary** – A stream or river that flows into a larger stream or main stem (or parent) river or a lake.
- **Trough** – A structure employed to hold or transport fluids.
- **Tsunami** – A sea wave caused by an underwater seismic disturbance such as sudden faulting, a landslide, or volcanic activity.
- **Turbid water** – Water that is cloudy due to fine particles in suspension.
- **Turbidity** – A measure of the fine particles suspended in a fluid; measured as the proportion of light passed through a sample that is refracted by suspended particles in a water column.
- **Ultrafiltration** – The separation of a solute with a specific molecular size and shape from a solution by applying pressure to force the solvent to flow through a membrane.

- **Underground sources of drinking water** – The aquifers that are currently being employed as a source of drinking water, and those that are capable of supplying a public water system.
- **Unconfined aquifer** – An aquifer whose upper surface is free of a confining layer and thus is able to fluctuate under atmospheric pressure.
- **Upstream** – The regions of a river system located in the direction opposite to the flow of a stream from a given position; a section of a river system that is hydraulically higher, if gravity transports the water away from the given location.
- **Uptake** – The act of taking up, drawing up, or absorbing.
- **Urban runoff** – The storm water from city streets and adjacent domestic or commercial properties.
- **Utility** – Public water service provider.
- **Vapor pressure** – The pressure exerted by a vapor in equilibrium with a liquid at a given temperature.
- **Virgin material** – A raw, unused material.
- **Wake** – The visible trail of turbulence left behind a moving stream.
- **Wash solvent** – A liquid added to a liquid–liquid extraction to wash or enrich the purity of the solute used in the extraction process.
- **Wash water** – The water employed to wash equipment.
- **Wastewater** – The water used to carry liquid waste material, consisting of dissolved and suspended solids, organics, and nutrients from homes, businesses, institutions, and industries to wastewater treatment plants for contaminant removal prior to release to surface water.
- **Wastewater operations and maintenance** – The actions taken after construction of wastewater treatment facilities to assure that the facilities will be properly operated and maintained.
- **Wastewater treatment** – A series of processes in which wastewater is treated in order to remove or alter its objectional constituents to a degree that renders it less harmful or dangerous.
- **Wastewater treatment plant** – A series of unit operations including screening, sedimentation, digestion, stabilization, dewatering, disinfection, and other processes for removing pollutants from wastewater before it discharges into the environment.
- **Wastewater treatment unit** – A device that is part of a wastewater treatment facility which is subject to regulation.
- **Water** – A colorless, odorless, tasteless liquid composed of the elements hydrogen and oxygen.
- **Water audit** – Program involving sending trained water auditors to participating family homes, free of charge, to identify water conservation

opportunities such as repairing leaks and installing low-flow plumbing and to recommend changes in water use practices to reduce home water use.

- **Water conditioning** – The treatments, excluding disinfection, that are intended to produce a water which is free of taste, odor, and undesirable contaminants.
- **Water conservation** – Activities designed to reduce the demand for water, improve efficiency of use, and reduce losses and waste of water in a potable water system.
- **Water consumption** – The quantity of water supplied in a municipality or district for a variety of uses during a given period.
- **Water main** – The water pipe, located in the street, from which domestic water supply is delivered to specific premises.
- **Water pollution** – The contamination of fresh or saltwater with materials that are toxic, noxious, or otherwise harmful to fish, man, or other animals.
- **Water purification** – Any process in which water is treated in such a way as to remove or reduce undesirable impurities.
- **Water quality** – The chemical, physical, or biological characteristics of water with respect to its suitability for a particular purpose.
- **Water quality criteria** – The specific levels of water quality that, if reached, are expected to render a body of water suitable for its designated beneficial use.
- **Water quality standards** – The state-adopted, EPA-approved, allowable numeric ambient water quality concentrations for surface water bodies defined for each beneficial use.
- **Water recycling** – Reuse of water for the same application for which it was originally used.
- **Water reuse** – Using treated wastewater (reclaimed water) for some beneficial purpose rather than discharging it to surface water; the deliberate use of reclaimed wastewater must be in compliance with applicable rules for a beneficial purpose (landscape irrigation, agricultural irrigation, aesthetic uses, ground water recharge, industrial uses, or fire protection).
- **Water rights** – The rights acquired under the law to use surface or groundwater for a specified purpose, in a given manner, and usually within the limits of a given period.
- **Water softening** – A treatment process designed to completely or partially remove hardness-producing ions, Ca^{2+} or Mg^{2+}, from a potable water.
- **Water solubility** – A measure of the maximum concentration of a chemical compound that can result when it is dissolved in water.

- **Water supplier** – A person who owns or operates a water supply system; they can be public or private.
- **Water supply system** – The collection, treatment, storage, and distribution of potable water from source to consumer.
- **Water surcharge** – Imposition of an increased cost due to excessive water use.
- **Water table** – The upper level of the groundwater below which the ground is saturated with water.
- **Water treatment** – The purification of water to make it suitable for drinking or other beneficial uses.
- **Waterlog** – Occurs when water is added to land faster than it can drain.
- **Watershed** – The area surrounding a stream that supplies it with runoff.
- **Waterway** – Any body of water, other than the open sea, that is or can be employed by boats as a means of travel.
- **Well** – A bored, drilled, or driven shaft, or a dug hole, whose depth is greater than the largest surface dimension and whose purpose is to reach underground water supplies, or to store or bury fluids below ground.
- **Wetland** – An area covered or saturated permanently, occasionally, or periodically by fresh or salt water.
- **WHO** – World Health Organization of the United Nations, based in Geneva (www.who.int).

Reference

Theodore, L., Reynolds, J., and Morris, K. (1997). *Concise Dictionary of Environmental Terms.* Amsterdam, The Netherlands: Gordon and Breach Science Publishers.

3

Engineering Principles

3.1 Introduction

This chapter introduces the reader to the general subject of engineering principles. Seven topics are addressed in the chapter that include: the metric system, SI multiples and prefixes, conversion constants, dimensional analysis, flow diagrams, significant figures and approximate numbers, and generic problem-solving techniques. A series of illustrative examples are provided to highlight concepts presented throughout the chapter.

3.2 The Metric System

The need for a single worldwide coordinated measurement system was recognized over 300 years ago. Gabriel Mouton, Vicar of St. Paul in Lyons, proposed a comprehensive decimal measurement system in 1670 based on the length of one minute of arc of a great circle of the Earth. In 1671 Jean Picard, a French astronomer, proposed the length of a pendulum beating seconds as the unit of length. (Such a pendulum would have been easily reproducible, thus facilitating the widespread distribution of uniform standards). Other proposals were made, but over a century elapsed before any action was taken.

In 1790, in the midst of the French Revolution, the National Assembly of France requested the French Academy of Sciences to "deduce an invariable standard for all the measures and weights." The Commission appointed by the Academy created a system that was, at once, simple and scientific. The unit of length was to be a portion of the Earth's circumference. Measures for capacity (volume) and mass (weight) were to be derived from the unit of length, thus relating the basic units of the system to each other and to nature. Furthermore, the larger and smaller versions of each unit were to be created by multiplying or dividing the basic units by 10 and its multiples. This feature provided great convenience to users of the system by eliminating the need for such

Introduction to Desalination: Principles, Processes, and Calculations, First Edition. Louis Theodore and R. Ryan Dupont.

calculations and divisions by 16 (to convert ounces to pounds) and by 12 (to convert inches to feet). Similar calculations in the metric system could be performed simply by shifting the decimal point. Thus, the metric system is a *base-10* or decimal system.

The Commission assigned the name metre (which is now spelled *meter*) to the unit of length. This name was derived from the Greek word metron meaning "a measure." The physical standard representing the meter was to be constructed so that it would equal one ten-millionth of the distance from the north pole to the Equator along the meridian of the Earth running near Dunkirk in France and Barcelona in Spain.

The metric unit of mass, called the *gram*, was defined as the mass of one cubic centimeter (a cube that is 1/100 of a meter on each side) of water at its temperature of maximum density. The cubic decimeter (a cube 1/10 of a meter on each side) was chosen as the unit of fluid capacity. This measure was given the name *liter*.

Although the metric system was not accepted with enthusiasm at first, adoption by other nations occurred steadily after France made its use compulsory is 1840. The standardized character and decimal features of the metric system made it well suited to scientific and engineering work. Consequently, it is not surprising that the rapid spread of the system coincided with an age of rapid technological development. In the United States, by Act of Congress in 1866, it was made "lawful throughout the United States of America to employ the weights and measures of the metric system in all contracts, dealings, or court proceedings."

By the late 1860s, even better metric standards were needed to keep pace with scientific advances. In 1875, the international "Treaty of the Meter," set up well-defined metric standards for length and mass and established permanent machinery to recommend and adopt further refinements in the metric system. This treaty, known as the Metric Convention, was signed by 17 countries, including the United States. Because of the Treaty, metric standards were constructed and distributed to each nation that ratified the Convention. Since 1893, the internationally agreed-to metric standards have served as the fundamental weights and measures standards of the United States.

A total of 35 nations – including the major nations of continental Europe and most of South America – had officially accepted the metric system by 1900. Today, except for the United States and a few small countries, the entire world is using predominantly the metric system or is committed to its use. In 1971 the Secretary of Commerce, in transmitting to Congress the results of a 3-year study authorized by the Metric Study Act of 1968, recommended a program that the United States change to predominant use of the metric system through a coordinated national program.

The International Bureau of Weights and Measures (located at Sevres France) serves as a permanent secretariat for the Metric Convention, coordinating the exchange of information about the use and refinement of the metric system. As measurement science develops more precise and easily reproducible ways of defining the measurement units, the General Conference of Weights and Measures – the diplomatic organization made up of adherents to the Convention – meets periodically to ratify improvements in the system and the standards.

The aforementioned General Conference adopted an extensive revision and simplification of the system in 1960. The name Le Systeme International d'Unites (International System of Units), with the international abbreviation SI, was adopted for this modernized metric system. Further improvements in and additions to SI were made by the General Conference in 1964, 1968, and 1971.

The basic units in the SI system are the *meter* (length), *kilogram* (mass), *second* (time), *ampere* (electric current), *Kelvin* (temperature), *mole* (amount of a substance), and *candela* (the unit of luminous intensity). All are commonly used by scientists and engineers. The Celsius scale of temperature (0°C, 273.15 K) is commonly used with the absolute Kelvin scale. The important derived units are the *newton* (SI unit of force), the *joule* (SI unit of energy), the *watt* (SI unit of power), the *pascal* (SI unit of pressure), the *hertz* (unit of frequency). There are also a number of electrical units: *coulomb* (charge), *farad* (capacitance), *henry* (inductance), *volt* (potential), and *weber* (magnetic flux). As noted, one of the major advantages of the metric system is that larger and smaller units are given in powers of ten. In the SI system, a further simplification is introduced by recommending only those units with multipliers of 10^3. Thus, for lengths in engineering, the *micrometer* (previously micron), *millimeter*, and *kilometer* are recommended, and the *centimeter* is generally avoided. A further simplification is that the decimal point may be substituted by a comma (as in France, Germany, and South Africa), while the other number, before and after the comma, will be separated by spaces between groups of three, i.e. one million dollars will be $1 000 000,00. More details are provided in a later section. Details on the seven base units are provided shortly.

3.2.1 Length – Meter (m)

The meter (common international spelling, *metre*) is defined as 1,670,763.00 wavelengths in vacuum of the orange-red line of the spectrum of krypton-86. The SI unit of area is the square meter (m^2). The SI unit of volume is the cubic meter (m^3). The *liter* (0.001 cubic meter), although not an SI unit, is commonly used to measure fluid volume.

3.2.2 Mass – Kilogram (kg)

The standard for the unit of mass, the *kilogram*, is a cylinder of platinum-iridium alloy kept by the International Bureau of Weights and Measures in Paris. A duplicate in the custody of the National Bureau of Standards serves as the mass standard for the United States. This is the only base unit defined by an artifact. The SI unit of force is the *newton* (N). One newton is the force which, when applied to a 1 kilogram mass, will give the kilogram mass an acceleration of 1 (meter per second) per second, i.e.

$$1\ \text{N} = 1\ \text{kg-m/s}^2 \tag{3.1}$$

The SI unit for pressure is pascal (Pa). One pascal equals 1 N/m^2. The SI unit for work and energy of any kind is the *joule* (J). One joule is equal to 1 N-m. The SI unit for power of any king is the *watt* (W). One watt is equal to 1 J/s.

3.2.3 Time – Seconds (s)

The *second* is defined as the duration of 9,192,632,770 cycles of the radiation associated with a specified transition of the cesium-133 atom. It is realized by tuning an oscillator to the resonance frequency of cesium-133 atoms as they pass through a system of magnets and a resonant cavity into a detector. The number of periods or cycles per second is called *frequency*. The SI unit for frequency is the *hertz* (Hz). One hertz equals one cycle/s The SI unit for speed is the *meter per second* (m/s). The SI unit for acceleration is the *meter per second per second* (m/s^2).

3.2.4 Electrical Current – Ampere (A)

The *ampere* is defined as that current which, if maintained in each of two long parallel wires separated by 1 meter in free space, would produce a force between the two wires (due to their magnetic fields) of 2×10^{-7} N/m length. The SI unit of voltage is the *volt* (V), where one volt equals 1 W/A. The unit of electrical resistance is the *ohm* (Ω). One ohm equals 1 V/A.

3.2.5 Temperature – Kelvin (K)

The *Kelvin* is defined as the fraction 1/273.15 of the thermodynamic temperature of the triple point of water. The temperature 0 K is called absolute zero. On the commonly used Celsius temperature scale, water freezes at ambient conditions at about 0°C and boils at about 100°C. The C is defined as an interval of 1 K, and the Celsius temperature 0°C is defined as 273.15 K. On the Fahrenheit scale, 1.8°Fahrenheit is equal to 1.0°C or 1.0 K; the Fahrenheit scale uses 32°F as a temperature corresponding to °C.

3.2.6 Amount of a Substance – Mole (mol)

The *mole* is the amount of substance of a system that contains as many elementary entities as there are atoms and is 0.012 kg of carbon-12. When the mole is used, the elementary entities must be specified and may be atoms, molecules, ions, electrons, other particles, or specified groups of such particles. The SI unit of concentration of a substance is the *mole per cubic meter* (mol/m^3).

3.2.7 Luminous Intensity – Candela (cd)

The *candela* is defined as the luminous intensity of 1/600,000 of a square meter of a black body at the temperature of freezing platinum (2045 K). The SI unit of light flux is the *lumen* (lm). A source having an intensity of 1 candela in all directions radiates a light flux of 4π lumens.

3.2.8 Supplementary Units

Two supplementary units complement these seven primary units. They include:

1. Phase angle – radian (rad). The *radian* is the plane angle with its vertex at the center of a circle that is subtended by an arc equal in length to the radius.
2. Solid angle – steradian (sr). The *steradian* is the solid angle with its vertex at the center of a sphere that is subtended by an arc of the spherical surface equal to that with sides equal in length to the radius.

3.3 SI Multiples and Prefixes

Important SI multiples and prefixes are provided in Table 3.1.

3.4 Conversion Constants

A significant amount of a practicing engineer's time is spent in converting data and equations from one set of units to another. As long as scientists provide data and information in grams and centimeters while engineers calculate in terms of pounds and feet, this confusion in terminology will continue to exist.

Table 3.1 SI multiples and prefixes.

Multiples submultiples	Prefixes	Symbols	
1 000 000 000 000	10^{12}	tera	T
1 000 000 000	10^{9}	giga	G
1 000 000	10^{6}	mega	M
1000	10^{3}	kilo	k
100	10^{2}	hecto	h
10	10^{1}	deka	da
Base unit 1	10^{0}		
0.1	10^{-1}	deci	d
0.01	10^{-2}	centi	c
0.001	10^{-3}	milli	m
0.000 001	10^{-6}	micro	μ
0.000 000 001	10^{-9}	nano	n
0.000 000 000 001	10^{-12}	pico	P
0.000 000 000 000 001	10^{-15}	femto	f
0.000 000 000 000 000 001	10^{-18}	atto	a

For the most part, the units in this text are consistent with those adopted by the engineering profession in the United States.

The physical, biological, and chemical properties of materials are important considerations in any study of desalination. To promote a better understanding of such properties, many of which are covered in Chapter 4, some definitions are briefly introduced in this section. Equations used in engineering as well as in the physical sciences are dimensional equations. The choice of the fundamental system of units determines the magnitude of the quantity described. Since the choice is optional, it is usually made to suit the convenience of a certain segment of the scientific community. For example, astronomers measure distance in light years, nuclear physicists employ angstroms, engineers use feet, etc. Engineers in the United States generally employ English units, but technical individuals need to be able to convert from one system of units to another.

Information on conversion constants (sometimes referred to as conversion factors) for length, mass, time, force, pressure, volume, energy, power, concentration, viscosity, and heat capacity are provided in Table 3.2. Commonly used water and water-related conversion constants, as they relate to volume and volumetric flowrate, are provided in Tables 3.3 and Table 3.4, respectively.

Table 3.2 Conversion constants.

To convert from	To	Multiply by
Length		
m	cm	100
m	mm	1000
m	microns (μm)	10^6
m	angstroms (Å)	10^{10}
m	in	39.37
m	ft	3.281
m	mi	6.214×10^{-4}
ft	in	12
ft	m	0.3048
ft	cm	30.48
ft	mi	1.894×10^{-4}
Mass		
kg	g	1000
kg	lb	2.205
kg	oz	35.24
	ton	2.268×10^{-4}
kg	grains	1.543×10^4
lb	oz	16
lb	ton	5×10^{-4}
lb	g	453.6
lb	kg	0.4536
lb	grains	7,000
Time		
s	min	0.01667
s	hr	2.78×10^{-4}
s	day	1.157×10^{-7}
s	week	1.653×10^{-6}
s	yr	3.171×10^{-8}
Force		
N	kg-m/s^2	1
N	dynes	10^5
N	g-cm/s^2	10^5

(*Continued*)

Table 3.2 (Continued)

To convert from	To	Multiply by
N	lb_f	0.2248
N	$lb\text{-}ft/s^2$	7.233
lb_f	N	4.448
lb_f	dynes	4.448×10^5
lbf	$g\text{-}cm/s^2$	4.448×10^5
lbf	$lb\text{-}ft/s^2$	32.17
Pressure		
atm	$N/m^2 = Pa$	1.013×10^5
atm	kPa	101.3
atm	bars	1.013
atm	$dynes/cm^2$	1.013×10^6
atm	$lb_f/in^2 = psi$	14.696
atm	mm Hg at 0°C (torr)	760
atm	in Hg at 0°C	29.92
atm	ft H_2O at 4°C	33.9
atm	in H_2O at 4°C	406.8
psi	atm	6.80×10^{-2}
psi	mm Hg at 0°C (torr)	51.71
psi	in H_2O at 4°C	27.70
in H_2O at 4°C	atm	2.458×10^{-3}
in H_2O at 4°C	psi	0.0361
in H_2O at 4°C	mm Hg at 0°C (torr)	1.868
Volume		
m^3	L	1,000
m^3	cm^3 (cc, mL)	10^6
m^3	ft^3	35.31
m^3	gal (U.S.)	264.2
m^3	qt	1,057
ft^3	in^3	1,728
ft^3	gal (U.S.)	7.48
ft^3	m^3	0.02832
ft^3	L	28.32

(Continued)

Table 3.2 (Continued)

To convert from	To	Multiply by
Energy		
J	N-m	1
J	erg	10^7
J	dyne-cm	10^7
J	kW-hr	2.778×10^{-7}
J	cal	0.2390
J	ft-lb_f	0.7376
J	Btu	9.486×10^{-4}
cal	J	4.186
cal	Btu	3.974×10^{-3}
cal	ft-lb_f	3.088
Btu	ft-lb_f	778
Btu	hp-hr	3.929×10^{-4}
Btu	cal	252
Btu	kW-hr	2.93×10^{-4}
ft-lb_f	cal	0.3239
ft-lb_f	J	1.356
Power		
W	J/s	1
W	cal/s	0.2390
W	ft-lb_f/s	0.7376
W	kW	10^{-3}
kW	Btu/s	0.949
kW	hp	1.341
hp	ft-lb_f/s	550
hp	kW	0.7457
hp	cal/s	178.2
hp	Btu/s	0.707
Concentration		
$\mu g/m^3$	lb/ft^3	6.243×10^{-11}
$\mu g/m^3$	lb/gal	8.346×10^{-12}
$\mu g/m^3$	gr/ft^3	4.370×10^{-7}

(Continued)

Table 3.2 (Continued)

To convert from	To	Multiply by
gr/ft^3	$\mu g/m^3$	2.288×10^6
gr/ft^3	g/m^3	2.288
lb/ft^3	$\mu g/m^3$	1.602×10^{10}
lb/ft^3	μg/L	1.602×10^7
lb/ft^3	lb/gal	7.48
Viscosity		
P (poise)	g/cm-s	1
P	cP (centipoise)	100
P	kg/m-hrv	360
P	lb/ft-s	6.72×10^{-2}
P	lb/ft-hr	241.9
P	lb/m-s	5.6×10^{-3}
lb/ft-s	P	14.88
lb/ft-s	g/cm-s	14.88
lb/ft-s	kg/m-hr	5.357×10^3
lb/ft-s	lb/ft-hr	3,600
Volume		
m^3	L	1,000
m^3	cm^3 (cc, mL)	10^6
lb/ft-s	kg/m-hr	5.357×10^3
lb/ft-s	lb/ft-hr	3,600
Heat Capacity		
cal/g-°C	Btu/lb-°F	1
cal/g-°C	kcal/kg-°C	1
cal/g-°C	cal/gmol-°C	molecular weight
cal/gmol-°C	Btu/lbmol-°F	1
J/g-°C	Btu/lb-°F	0.2389
Btu/lb-°F	cal/g-°C	1
Btu/lb-°F	J/g-°C	4.186
Btu/lb-°F	Btu/lbmol-°F	molecular weight

Table 3.3 Conversion constants for volume.

Units	Conversion constants				
	L	gal	ft^3	m^3	ac-ft
Liters (L)	1.0	0.2642	3.531 x 10^{-2}	0.0001	8.106 x 10^{-7}
Gallons (gal)	3.785	1.0	0.1337	3.785 x 10^{-3}	3.068 x 10^{-6}
Cubic feet (ft^3)	28.32	7.481	1.0	2.832 x 10^{-3}	2.296 x 10^{-5}
Cubic meters (m^3)	1,000	264.2	35.31	1.0	8.106 x 10^{-4}
Acre feet (ac-ft)	1.223 x 10^6	3.259 x 10^5	4.356 x 10^4	1,233	1.0

Table 3.4 Conversion constants for volumetric flowrate.

Units	Conversion constant				
	gpm	L/s	ac-ft/d	cfs	m^3/d
Gallons/min (gpm)	1.0	6.309 x 10^{-2}	4.419 x 10^{-3}	2.228 x 10^{-3}	5.45
Liters/sec (L/s)	15.85	1.0	7.005 x 10^{-2}	3.351 x 10^{-2}	86.4
Acre feet/day (ac-ft/d)	226.3	14.28	1.0	0.5042	1,234
Cubic feet/sec (ft^3/s)	448.8	28.32	1.983	1.0	2.447
Cubic meters/day (m^3/d)	1.369 x 10^9	8.64 x 10^7	6.051 x 10^6	3.051 x 10^6	1.0

3.5 Dimensional Analysis

A unit is defined as a measure of a physical extent, while a dimension is a description of the physical extent. Units, unlike physical laws, can be considered as either derived or basic. There is a certain latitude in choosing the basic units, and, unfortunately, this free choice has resulted in the aforementioned mild form of confusion. As described earlier, two systems of units have arisen: metric, the cgs, or centimeters-gram-second system and the English, the fps, or foot–pound–second system of engineering.

There are hundreds of conversion constants employed by engineers and scientists. Some of the more common "conversion constants" are provided in Tables 3.2 to Table 3.4.

Conversion of units can be accomplished by the multiplication of the quantity to be converted by appropriate unit ratios, i.e. the conversion constants. For example, suppose an energy of 50 *Btu* must be converted to units of *ft-lb*$_f$. From the energy section in Table 3.2, one notes that to convert from *Btu* to *ft-lb*$_f$, on simply multiplies by 778, Therefore:

$$1Btu = 778 ft\text{-}lb_f \tag{3.2}$$

The conversion constant, or unit ratio, is:

$$\frac{778 ft\text{-}lb_f}{1Btu} = 1 \tag{3.3}$$

The 50 *Btu* may then be multiplied by the earlier conversion constant without changing its value. Therefore:

$$(50Btu)\left(\frac{778 ft\text{-}lb_f}{1Btu}\right) = 38{,}900 ft\text{-}lb_f \tag{3.4}$$

with the *Btu* units cancelling, just like numbers.

Problems are frequently encountered in environmental studies and other engineering work that involve several variables. Engineers and scientists are generally interested in developing functional relationships (equations) between these variables. When these variables can be grouped together in such a manner that they can be used to predict the performance of similar pieces of equipment, independent of the scale or size of the operation, something very valuable has been accomplished. In addition, one of the properties of equations that has a rational basis and is deduced from general relations is that they must be dimensionally homogeneous, or consistent. Dimensional analysis is a relatively "compact" technique for reducing the number and the complexity of the variables affecting a given phenomenon, process, or calculation. It can help obtain not only the most out of experimental data but also scale-up data from a model to a prototype. To do this, one must achieve similarity between the prototype and the model. This similarity may be achieved through dimensional analysis by determining the important dimensionless numbers, and then designing the model and prototype such that the important dimensionless numbers are the same in both.

3.6 Flow Diagrams

The complete design specification for a medium-sized chemical process would cover several hundred pages. It would include diagrams, tables, and discussion of all aspects of the plant, including chemical, environmental, mechanical, electrical, metallurgical, and civil engineering considerations. To understand a complete design, an engineer or scientist must have training and experience in this area.

Any plant whether for the production of canned fruit, vacuum cleaners, sulfuric acid, or potable water may be visualized as a box into which raw materials and energy are fed and from which useful products, waste, and energy emerge. Ordinarily the manufacturing process involves a number of consecutive operations or steps through which the materials in the process pass.

Figure 3.1 is a schematic representation of materials passing into and out of an unidentified process. In this hypothetical three-step process, raw materials *A* and *B* are fed into Step 1; material *C* is drawn off while material *D* is passed to Step 2 for further processing. One cannot tell whether *C* is a useful product or waste from the sketch. In Step 2, it is necessary to combine raw material *E* with *D* in order to produce *F*. In Step 3, material *F* is separated into *G*, *H*, and *I*, ending the process. Again, the figure does not show which of these last three items are useful products. The arrows lettered *A*, *B*, *C*, etc., represent *material streams*, and the sketch is known as a *flow diagram*.

Although nothing has been said about the specific nature of the process, this flow diagram nevertheless conveys a great deal of information; namely, that in this three-step process, three raw materials *A*, *B*, and *E*, are required to produce four products *C*, *G*, *H*, and *I*, and that these seven materials enter and leave the manufacturing process at the points shown on the diagram (Reynolds 1992).

As one might expect, a process flow diagram for a chemical or environmental plant is usually significantly more complex than that for a simple facility. For the latter case, the flow sequence and determinations often reduce to an approach that employs a "railroad" or sequential type of calculation that does not require iterative calculations (Reynolds 1992).

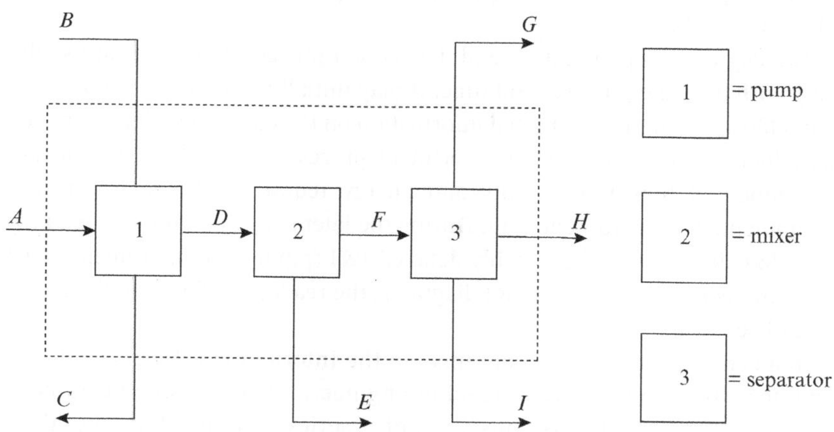

Figure 3.1 Flow diagram for a three-step manufacturing process.

To the environmental engineer, but particularly the chemical engineer, the proceeding flowchart is the key instrument for defining, refining, and documenting a process. The process flow diagram is the authorized process blueprint, the framework for specifications used in equipment designation and design; it is the single, authoritative document employed to define, construct, and operate the process (Kauffman 1992).

There are several essential constituents to a detailed process flowchart beyond equipment symbols and process stream flow lines. These include equipment identification numbers and names; temperature and pressure designations; utility designations; mass, molar, and volumetric flow rates for each process stream; and a material balance table pertaining to process flow lines. The process flow diagram may also contain additional information such as energy requirements, major instrumentation, environmental equipment (and concerns), and physical properties of the process streams. When properly assembled and employed, a process schematic provides a coherent picture of the overall process; it can pinpoint some deficiencies in the process that may have been overlooked earlier in a study, e.g. instrumentation overkill, byproducts (undesirable or otherwise), and recycle needs. Basically, the flowchart symbolically and pictorially represents the interrelation among the various flow streams and equipment, and permits easy calculation of material and energy balances.

Various symbols are universally employed to represent equipment, parts, valves, piping, etc. Some of these are depicted in the schematic in Figure 3.2. Although a significant number of these symbols are used to describe chemical and petrochemical processes, only a few are needed for simpler facilities. These symbols obviously reduce, and in some instances replace, detailed written descriptions of the process. Note that many of the symbols are pictorial, which helps in better describing process components, units, and equipment (Theodore 2014).

The degree of sophistication and details of a process flow diagram usually vary with both the preparer and time. It may initially consist of a simple freehand block diagram with limited information on the equipment; later versions may include line drawings with pertinent process data such as overall and componential flow rates, utility and energy requirements, environmental equipment, and instrumentation. During the later stages of a project, the process flow diagram will be a highly detailed P&I (piping and instrumentation) diagram. For information on P&I diagrams, the reader is referred to the literature (Theodore 2014).

In a sense, process flow diagrams are the international language of the engineer, particularly the practicing engineer. Chemical engineers conceptually view a plant as consisting of a series of interrelated building

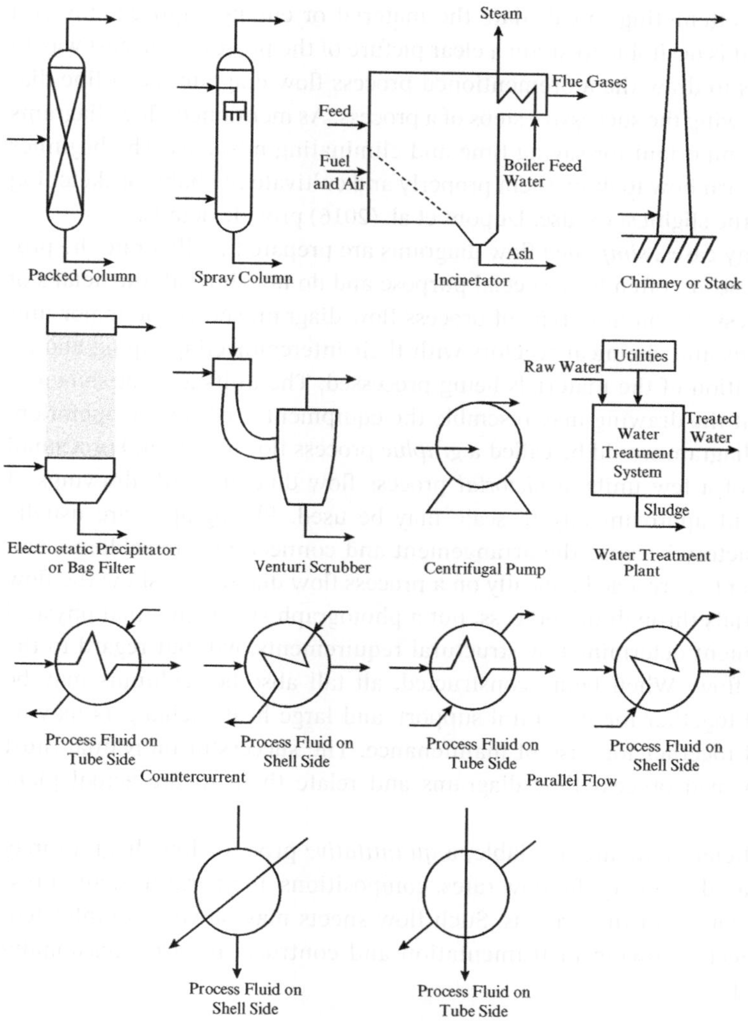

Figure 3.2 Common process flow diagram symbols.

blocks that are defined as *units* or *unit operations* (Abulencia and Theodore 2007; Theodore 2011; Theodore and Ricci 2010). A plant ties together the various pieces of equipment that make up the process. Process flow diagrams follow the successive steps of a process by indicating where the pieces of equipment are located, and the material streams entering and leaving each unit (Felder and Rousseau 1986; Green and Perry 2008; McCabe et al. 1993).

Before attempting to calculate the material or energy requirements of a process, it is desirable to attain a clear picture of the process. The best way to do this is to draw the aforementioned process flow diagram, i.e. a line diagram showing the successive steps of a process. As mentioned, flow diagrams are very important for saving time and eliminating mistakes. The beginner should learn how to draw them properly and cultivate the habit of sketching them at the slightest excuse. Dupont et al. (2016) provide details.

In many cases, *simplified* flow diagrams are prepared to illustrate the process. These are often for a special purpose and do not show all the details of the process. A common type of process flow diagram shows the major unit operations and chemical reactors with their interconnecting piping and an identification of the materials being processed. The units are not shown to scale, but the drawing may resemble the equipment used for the operation. Such a diagram would be called a *graphic* process flow diagram. For a small process of a few units, a *pictorial* process flow diagram with drawings of equipment approximately to scale may be used. Photographs are usually unsatisfactory because the arrangement and connections are seldom clear. Equipment is arranged logically on a process flow diagram to show the flow of materials through the process, but a photograph shows the final physical arrangement determined by structural requirements, without regard to the process flow. When being constructed, all tall absorber columns may be grouped together for structural support, and large heat exchangers may be grouped together for ease of maintenance. The successful individual must learn to read process flow diagrams and relate them to the actual plant layout.

If sufficient data are available, a *quantitative* process flow diagram may be prepared showing the flow rates, compositions, temperatures, and pressures throughout the process. Such flow sheets may be very complicated. Flow sheets showing instrumentation and controls are also occasionally prepared.

3.7 Significant Figures and Approximate Numbers

Significant figures provide an indication of the *precision* with which a quantity is measured or known. The last digit represents, in a qualitative sense, some degree of uncertainty or error. For example, a measurement of 8.32 inches implies that the actual quantity is somewhere between 8.315 and 8.325 inches. This applies to calculated and measured quantities; quantities that are known exactly (e.g. pure integers) have an infinite number of significant figures. Note, however, that there is an upper limit to the accuracy with which physical measurements can be made.

The significant digits of a number are the digits from the first nonzero digit on the left to either (a) the last digit (whether it is nonzero or zero) on the right if there is a decimal point, or (b) the last nonzero digit of the number if there is no decimal point. For example:

370	has 2 significant figures
370.	has 3 significant figures
370.0	has 4 significant figures
28,070	has 4 significant figures
0.037	has 2 significant figures
0.0370	has 3 significant figures
0.02807	has 4 significant figures

Whenever quantities are combined by multiplication and/or division, the number of significant figures in the result should equal the lowest number of significant figures of any of the quantities. In *long* calculations, the final result should be rounded off to the correct number of significant figures. When quantities are combined by addition and/or subtraction, the final result cannot be more precise than any of the quantities added or subtracted. Therefore, the position (relative to the decimal point) of the last significant digit in the number that has the lowest degree of precision is the position of the last permissible significant digit in the result. For example, the sum of 3702., 370, 0.037, 4, and 37 should be reported as 4110 (without a decimal). The least precise of the five numbers is 370, which has its last significant digit in the tens position. The answer should also have its last significant digit in the tens position.

Unfortunately, engineers and scientists rarely concern themselves with significant figures in their calculations. However, it is recommended that the reader attempt to follow the calculation procedure set forth in this section.

In the process of performing engineering calculations, very large and very small numbers are often encountered. A convenient way to represent these numbers is to use scientific notation. Generally, a number represented in scientific notation is the product of a number ($<$ 10 but $>$ or $=$ 1) and 10 raised to an integer power. For example,

$$28{,}070{,}000{,}000 = 2.807 \times 10^{10}$$

and

$$0.000002807 = 2.807 \times 10^{-6}$$

A positive feature of using scientific notation is that only the significant figures need to appear in the number.

When approximate numbers are added or subtracted, the results should also be represented in terms of the least precise number. Since this is a relatively simple rule to master, note that the answer follows the aforementioned rule of precision. Therefore,

$$6.04\text{ L} + 2.8\text{ L} - 4.173\text{ L} = 4.7\text{ L}$$

These numbers have two, one, and three decimal places, respectively. The least precise number (least number of decimal places) is 2.8, a value carried only to the tenths position. Therefore, the answer must be calculated to the tenths position only. Thus, the correct answer is 4.7 L (the last 6 and 7 are dropped from the full answer 4.667, and the result is rounded up to provide 4.7 L).

In multiplication and division of approximate numbers, finding the number of significant digits in the appropriate numbers is used to determine how many digits to keep (i.e. where to truncate). One must understand *significant digits* in order to determine the correct number of digits to keep or remove in multiplication and division problems. As noted earlier in this section, *the digits 1 through 9 are considered to be significant*. Thus, the numbers 123, 53, 7492, and 5 contain three, two, four, and one significant digit, respectively. The digit zero must be considered separately.

Zeros are significant when they occur *between* significant digits. In the following examples, all zeros are significant: 10001, 402, 1.1001, and 500.09 (five, three, five, and four significant figures, respectively). *Zeroes are not significant when they are used as place holders*. When used as a holder, a zero simply identifies where a decimal is located. For example, each of the following numbers has only one significant digit: 1000, 500, 60, 0.09, and 0.0002. In the numbers 1200, 540, and 0.0032, there are two significant digits, and the zeroes are not significant. Once again, when *zeros follow a decimal and are preceded by a significant digit, the zeros are significant*. In the following examples, all zeroes are significant: 1.00, 15.0, 4.100, 1.90, 10.002, and 10.0400. For 10.002, the zeroes are significant because they fall between two significant digits. For 10.0400, the first two zeroes are significant because they fall between two significant digits; the last two zeroes are significant because they follow a decimal and are preceded by a significant digit. Finally, when approximate numbers are multiplied or divided, the result is expressed as a number having the same number of significant digits as the expression in the problem having the least number of significant digits.

When truncating (removing final, unwanted digits), rounding is normally applied to the last digit to be kept. Thus, *if the value of the first digit to be discarded is less than 5, one should retain the last kept digit with no change. If the first digit to be discarded is 5 or greater, increase the last kept digit's value by one.* Assume, for example, only the first two decimal places are to be kept for 25.0847 (the 4 and 7 are to be dropped). The number is then 25.08. Since the first digit to be discarded (4) is less than 5, the 8 is not rounded up. If only the first two decimal places are to be kept for 25.0867 (the 6 and the 7 are to be dropped), it should be rounded to 25.09. Since the first digit to be discarded (6) is 5 or more, the 8 is rounded up to 9.

In some desalination studies, a measurement in gallons or liters may be required. Although a gallon or liter may represent an exact quantity, the measurement instruments that are used are capable of producing approximations only. Using a standard graduated flask in liters as an example, can one determine whether there is exactly 1 liter? Likely not. In fact, one would be pressed to verify that there was a liter within the ± 1/100 of a liter. Therefore, depending upon the instruments used, the precision of a given measurement may vary.

If a measurement is given as 16.0 L, the zero after the decimal indicates that the measurement is precise to within 1/10 L. Given a measurement of 16.00 L, one has precision to the 1/100 L. As noted, the digits following the decimal indicate how precise the measurement is. Thus, precision is used to determine where to truncate when approximate numbers are added or subtracted.

3.8 Generic Problem Solving Techniques

Certain methods of logic and techniques of calculation are fundamental to the solution of many problems, and there is a near-infinite number of methods that can be used for technical problem solving. Words such as creativity, ingenuity, originality, etc., appear in all these approaches. What do they all have in common? They provide a systematic, logical approach to solving problems, and what follows is the authors' definition of a generic approach.

The methodology of solving problems has been discussed by most mathematicians and logicians since the days of Aristotle. Heuristic ("serving to discover") is the term often given to this study of the methods and rules of solving problems. Nearly always, a stepwise approach to the solution is desirable. The broad steps are:

1. Understand the problem.
2. Devise a plan for the solution.
3. Carry out the plan.
4. Look back and check the problem solution.

Details on each of the four points follows.

3.8.1 Understanding the Problem

The first step in the problem solving approach is to understand the problem. Get acquainted with the problem. Read it and attempt to visualize it in its entirety without concerning yourself with details. Be particularly sure that you understand what is known and what is unknown.

Next, organize the available data. Engineering and science data are often insufficient, redundant, or contradictory. The units may have to be converted. The accuracy of the data must also be verified. Sketch a flow sheet. This step will usually facilitate the "get acquainted" operation. If the problem is a simple one, this sketch may consist of nothing more than a small box with arrows. For chemical processes, the diagrams should indicate the amount and composition of all streams entering or leaving the process.

Finally, introduce suitable and consistent symbols for the unknowns. It is usually helpful to adopt a notation which will serve to identify the unknown. This is particularly important when a problem has many unknowns.

3.8.2 Devising a Plan for the Solution

Once the problem is understood, the next step in the problem solution process is to develop a plan to solve it. Memory plays an important part in even the most creative thinking. It is almost impossible to generate ideas or plans with little to no background knowledge. Although mere memory is not enough to create new ideas, such ideas must be based on experience and acquired knowledge. When devising a plan to solve a problem, here are some things to consider:

1. Have I seen a similar or slightly different problem before?
2. Is there a very general equation or theorem which applies to this type of problem?
3. Can the problem be restated or simplified?
4. If I cannot solve the problem, can I solve a related problem?
5. Can part of the problem be solved? What data are lacking to solve the rest of the problem? Did I use all of the data given?

3.8.3 Carrying Out the Plan

Once a plan for the solution has been developed, carry out the plan. Choose a basis of calculation, if applicable. During various stages in the solution of a complex problem it may be necessary to use more than one basis. Also remember that the most brilliant work is useless if it cannot be conveyed clearly to other persons by written or oral presentation.

3.8.4 Looking Back and Checking the Problem Solution

The amount of time it takes to check a problem is usually only a small fraction of the time it takes to solve it. Nearly all solutions to physical problems are amendable to verification. Thus, one should always:

1. Look for order-of-magnitude errors. Can the answer be checked by an approximate solution? Does the answer make sense?
2. Attempt to solve the problem by an alternate method.
3. Can the results be verified?

3.9 Illustrative Examples

Five illustrative examples complement the material presented in this chapter.

3.9.1 Illustrative Example 1

Convert the following temperatures.

20°F to °C, K, and °R

Solution. The following key equations are employed:

$$T(°\text{F}) = 1.8T(°\text{C}) + 32 \tag{3.5}$$

$$T(\text{K}) = T(°\text{C}) + 273 \tag{3.6}$$

$$T(°\text{R}) = T(°\text{F}) + 460 \tag{3.7}$$

$$T(°\text{R}) = 1.8\ T(\text{K}) \tag{3.8}$$

Thus:

Equations 3.5, 3.6, and 3.7 are used as follows:

$$(20°\text{F}) = 1.8\ T(°\text{C}) + 32; (20°\text{F} - 32)/1.8 = T(°\text{C}) = -6.7°\text{C}$$

$$T(\text{K}) = (-6.7°\text{C}) + 273 = 266.3\ \text{K}$$

$$T(°\text{R}) = (20°\text{F}) + 460 = 480°\text{R}$$

3.9.2 Illustrative Example 2

The height of a liquid column of mercury is 2.493 ft. Assume the density of mercury is 848.7 lb/ft^3 and atmospheric pressure is 2116 lb$_f$/ft^2 absolute. Calculate the gauge pressure in lb$_f$/ft^2 and the absolute pressure in lb$_f$/ft^2, psia, mm Hg, and in H_2O.

Solution. Expressed in various units, the standard atmosphere is equal to:

1.0	Atmospheres (atm)
33.91	Feet of water (ft H_2O)
14.7	Pounds force per square inch absolute (psia)
2116	Pounds force per square foot absolute (psfa)
29.92	Inches of mercury (in Hg)
760.0	Millimeters of mercury (mm Hg)
1.013 x 10^5	Newtons per square meter (N/m^2)

The density equation is describing the gauge pressure in terms of the column height and liquid density is:

$$P_g = \rho gh / g_c \tag{3.9}$$

where Pg = gauge pressure, ρ = liquid density, g = acceleration of gravity, h = column height, and g_c = conversion constant. Thus,

$$P_g = \left(848.7 lb / ft^3\right)\left(1\frac{lb_f}{lb}\right)\left(2.493\, ft\right) = 2116\, {}^{lb_f}\!\!\diagup_{ft^2} \text{ gauge}$$

The pressure in lb$_f$/ft^2 absolute is:

$$P_{absolute} = P_g + P_{atmospheric} = 2{,}116 \text{ lb}_f/\text{ft}^2 + 2{,}116 \text{ lb}_f/\text{ft}^2 = 4{,}232 \text{ lb}_f/\text{ft}^2 \text{ absolute}$$

The pressure in psia is;

$$P(\text{psia}) = \left(4{,}232 \text{ lb}_f/\text{ft}^2\right)\left(1 \text{ ft}^2/144 \text{ in}^2\right) = 29\ .4 \text{ psia}$$

The corresponding gauge pressure in psi is:

$$P(\text{psig}) = 29.4 - 14.7 = 14.7 \text{ psig}$$

The pressure in mm Hg is:

$$P(\text{mm Hg}) = (29.4 \text{ psia})(760 \text{ mm Hg})/(14.7 \text{ psia}) = 1{,}520 \text{ mm Hg}$$

Finally, the pressure in in H_2O is:

$$P(\text{in } H_2O) = \left(\frac{29.4 \text{ psia}}{14.7 \text{ psia / atm}}\right)\left(\frac{33.91 \text{ ft } H_2O}{1 \text{ atm}}\right)\left(\frac{12 \text{ in}}{1 \text{ ft}}\right) = 813.8 \text{ in } H_2O$$

The reader should note that absolute and gauge pressures are usually expressed with units of atm, psi, or mm Hg. This statement also applies to partial pressures. One of the most common units employed to describe pressure drop is inches of H_2O, with the notation in H_2O or IWC (inches of water column).

3.9.3 Illustrative Example 3

Given the following data for liquid methanol, convert the heat capacity, thermal conductivity, and viscosity from SI units to English units:

$$\text{Density of reference substance (water)} = 62.4 \text{ lb/ft}^3 \text{@ } 60°\text{F}$$

$$\text{Heat capacity} = 0.61 \text{ cal/(g-°C) @ } 60°\text{F}$$

$$\text{Thermal conductivity} = 0.0512 \text{ cal/(m-s-°C) @ } 60°\text{F}$$

$$\text{Viscosity} = 0.64 \text{ cP @ } 60°\text{F}$$

Solution. Convert the heat capacity of methanol from units of cal/(g-°C) to corresponding English units of Btu/(lb-°F):

$$\left(\frac{0.61 \text{ cal}}{\text{g-°C}}\right)\left(\frac{454 \text{ g}}{1 \text{ lb}}\right)\left(\frac{1 \text{ Btu}}{252 \text{ cal}}\right)\left(\frac{°\text{C}}{1.8°\text{F}}\right) = 0.61 \text{Btu/(lb-°F)}$$

Note that 1.0 Btu/(lb-°F) is equivalent to 1.0 cal/(g-°C). This also applies on a mole basis, i.e. 1 Btu/(lbmol-°F) = 1 cal/(gmol-°C). The usual notation for heat capacity is C_p. In this book, C_p represents the heat capacity on a mole basis, while *cp* indicates heat capacity on a mass basis.

Convert the thermal conductivity of methanol from units of cal/(m-s-°C) to corresponding English units of Btu/(ft-hr-°F):

$$\left(\frac{0.051 \text{ cal}}{\text{m-s-°C}}\right)\left(\frac{1 \text{ Btu}}{252 \text{ cal}}\right)\left(\frac{0.3048 \text{ m}}{1 \text{ ft}}\right)\left(\frac{3{,}600 \text{ s}}{1 \text{ hr}}\right)\left(\frac{°\text{C}}{1.8°\text{F}}\right) = 0.124 \text{Btu/(ft-hr-°F)}$$

The usual notation for thermal conductivity is *k*.

Convert viscosity of methanol from centipoise to lb/(ft-s):

$$(0.64 \text{ cP})\left(\frac{6.72\text{x}10^{-4} \text{ lb / (ft-s)}}{1 \text{ cP}}\right) = 4.3 \text{ x}10^{-4} \text{lb/(ft-s)}$$

The notation for viscosity is typically μ. The kinematic viscosity, ν, is defined by the ratio of viscosity to density, i.e. $\nu = \mu / \rho$ with units of length^2/time. Finally, note that the above physical properties are strong functions of the temperature but weak functions of the pressure. Interestingly, the viscosity of a gas increases with increasing temperature, while the viscosity of a liquid decreases with an increase in temperature. These properties will be revisited in Chapter 4.

3.9.4 Illustrative Example 4

Consider the manufacture of ammonia from natural gas, which can be considered to be pure methane. In the first stage of the process, natural gas is partially oxidized in the presence of steam to produce a synthesis gas containing hydrogen, nitrogen, and carbon dioxide. The carbon dioxide is removed. The remaining mixture of hydrogen and nitrogen is reacted over a catalyst at high pressure to form ammonia. The two reactions for the process, neglecting side reactions and impurities, are as follows:

$$7CH_4 + 10\,H_2O + 8\,N_2 + 2\,O_2 \rightarrow 7\,CO_2 + 8\,N_2 + 24\,H_2$$

$$N_2 + 3\,H_2 \rightarrow 2\,NH_3$$

Sketch a flow diagram for this process.

Solution. See Figure 3.3 for a simplified process flow diagram for this ammonia manufacturing process.

3.9.5 Illustrative Example 5

All measurements and calculations are subject to two broad classes of errors: *determinate* and *indeterminate*. Any error that is discovered and allowed for in magnitude and sign in the form of a correction allowing for its effect is a

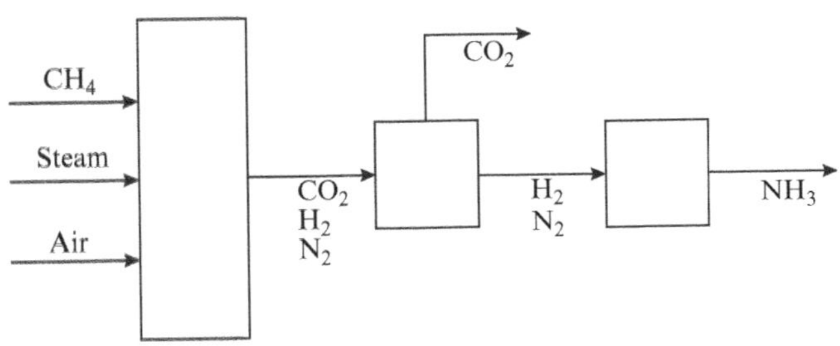

Figure 3.3 Simplified process flow diagram for Illustrative Example 4.

"determinate error." All errors that either cannot be or are not properly allowed for in magnitude and in sign are known as "indeterminate errors." A particularly important class of indeterminate errors is that of accidental errors. Qualitatively discuss these two error types.

Solution. Determinate errors are systematic errors that can be discovered and corrected. Determinate errors include *instrument errors*, i.e. calibration errors, power fluctuations, *method errors*, i.e. sampling problems, inconsistent recovery, and *personal errors*, i.e. errors that occur when measurements require judgment, color interpretation, etc. Determinate errors can be corrected via proper instrument calibration, method development, or training. To illustrate the nature of indeterminate errors, consider the very simple and direct measurement of temperature. Suppose that several independent readings are made and that the temperature varies by 0.1°F. When the results of the different readings are compared, it may be found that even though they have been performed very carefully, they may differ from each other by several tenths of a degree. Experience has shown that such deviations are inevitable in all measurements and that they result from small unavoidable random errors of observation due to the sensitivity of measuring instruments and the keenness of the sense of perception. Such errors are due to the combined effect of a large number of undetermined causes and they can be defined as accidental or indeterminate errors (Prochaska and Theodore 2018; Theodore and Behan 2018).

3.10 Problems

1. The suspended particulate concentration in a brackish water source has been determined to be 27.6 mg/L. Convert this value to units of μg/L, g/L, lb/ft^3, and lb/gal.
2. As part of a plant's proposed desalination process cold seawater is mixed with recycled water from a dehumidifier (D), and the mixture then passes through a waste heat boiler (W), which uses air (A) as the heat transfer medium. The hot air and water mixture then pass through a humidifier (H), in which humid air is generated and then passed through the dehumidifier (D) and condensed to form potable water. Prepare a simplified process flow diagram for this process.
3. Precision and accuracy are two terms used to describe measurements and measurement error. Discuss the difference between these two terms in the context of quantitative parameter measurements.
4. A common English volumetric flowrate unit in water and wastewater treatment applications in the United States is million gallons per day, MGD. A desalination plant is being designed for an average flow of 25

MGD. Determine the plant's capacity in other common volumetric flow units shown in Table 3.4 of gpm, cfs, and m^3/d.

5. The concentration of many organic chemicals at which they are potentially toxic or hazardous to human health or the environment is often very low. Many emerging contaminants, for example PFAS (per- and polyfluoroalkyl substances) are very persistent in the environment and in the human body, and there is evidence that exposure to PFAS can lead to adverse human health effects. The drinking water health advisory for two of these PFAS compounds, PFOA and PFOS, is 70 parts/trillion or 70 ng/L. For these two PFAS compounds, determine the following: (1) what is their daily dose at the health advisory limit if the average adult water consumption rate is 1 L/d? (2) if a desalination plant is treating combined brackish water and recharged treated wastewater at a ratio of 50 to 1 on a volume basis, the concentration of PFOA + PFOS in the treated wastewater is 8 ng/L, and the plant is expected to provide 98% PFAS removal, what is the margin of exposure (MOE = NOEL/Actual Dose, g/d) if the health advisory limit dose from (1) is assumed to be the no-observed-adverse-effect level (NOEL)? (3) If the MOE is recommended to be > 100, comment on the MOE for this facility and indicate any corrective action or modified operating conditions that should be implemented to ensure the safety of this produced water supply.

References

Abulencia, P. and Theodore, L. (2007). *Fluid Flow for the Practicing Chemical Engineer*. Hoboken, NJ: John Wiley & Sons.

Dupont, R.R., Ganesan, K., and Theodore, L. (2016). *Pollution Prevention, Industrial Ecology, Green Science and Engineering, and Sustainability*. Boca Raton, FL: CRC Press/Taylor and Francis Group.

Felder, R. and Rousseau, R. (1986). *Elementary Principles of Chemical Processes*, 2nd Edition. Hoboken, NJ: John Wiley & Sons.

Green, D.W. and Perry, R.H. (2008). *Perry's Chemical Engineer's Handbook*, 8th Edition. New York, NY: McGraw-Hill.

Kauffman, D. (1992). *Process Synthesis and Design, a Theodore Tutorial*. East Williston, NY: Theodore Tutorials. Originally published Research Triangle Park, NC: USEPA/APTI.

McCabe, W., Smith, J., and Harriot, P. (1993). *Unit Operations of Chemical Engineering*, 5th Edition. New York, NY: McGraw-Hill.

Prochaska, C. and Theodore, L. (2018). *Introduction to Mathematical Methods for Environmental Engineers and Scientists*. Salem, MA: Scrivener/Wiley.

Reynolds, J. (1992). *Material and Energy Balances, a Theodore Tutorial*. East Williston, NY: Theodore Tutorials. Originally published Research Triangle Park, NC: USEPA/APTI.

Theodore, L. (2011). *Heat Transfer for the Practicing Chemical Engineer*. Hoboken, NJ: John Wiley & Sons.

Theodore, L. (2014). *Chemical Engineering: The Essential Reference*. New York, NY: McGraw-Hill.

Theodore, L. and Behan, K. (2018). *Introduction to Optimization for Chemical and Environmental Engineers*. Boca Raton, FL: CRC Press/Taylor and Francis Group.

Theodore, L. and Ricci, F. (2010). *Mass Transfer Operations for the Practicing Engineer*. Hoboken, NJ: John Wiley & Sons.

4

Physical, Chemical, and Biological Properties of Materials

4.1 Introduction

Physical, chemical, and biological properties of materials are important considerations in any study of desalination. A substance may exhibit certain characteristics under one particular set of reaction or process conditions of temperature, pressure, and composition. However, these conditions are often changed when an attempt is made to implement a desalination strategy. To promote a better understanding of such effects, many of which are covered in this chapter, certain key physical, chemical, and biological properties are briefly reviewed in this chapter. The last two sections examine the properties of mixtures and several key process concepts as they relate to material properties. An application section provides seven illustrative examples related to physical, chemical, and biological properties of materials relevant to desalination processes.

4.2 Fundamental Definitions

This section is concerned with key terms and definitions that are used in engineering practice. Since desalination is an important subject that finds application in engineering, the understanding of fundamental definitions is important to the practicing engineer. Four terms are briefly reviewed before proceeding to specific properties of materials. It should also be noted that the same substance in its different phases may have various properties that have different orders of magnitude. As an example, heat capacity values are low for solids, high for liquids, and usually intermediate for gases.

4.2.1 Fluids

For the purpose of this desalination text, a fluid may be defined as a substance that does not permanently resist distortion. An attempt to change the shape of

Introduction to Desalination: Principles, Processes, and Calculations, First Edition. Louis Theodore and R. Ryan Dupont.

a mass of fluid will result in layers of fluid sliding over one another until a new shape is attained. During the change in shape, shear stresses (forces parallel to a surface) will result, the magnitude of which depends upon the viscosity (to be discussed shortly) of the fluid and the rate of sliding. However, when a final shape is reached, all shear stresses will have disappeared. Thus, a fluid at equilibrium is free from shear stresses. This definition applies for both liquids and gases.

4.2.2 Temperature

Whether in the gaseous, liquid, or solid state, all molecules possess some degree of kinetic energy; i.e. they are in constant motion-vibrating, rotating, or translating. The kinetic energies of individual molecules cannot be measured, but the combined effect of these energies in a very large number of molecules can be. This measurable quantity is known as temperature; it is a macroscopic concept only (i.e. it does not exist on the molecular level).

Temperature can be measured in many ways; the most common method makes use of the expansion of mercury with increasing temperature. (On process equipment, however, thermocouples or thermistors are more commonly employed.) The two most commonly used temperature scales are the Celsius and Fahrenheit scales. The Celsius scale is based on the boiling and freezing points of water at 1 atm pressure; to the former, a value of 100°C is assigned, and to the latter, a value of 0°C is assigned. On the older Fahrenheit scale, the corresponding temperatures are 212°F and 32°F, respectively. Equations 4.1 and 4.2 show the conversion from one scale to the other.

$$^{\circ}\mathrm{F} = 1.8\left(^{\circ}\mathrm{C}\right) + 32 \tag{4.1}$$

$$^{\circ}\mathrm{C} = \left(^{\circ}\mathrm{F} - 32\right)/1.8 \tag{4.2}$$

The volume of a gas would theoretically be zero at a temperature of approximately −273°C or −460°F. This temperature, which has become known as absolute zero, is the basis for the definition of two absolute temperature scales, the Kelvin (K) and Rankine (°R) scales. The former scale is defined by shifting the Celsius scale by 273°Celsius so that 0 K is equal to −273°C. Equation 4.3 shows this relationship.

$$\mathrm{K} = ^{\circ}\mathrm{C} + 273 \tag{4.3}$$

The Rankine scale is defined by shifting the Fahrenheit scale 460°Fahrenheight so that (Equation 4.4):

$$^{\circ}\mathrm{R} = ^{\circ}\mathrm{F} + 460 \tag{4.4}$$

As one might expect, temperature is an important parameter in desalination applications.

4.2.3 Pressure

In the gaseous state, molecules possess a high degree of translational kinetic energy, which means that they are able to move quite freely throughout the body of a gas. If the gas is in a container, the molecules are constantly bombarding the container walls. The macroscopic effect of this bombardment by a tremendous number of molecules, enough to make the effect measurable, is called pressure. The basic units of pressure are force per unit area.

A number of units are used to express a pressure measurement. Some are based on a force per unit area: for example, pound (force) per square inch (psi) or dyne per square centimeter (dyne/cm^2). Others are based on fluid height such as inches of water (in H_2O) or millimeters of mercury (mm Hg). Units such as these are convenient when the pressure is indicated by a difference between levels of a liquid, as in a manometer or barometer. Barometric pressure is a measure of the ambient air pressure. Standard barometric pressure is 1 atm and is equivalent to 14.696 psi and 29.921 mm Hg.

Measurements of pressure by most gauges indicate the difference in pressure either above or below that of the surrounding atmosphere. The system pressure is greater than atmospheric if the gauge (reading) is positive; if the gauge (reading) is negative, it is less than atmospheric pressure. The term vacuum designates this negative gauge pressure. Gauge pressures are usually identified by the letter g after the pressure unit (e.g. psig). Since gauge pressure is the pressure relative to the prevailing atmospheric pressure, the sum of the two gives the absolute pressure and is often indicated by the letter a after the unit (e.g. psia) as indicated in Equation 4.5:

$$P_a = P + P_g \tag{4.5}$$

A pressure of zero psia is the lowest possible pressure theoretically achievable, i.e. a perfect vacuum. Pressure, like temperature, can have a significant impact on any process change.

4.2.4 Moles and Molecular Weight

An atom consists of protons and neutrons in a nucleus surrounded by electrons. An electron has such a small mass relative to that of the proton and neutron that the weight of the atom, the atomic weight, is approximately equal to the sum of the weights of the particles in its nucleus. Atomic weight may be expressed in atomic mass units per atom or in grams per gram-atom. The number of atoms in 1 gram-atom, called Avogadro's number, is 6.02×10^{23}.

The molecular weight (MW) of a compound is the sum of the atomic weights of the atoms that make up one molecule of the material. Units of grams per gram-mole (g/gmol) are used for molecular weight. One gmol contains 6.02 x 10^{23} molecules. In the English system, one pound-mole (lbmol) contains $454 \times 6.02 \times 10^{23}$ molecules.

4.3 Physical Properties

4.3.1 Density

The *density* of a substance is the ratio of its mass to its volume and may be expressed in units of pounds per cubic foot (lb/ft^3) or kilograms per cubic meter (kg/m^3). For solids, density can be determined easily by placing a known mass of a substance in a liquid and measuring the displaced volume. The density of a liquid can be measured by weighing a known volume of the liquid in a graduated cylinder. For gases, the ideal gas law (to be discussed later) can be used to calculate the density from the temperature, pressure, and molecular weight of the gas.

The specific gravity (SG) (Equation 4.6) is the ratio of the density of a substance (ρ) to the density of a reference substance at a specific condition, i.e.

$$\text{SG} = \rho / \rho_{ref} \tag{4.6}$$

The reference most commonly used for solids and liquids is water. At its maximum density, which occurs at 4°C, this reference density is 1.000 g/cm^3, 1000 kg/m^3, or 62.4 lb/ft^3.

4.3.2 Viscosity

Viscosity is a property associated with a fluid's resistance to flow; more precisely, this property accounts for the energy losses resulting from shear stresses that occur between different portions of the fluid, moving at different velocities. The absolute viscosity μ has units of mass per length-time; the fundamental unit is the poise, which is defined as 1 g/cm-s. Viscosities are frequently given in centipoises (0.01 poise), the abbreviation for which is cP. In English units, the absolute viscosity is expressed as pounds (mass) per foot-second (lb/ft-s). The absolute viscosity depends primarily on temperature and, to a lesser degree, on pressure. The kinematic viscosity, υ, the absolute viscosity divided by the density of the fluid, is useful in certain fluid flow applications. The units for this quantity are length squared per time (e.g. ft^2/s). Because fluid viscosity changes rapidly with temperature, a numerical value of viscosity has no significance unless the temperature is specified.

4.3.3 Heat Capacity

The *heat capacity* of a substance is defined as the quantity of heat required to raise the temperature of a unit mass by 1°. The term *specific heat* is frequently used in place of *specific heat capacity*. This is not strictly correct because, traditionally, specific heat has been defined as the ratio of the heat capacity of a substance to the heat capacity of water. However, since the specific heat of water is approximately 1 cal/g-°C or 1 Btu/lb-°F, the term specific heat has come to imply heat capacity. For gases, the addition of heat to cause a 1° temperature rise may be accomplished either at constant pressure or at constant volume. Since the amounts of heat necessary are different for the two cases, subscripts are used to identify which heat capacity is being used – C_p for constant pressure, C_v, for constant volume. This distinction does not have to be made for liquids and solids, since there is little difference between the two. Values of heat capacity are available in the literature (e.g. Theodore and Reynolds 1992; Theodore 2014) or online, e.g. The Engineering Toolbox, www.EngineeringToolBox.com.

4.3.4 Thermal Conductivity

Experience has shown that when a temperature difference exists across a solid body, heat energy will flow from the region of high temperature to that of low temperature until thermal equilibrium is reached. This mode of heat transfer, in which vibrating molecules pass kinetic energy through a solid, is called conduction. Liquids and gases also transport heat in this fashion. Thermal conductivity provides a measure of how fast heat flows through a substance, and is defined as the amount of heat that flows in a unit of time through a unit surface area of unit thickness as a result of a unit difference in temperature. Typical units for thermal conductivity are Btu/h-ft-°F.

4.3.5 Diffusivity

The *diffusion coefficient*, or *diffusivity*, D, is a measure of the rate of transfer of one substance through another by molecular diffusion. The diffusivity for a component A migrating through a solute B, for example, is given in Equation 4.7 by the ratio of the flux J_A (mass transferred per unit time through a unit area) to its concentration gradient dC_A / dz,

$$D = -J_A / \left(dC_A / dz\right) \tag{4.7}$$

where C_A is the concentration of Component A, and z is the diffusion length. The negative sign indicates that diffusion occurs in the direction of decreasing concentration. Diffusivity is a function of temperature, pressure, concentration, phase, and the number and nature of the other components within a mixture.

4.3.6 Vapor Pressure

Vapor pressure is an important property of liquids and, to a much lesser extent, of solids. If a liquid is allowed to evaporate in a confined space, the pressure of the vapor phase increases as the amount of vapor within the space increases. If there is sufficient liquid present, the pressure in the vapor space eventually comes to equal exactly the pressure exerted by the liquid at its own surface. At this point, a dynamic equilibrium exists in which vaporization and condensation take place at equal rates, and the pressure in the vapor space remains constant. This gas pressure exerted at equilibrium is called the vapor pressure of the liquid. Solids, like liquids, also exert vapor pressure. Evaporation of solids (sublimation) is noticeable only for the few solids characterized by materials with an appreciable vapor pressure.

4.3.7 Boiling Point

The *boiling point* is the temperature at which the vapor pressure of a liquid is equal to its surrounding pressure. Since the vapor pressure remains constant during boiling, so does the liquid temperature. The higher the system pressure, the higher the temperature must be to induce boiling. The boiling point is specific to each individual substance and is a strong function of pressure.

4.3.8 Freezing Point

The *freezing point* is the temperature at which a liquid and solid state can coexist in equilibrium at a given pressure. At this point, the rate at which the substance leaves the solid state equals the rate at which it leaves the liquid state.

4.3.9 Surface Tension

An interface forms when a liquid comes in contact with another fluid. At the surface, the molecules are more densely packed than those within the fluid. This results in surface tension effects, an interfacial phenomenon. The surface tension coefficient, σ, is the force per unit length of the circumference of the interface, or the energy per unit area of the interface area. The surface tension for water is provided in the literature (Santoleri et al. 2002).

4.4 Chemical Properties

4.4.1 Flammability (or Explosion) Limit

Flammability limits for a flammable gas define the concentration range of a gas–air mixture within which an ignition source can start a self-propagating reaction. The minimum and maximum fuel concentrations in air that will produce a self-sustaining reaction under given conditions are called the lower flammability limit (LFL) and the upper flammability limit (UFL), respectively. The abbreviations LEL and UEL, for lower and upper explosive limits, are sometimes used. The range of flammability limits becomes greater when the ignition energy is higher, moving the UFL to a higher concentration. The flammability limits also increase when the initial pressure and/or initial temperature at which the ignition source is activated increases. In addition, the increase in pressure and temperature will increase the rate at which a flame propagates through a gas. Changing the oxygen content or adding an inert gas to the gas-air mixture can significantly alter flammability limits. The heat capacity of the diluent, or inert gas, plays a role in flammability because the diluent will act as a heat sink. Thus, carbon dioxide is a better diluent than nitrogen because it has a higher heat capacity. The flammability limits also increase for drier mixtures. Tables of flammability limits of flammable gases are available in the literature, for example, Theodore et al. (1988) or Theodore and Dupont (2012) or online, e.g. at The Engineering Toolbox, www.EngineeringToolBox.com.

The flammability of a gas mixture (Equations 4.8 and 4.9) can be calculated by using Le Chatelier's law given the flammability of the gas components:

$$LFL_{mix} = 1/\left[\left(f_1/LFL_1\right)+\left(f_2/LFL_2\right)+\ldots++\left(f_n/LFL_n\right)\right] \tag{4.8}$$

$$UFL_{mix} = 1/[\left(f_1/UFL_1\right)+\left(f_2/UFL_2\right)+\ldots++\left(f_n/UFL_n\right) \tag{4.9}$$

where f_1, f_2, ..., f_n = volume or mole fraction of each of the n components; LFL_{mix}, UFL_{mix} = mixture lower and upper flammability limits respectively, in volume or mole fraction; LFL_1, LFL_2,...., LFL_n = component lower flammability limits in volume or mole fraction; and UFL_1, UFL_2,...., UFL_n = component upper flammability limits in volume or mole fraction.

If data are not available for a particular gas mixture, it is possible to estimate the flammability limit by taking data for a similar material and applying Equation 4.10:

$$LFL_A = \left(M_B/M_A\right) LFL_B \tag{4.10}$$

where M_A and M_B are the relevant molecular weights of Components A and B, respectively.

4.4.2 Flash Point

The *flash point* of a flammable liquid is defined as the temperature at which the vapor of the liquid is the same as the vapor pressure corresponding to the lower flammability limit concentration. The three major methods of measuring the flash point are:

1. The Cleveland open cup method,
2. The Penskey-Martens closed cup tester, and
3. The tag closed cup method.

These are experimental tests that measure the lowest temperature at which application of a test flame causes the vapor overlying the sample to ignite. Tables of flash points for selected substances are available in the literature (e.g. Theodore and Dupont 2012) or on line, e.g. The Engineering Toolbox, www.EngineeringToolBox.com.

Since liquids are condensed and fuel vapor (not the liquid fuel) is the ignition participant, the ignition concepts developed for gases often apply to liquids as well. In general, liquid reactions are inclined to be more sluggish and to have longer time lags than gaseous reactions. Time is required for the liquid to vaporize so that it can mix with air and become ignited.

The physical properties of a flammable solid, such as hardness, texture, waxiness, particle size, melting point, plastic flow, thermal conductivity, and heat capacity, impart a wide range of characteristics to the flammability of solids. A solid ignites by first melting and then producing sufficient vapor, which in turn combines with air to form a flammable mixture.

4.4.3 Autoignition Temperature (AIT)

The *autoignition temperature (AIT)*, or the maximum spontaneous ignition temperature, is defined as the minimum temperature at which combustion occurs in a bulk gas mixture when the temperature of a flammable gas-air mixture is raised in a uniformly heated apparatus. The AIT represents a threshold below which chemicals and combustibles can be handled safely. The AITs of selected substances are available in the literature (Theodore and Dupont 2012; Theodore et al. 1988) or online, e.g. The Engineering Toolbox, www.engineeringtoolbox.com/fuels-ignition-temperatures-d_171.html.

The AIT is strongly dependent on the nature of hot surfaces. The AIT may be reduced by as much as 100 to 200°C when surfaces are contaminated by dust.

When the temperature of a flammable mixture is raised to or above the AIT, ignition is not instantaneous. Most notably in liquids, there is a finite delay before ignition takes place, i.e. a lapse between the time a flammable mixture reaches its flame temperature and the first appearance of a flame.

4.4.4 Heat (Enthalpy) of Reaction

Many chemical reactions evolve or absorb heat. When applying energy balances in technical calculations, the *heat (enthalpy) of reaction* is often indicated in mole units so that they can be directly applied to show its chemical change.

To simplify the presentation that follows, examine Equation 4.11:

$$aA + bB \rightarrow cC + dD \tag{4.11}$$

The heat (enthalpy) of reaction, ΔH, is given by Equation 4.12:

$$\Delta H = c\left(\Delta H_f\right)_C + d\left(\Delta H_f\right)_D - a\left(\Delta H_f\right)_A - b\left(\Delta H_f\right)_B \tag{4.12}$$

Thus, the heat (enthalpy) of a reaction, $\Delta H°$, is obtained by taking the difference between the heat (enthalpy) of formation, ΔH_f, of products and reactants. If the heat (enthalpy) of reaction is negative (exothermic), as is the case of most combustion reactions, then energy is liberated due to chemical reaction. Energy is absorbed if ΔH is positive (endothermic). The standard heat of reaction, $\Delta H°$, is given by Equation 4.13:

$$\Delta H° = c\left(\Delta H_f°\right)_C + d\left(\Delta H_f°\right)_D - a\left(\Delta H_f°\right)_A - b\left(\Delta H_f°\right)_B \tag{4.13}$$

Thus, the standard heat (enthalpy) of a reaction, $\Delta H°$, is obtained by taking the difference between the standard heat (enthalpy) of formation, $\Delta H_f°$, of products and reactants. Once again, if the standard heat (enthalpy) of reaction of formation is negative (exothermic), as is the case of most combustion reactions, then energy is liberated due to chemical reaction. Energy is absorbed if $\Delta H°$ is positive (endothermic).

Tables of standard enthalpies of formation combustion and reaction are available in the literature (e.g. Theodore and Reynolds 1992; Theodore et al. 2009) or online, e.g. The Engineering Toolbox, www.EngineeringToolBox.com, for a wide variety of compounds. It is important to note that these are useless unless the stoichiometric equation and the state of reactants and products are included.

Enthalpy of reaction and standard enthalpy of reaction are not always employed in engineering reaction/combustion calculations. The two other terms that have been used are the gross (or higher) heating value and the net (or lower) heating value. These are discussed later in this section.

4.4.5 Heat (Enthalpy) of Combustion

The *heat (enthalpy) of combustion* at temperature T is defined as the enthalpy change during the chemical reaction where 1 mole of material is burned in oxygen, where all reactants and products are at temperature T. This quantity finds extensive application in calculating enthalpy changes for incineration reactions and is often given in the literature for a standard state temperature of 60°F (16°C). Although much of the literature data on standard heats of reaction are given for 25°C, there is little sensible enthalpy change between these two temperatures, and the two sets of data may be considered compatible. Thus, the standard heat enthalpy of combustion is obtained by taking the difference between the standard heat of formation, ΔH_f°, of the products and that of the reactants. If the standard heat of combustion is negative (exothermic), as is the case with most combustion/incineration reactions, energy is liberated as a result of the reaction. Energy is absorbed if enthalpy of combustion is positive (endothermic). Extensive standard heat of formation and standard enthalpy of combustion data at 298 K are provided in the literature (e.g. Theodore and Reynolds 1992; Theodore et al. 2009) or online, e.g. UC Davis ChemWiki, http://chemwiki.ucdavis.edu/Reference.

4.4.6 Gross Heating Value

The heating value of a waste and/or fuel is one of its most important chemical properties. This value represents the amount of heat evolved in the complete reaction (usually combustion) of a given quantity of the waste and/or fuel.

The *gross heating value* or *higher heating value (HHV)* represents the enthalpy change or heat released when a compound is stoichiometrically combusted at 60°F, with the final (flue) products at 60°F and any water present in the liquid state. Stoichiometric combustion requires that no oxygen be present in the flue gas following combustion of the starting fuel.

4.4.7 Net Heating Value

Due to the varying amounts of water that may be produced in a combustion reaction, a second method for expressing heating value is also in use, the *net heating value*. The net heating value is similar to the gross heating value except that it is calculated with the produced water in the vapor state. The net heating value is also known as the *lower heating value (LHV)*. The *LHV* is equal to the *HHV* minus the heat of vaporization of the water in the final combustion gas.

4.4.8 Theoretical Adiabatic Flame Temperature (TAFT)

Most combustion reactions do not operate at stoichiometric or 0% excess air. Incomplete combustion and high carbon monoxide levels would result under this condition. If all the heat liberated by the reaction goes into heating up the products of combustion (the flue gas), the temperature achieved is defined as the *flame temperature*. If the reaction process is conducted adiabatically, with no heat transfer loss to the surroundings, the final temperature achieved in the flue gas is defined as the *adiabatic flame temperature*. If the combustion process is conducted with theoretical or stoichiometric air (0% excess air), the resulting temperature is defined as the *theoretical flame temperature*. Theoretical or stoichiometric air is defined as that exact amount of air required to completely react with a compound to produce oxidized end products. Any air in excess of this stoichiometric amount is defined as excess air.

If the reaction is conducted both adiabatically and with stoichiometric air, the resulting temperature is defined as the *theoretical adiabatic flame temperature (TAFT)*. It represents the maximum temperature the products of combustion (flue) can achieve if the reaction is conducted both stoichiometrically and adiabatically. For this condition, all the energy liberated from combustion at or near standard conditions ($\Delta H_C{}^{\circ}$ and/or $\Delta H^{\circ}{}_{298}$) appears as sensible heat in raising the temperature of the flue products, ΔH_P as is shown in Equation 4.14:

$$\Delta H_C{}^{\circ} + \Delta H_P = 0 \tag{4.14}$$

where $\Delta H_C{}^{\circ}$ = standard heat of combustion at 25°C; and ΔH_P = enthalpy change of the products as the temperature increases from 25°C to the theoretical adiabatic flame temperature.

4.5 Property Estimation

No discussion of materials properties would be complete without a brief discussion of methods to analytically estimate these physical/chemical properties. This section provides references detailing procedures that will allow the practitioner to estimate key physical and chemical properties of materials. Although the scientific community has traditionally resorted to experimental methods to accurately determine the aforementioned properties, that option may very well not be available when dealing with new materials.

Predictive methods, albeit traditional ones, may be the only option available to obtain a first estimate of these properties. It should be noted that significant errors may be involved since extrapolating (or extending) satisfactory estimation procedures at the macroscale level may not always be reasonable. Notwithstanding these concerns, procedures to estimate some of the key physical and chemical properties in thermodynamics that follow, are available in the literature (Chopey 1994; Bird, Stewart, and Lightfoot 2002; Smith, Van Ness, and Abbott 2005; Green and Perry 2008).

1. Vapor pressure
2. Latent enthalpy
3. Critical properties
4. Viscosity
5. Thermal conductivity
6. Heat capacity

References (Smith, Van Ness, and Abbott 2005; Green and Perry 2008) are somewhat complementary, but each provides extensive information on this topic. This includes equations and procedures for several other properties not listed earlier. The interested reader should check these references for more details.

Is proper estimation important? Absolutely. As indicated earlier, there are times and situations when experimental procedures cannot be implemented. For this scenario, one can turn to theoretical and semi-theoretical methods and equations to obtain first estimates of important property values.

One could argue that the present procedures available to estimate the properties of materials are based on questionable approaches. Nonetheless, the traditional methods available for property estimation either may be applicable or may suggest alternative theoretical approaches.

4.6 Biological Properties

4.6.1 Biochemical Oxygen Demand (BOD or BOD5)

Over the years, many different processes have been developed for the treatment of wastewater. Many, including most biological processes, are based on technology originally developed for the treatment of municipal wastewater. One of the key indicator parameters of a wastewater, whether municipal or industrial, is its oxygen demand, i.e. the amount of oxygen needed to convert the biodegradable organic materials in a wastewater to oxidized end products by the microorganisms in a wastewater treatment plant. *Biochemical Oxygen Demand (BOD)* is the amount of oxygen used by microbes to oxidize waste materials, usually measured for a standardized 5-day period, at 20°C, in the dark. BOD_5 is the designation used for this biochemical oxygen demand of waste expressed over a 5-day incubation period and has units of parts per million (ppm) or milligrams per liter (mg/L). The end products of this biochemical oxidation reaction are carbon dioxide, water, partially oxidized organic materials, and more microbes. The BOD_5 test is used for wastewater treatment plant process and compliance monitoring and is a primary variable in evaluating the quantity of biodegradable organic material in a sample.

4.6.2 Ultimate Oxygen Demand (UOD or BODu)

Ultimate oxygen demand (UOD or BODu) is a calculated oxygen demand that assumes oxidation of all species to their most highly oxidized, stable form, e.g. CO_2, H_2O, etc. without any remaining, partially oxidized organic materials. Thee BODu is generally defined as the cumulative oxygen consumption measured over a 20-day period, is also often noted as BOD_{20}, and has units of ppm or mg/L. For normal domestic wastewater, the BOD_5 represents approximately 68% of the BODu, and the more rapid BOD_5 measurement is used to predict the BODu that will occur once the waste is released into the environment.

4.6.3 Chemical Oxygen Demand (COD)

Chemical oxygen demand (COD) is the amount of oxygen, expressed as ppm or mg/L, consumed under high temperature, low pH and strong oxidant, potassium permanganate, conditions in the rapid (3 h) chemical reaction of organic material in a waste stream. While strictly not a biological property of a material, this number is particularly relevant to industrial wastes where toxicity to biological systems could be suspected. If a waste stream has a high COD and no BOD, waste toxicity should be investigated. The end products of this chemical oxidation reaction are carbon dioxide, water, and other inorganic species.

The COD should generally always be larger than the BODu because the COD test will chemically oxidize many organic chemicals that are not readily biodegradable. The BOD_5 of a waste is always much less than the COD, due both to the inability of the microbes to completely oxidize materials that are chemically oxidizable, and to due to the incorporation of some of the organic material into the microbial cell mass, which results in the production of cells and not a measured oxygen consumption.

4.7 Properties of Mixtures

4.7.1 pH

An important chemical property of an aqueous solution is its *pH* value, which indicates the acidity or basicity of the substance. In a neutral solution, i.e. pH = 7.0, such as pure water, the hydrogen (H^+) and hydroxyl (OH^-) ion concentrations are equal. At ordinary temperatures, this concentration can be described using Equation 4.15:

$$\left[H^{+}\right]=\left[OH^{-}\right]=10^{-7}\,g\text{-}ion\,/\,L \tag{4.15}$$

where $[H^+]$ = hydrogen ion concentration, and $[OH^-]$ = hydroxyl ion concentration. The notation g-ion is the unit for gram-ion, which represents Avogadro's number (6.02 x 10^{23}) of ions.

In all aqueous solutions, whether neutral, basic, or acidic, a chemical equilibrium or balance is established between these two concentrations, so that as shown in Equation 4.16:

$$K_{eq}=\left[H^{+}\right]\left[OH^{-}\right]=10^{-14} \tag{4.16}$$

where K_{eq} = equilibrium constant. The numerical value for K_{eq} given earlier holds for room temperature and only when the concentrations are expressed in g-ion/L. In acid solutions, $[H^+]$ is > $[OH^-]$ in basic solutions, $[OH^-]$ predominates.

The pH is a direct measure of the hydrogen ion concentration and is defined by Equation 4.17:

$$pH=-log[H^{+}] \tag{4.17}$$

Thus, an acidic solution is characterized by a pH below 7 (the lower the pH, the higher the acidity), a basic solution by a pH above 7, and a neutral solution by a pH of 7.

For example, for a pH of 1.0,

$$\text{pH}=1.0=-log\left[H^{+}\right];\left[H^{+}\right]=10^{-1}=0.1\text{ g-ion/L; }10^{-13}\text{g-ion/L}$$

It should be pointed out that the equations employed are not the exact definition of pH but are a close approximation of it. Strictly speaking, the *activity* of the hydrogen ion, α_{H^+}, rather than the ion concentration, $\left[H^{+}\right]$, belongs in the equations.

4.7.2 Partial Pressure

In engineering practice, mixtures of gases are more often encountered than single or pure gases. The ideal gas law is based on the number of molecules present in the gas volume; the nature of molecules is not of significance. This law applies equally well to mixtures and to pure gases. Since pressure is caused by gas molecules colliding with the walls of a container, it seems reasonable that the total pressure of a gas mixture is made up of pressure contributions from each of the component gases. These pressure contributions are called *partial pressures*. Dalton defined the partial pressure of a component as the pressure that would be exerted if the same mass of the component gas occupied the same total volume alone at the same temperature as the mixture. The sum of these partial pressures equals the total pressure as indicated in Equation 4.18:

$$P = P_1 + P_2 + P_3 + \ldots + P_n = \sum_{i=1}^{n} P_i \tag{4.18}$$

For ideal gases, partial pressure is defined by Equation 4.19:

$$P_i = y_i P \tag{4.19}$$

where y_i is the mole fraction of Component i.

4.7.3 Humidity

Curves showing the relative humidity (ratio of the mass of the water vapor in the air to the maximum water vapor the air can hold at that temperature, i.e. if the air were saturated) of humid air appear on psychrometric charts (Green and Perry 2008). The curve for 100% relative humidity is also referred to as the saturation curve. The abscissa of the humidity chart is air temperature, also known as the dry-bulb temperature (TDB). The wet-bulb temperature (TWB) is another measure of humidity. It is the temperature at which a thermometer with a wet wick wrapped around the bulb stabilizes. As water evaporates from the wick to the ambient air, the bulb is cooled; the rate of cooling depends on how humid the air is. No evaporation occurs if the air is saturated with water, hence at that point the TWB and TDB are the same.

The lower the humidity, the greater the difference between these wet and dry bulb temperatures. On a psychometric chart, constant wet-bulb temperature lines are straight with negative slopes. Finally, the value of TWB corresponds to the value of the abscissa at the point of intersection of this line with the saturation curve. Additional details of the humidity chart are provided in Chapter 5.

4.7.4 Concentration Terms

A variety of concentration terms can be used to describe the concentration of components in a mixture. Table 4.1 presents six concentration terms common in engineering practice related to water treatment and desalination.

Table 4.1 Commonly used concentration terms for mixtures.

Term	Application	Units
Density	$\frac{\text{Mass of solution}}{\text{Unit of volume}}$	kg/m^3
Volume fraction	$\frac{\text{Volume of solute}}{\text{Total volume of solution}}$	% by volume
Mass fraction	$\frac{\text{Mass of solute}}{\text{Combined mass of solute + solvent}}$	% by mass
Molality	$\frac{\text{Moles of solute}}{\text{1000 g of solvent}}$	mol/kg
Molarity	$\frac{\text{Moles of solute}}{\text{1 L of solution}}$	mol/L
Normality	$\frac{\text{Equivalents of solute}}{\text{1 L of solution}}$	eq/L

The concentration of trace quantities of substances is very often expressed in units of parts per million (ppm) or, if the amount is even smaller as parts per billion (ppb). These are usually represented on a volume basis for gases and Equations 4.20 and 4.21 apply:

$$\text{ppm} = 10^6 y_A \tag{4.20}$$

$$\text{ppb} = 10^3 (\text{ppm}) \tag{4.21}$$

where y_A = mole or volume fraction of a gas A.

4.8 Process Concepts

Five process concepts that find application in desalination studies are briefly reviewed.

4.8.1 Reynolds Number

The Reynolds number, *Re*, is a dimensionless number that indicates whether a moving fluid is flowing in the laminar or turbulent mode. Laminar flow is characteristic of fluids flowing slowly enough so that there are no eddies (whirlpools) or macroscopic mixing of different portions of the fluid. (Note: In any fluid, there is always molecular mixing due to the thermal activity of the molecules; this is distinct from macroscopic mixing due to the swirling motion of different portions of the fluid). In laminar flow, a fluid can be imagined to flow like a deck of cards, with adjacent layers sliding past one another. Turbulent flow is characterized by eddies and macroscopic currents. In practice, moving gases are generally in the turbulent region. For flow in a pipe, a Reynolds number above 2,100 is an indication of turbulent flow.

The Reynolds number (Equation 4.22) is dependent on the fluid velocity, density, viscosity, and some length characteristic of the system or conduit; for pipes, this characteristic length is the inside diameter:

$$\mathrm{Re} = DV\rho/\mu = DV/\upsilon \tag{4.22}$$

where D = inside diameter of a pipe, ft; V = fluid velocity, ft/s; ρ = fluid density, lb/ft^3; μ = fluid viscosity, lb/ft-s; and >υ = fluid viscosity, ft^2/s.

Any consistent set of units may be used with Equation 4.22. In addition, the reader should note that both υ and V have been employed in the literature to represent velocity while volume is solely designated as V.

4.8.2 Kinetic Energy

Consider a body of mass, m, that is acted upon by a force, F. If the mass is displaced a distance, dL, during a differential interval of time, dt, the kinetic energy, E_{KE} expended is given by Equation 4.23:

$$dE_{KE} = m\left(\frac{a}{g_c}\right)dL \tag{4.23}$$

Since the acceleration is given by $a = dV/dt$, Equation 4.24 can be written as:

$$dE_{KE} = \left(\frac{m}{g_c}\right)\left(\frac{dV}{dt}\right)dL = \left(\frac{m}{g_c}\right)\left(\frac{dL}{dt}\right)dV \tag{4.24}$$

Noting that $V = dL/dt$, the above expression becomes Equation 4.25:

$$dE_{KE} = m\left(\frac{V}{g_c}\right)dV \tag{4.25}$$

If this equation is integrated from V_1 to V_2 the change in kinetic energy is given by Equations 4.26 or 27:

$$\Delta E_{KE} = \left(\frac{m}{g_c}\right)\int_{V_1}^{V_2} V dV = \left(\frac{m}{g_c}\right)\left(\frac{V_2^2}{2} - \frac{V_1^2}{2}\right) \tag{4.26}$$

or

$$\Delta E_{KE} = \left(\frac{mV_2^2}{2g_c} - \frac{mV_1^2}{2g_c}\right) = \Delta\left(\frac{mV^2}{2g_c}\right) \tag{4.27}$$

This term is defined as the change in kinetic energy. The reader should note that for flow through pipes and tubes, the kinetic energy term can be retained as written if the velocity profile is uniform, i.e. the local velocities at all points in the cross section are the same. Ordinarily, there is a velocity gradient across the passage; this introduces an error in the calculation, the magnitude of which depends on the nature of the velocity profile and the shape of the cross section. For the usual case where the velocity is approximately uniform (e.g. turbulent flow), the error is not serious and, since the error tends to cancel because of the appearance of kinetic terms on each side of any energy balance equation, it is customary to ignore the effect of velocity gradients. When the error cannot be ignored, the introduction of a correction factor is needed (Theodore 2014).

4.8.3 Potential Energy

A body of mass m is raised vertically from an initial position z_1 to z_2. For this condition, an upward force at least equal to the weight of the body must be exerted on it, and this force must move through the distance z_2 to z_1. Since the weight of the body is the force of gravity on it, the minimum force required is given by Newton's law (Equation 4.28):

$$F = \left(\frac{ma}{g_c}\right) = m\left(\frac{g}{g_c}\right) \tag{4.28}$$

where g is the local acceleration of gravity. The minimum work required to raise the body is the product of this force and the change in vertical displacement, as shown in Equation 4.29:

$$dE_{PE} = F(z_2 - z_1) = m\left(\frac{g}{g_c}\right)(z_2 - z_1) = m\Delta\left(\frac{g}{g_c}z\right) \tag{4.29}$$

The term is defined as the change in the potential energy of the mass.

4.8.4 Ideal Gas Law

The Ideal Gas Law was derived from experiments in which the effects of pressure and temperature on gaseous volumes were measured over a moderate range of temperatures and pressures. As a general rule, this law works best when the molecules of the gas are far apart, i.e. when the pressure is low, and the temperature is high. Under these conditions, the gas is said to behave ideally. For engineering calculations, the Ideal Gas Law is almost always assumed to be valid since it generally works well for the temperature and pressure ranges used in most engineering applications.

The two precursors of the Ideal Gas Law are Boyle's Law and Charles's Law. Boyle found that the volume of a given mass of gas is inversely proportional to the absolute pressure if the temperature is kept constant (Equation 4.30):

$$P_1V_1 = P_2V_2 \tag{4.30}$$

where V_1 and V_2 are the volumes of gas at absolute pressures P_1 and P_2, respectively. Charles found that the volume of a given mass of gas varies directly with the absolute temperature at constant pressure (Equation 4.31):

$$V_1 / T_1 = V_2 / T_2 \tag{4.31}$$

where V_1 and V_2 are the volumes of gas at absolute temperatures T_1 and T_2, respectively. Boyle's and Charles's Laws may be combined into a single equation, Equation 4.32, in which neither temperature nor pressure need be held constant:

$$P_1V_1 / T_1 = P_2V_2 / T_2 \tag{4.32}$$

This equation indicates that for a given mass of a specific gas, PV/T has a constant value. Since, at the same temperature and pressure, volume and mass must be directly proportional, this statement may be extended to Equation 4.33:

$$PV / nT = R \tag{4.33}$$

where n is the number of moles. and R is the universal gas constant. Equation 4.33 is called the ideal gas law. Numerically, the value of R depends on the units of P, V, T, and n. Tabulated values of R are provided in Table 4.2.

4.8.5 Phase Equilibrium

The term *phase*, for a pure substance, indicates a state of matter, i.e. solid, liquid, or gas. For mixtures, however, a more stringent connotation must be used, since a totally liquid or solid system may contain more than one phase. A phase is characterized by uniformity or homogeneity, i.e. the same composition and properties must exist throughout the phase region. At most temperatures and pressures, a pure substance normally exists as a single phase. At

Table 4.2 Values of the ideal gas law constant, *R*, in various units.

R	Temperature scale	Units of *V*	Units of *n*	Units of *P*
10.73	°R	ft^3	lbmol	psia
0.7302	°R	ft^3	lbmol	atm
21.85	°R	ft^3	lbmol	in Hg
555.0	°R	ft^3	lbmol	mm Hg
297.0	°R	ft^3	lbmol	in H_2O
0.7398	°R	ft^3	lbmol	bar
545.0	°R	ft^3	lbmol	psfa
24.75	°R	ft^3	lbmol	ft H_2O
1.314	K	ft^3	lbmol	atm
998.9	K	ft^3	lbmol	mm Hg
19.32	K	ft^3	lbmol	psia
62.361	K	L	gmol	mm Hg
0.08205	K	L	gmol	atm
0.08314	K	L	gmol	bar
8314	K	L	gmol	Pa
8.314	K	m^3	gmol	Pa
82.057	K	cm^3	gmol	atm

certain temperatures and pressures, two or perhaps even three phases can coexist in equilibrium.

The most important equilibrium phase relationship is that between liquid and vapor. Raoult's and Henry's Laws theoretically describe liquid–vapor behavior and, under certain conditions, are applicable in practice. Raoult's Law, sometimes useful for mixtures of components of similar structure, states that the partial pressure of any component in the vapor is equal to the product of the vapor pressure of the pure component and the mole fraction of that component in the liquid (Equation 4.34):

$$P_i = P_i' x_i \tag{4.34}$$

where P_i = partial pressure of component i in the vapor; P_i' = vapor pressure of pure component i; and x_i = mole fraction of component i in the liquid. If the gas phase is ideal, Equation 4.34 becomes Equation 4.35:

$$y_i = (P_i'/P)x_i \tag{4.35}$$

where y_i = mole fraction of component i in the vapor; and P = total system pressure. Highly concentrated mixtures follow Raoult's Law. A more empirical relationship used for representing data for many dilute systems is Henry's Law, Equation 4.36, valid for low solution concentrations or low values of x_i:

$$P_i = H_i x_i \tag{4.36}$$

where H_i = Henry's Law constant for component i. Values and various units of Henry's Law constant can be found in the literature (Green and Perry 2008; Theodore and Ricci 2010).

4.9 Illustrative Examples

4.9.1 Illustrative Example 1

Specific gravity (Equation 4.37) is defined as the ratio of two densities, that of the subject of interest to that of a reference substance. The reference substance is normally water. If the specific gravity of a liquid (methanol) is 0.92 at 60°F, determine its density. Assume the density of water is 62.4 lb/ft^3 at 60°F.

Solution. Write the equation for the definition of specific gravity.

$$\text{Specific Gravity} = (\text{density of methanol})/(\text{density of water}) \tag{4.37}$$

Substitute the known quantities into the equation and solve for the unknown quantity (density of methanol).

$$\text{Density of methanol} = (0.92)\left(62.4\ \text{lb/ft}^3\right) = 57.4\ \text{lb/ft}^3$$

4.9.2 Illustrative Example 2

Briefly describe the American Petroleum Institute gravity unit.

Solution. As noted earlier, the reference material most commonly used for solid and liquid specific gravity values is water at its maximum density, which occurs at 4°C; this reference density is 1.000 g/cm^3, 1000 kg/m^3, or 62.43 lb/ft^3. Note that since specific gravity is a ratio of two densities, it is dimensionless. Therefore, any set of units may be employed for the two densities as long as they are consistent. The specific gravity of gases is used only rarely; and when it is, air at the same conditions of temperature and pressure as the gas is usually employed as the reference substance.

Another dimensionless quantity related to density is the API (American Petroleum Institute) gravity, which is often used to indicate densities of fuel oils. The relationship between the API scale and specific gravity is given by Equation 4.38:

$$degrees\ API = \frac{141.5}{SG(60/60°F)} - 131.5 \tag{4.38}$$

where *SG (60/60°F)* = specific gravity of the liquid at 60°F using water at 60°F as the reference.

4.9.3 Illustrative Example 3

If a 55-gal tank contains 20.0 lb of potable water, (1) how many lbmol of water does it contain? (2) how many gmol of water does it contain? and, (3) how many molecules of water does it contain?

Solution. The molecular weight of the water, H_2O, is:

$$\begin{aligned} \text{MW} &= 2(\text{MW H}) + 1(\text{MW O}) = 2(1.00\ \text{lb/lbmol}) + 1(16\ \text{lb/lbmol}) \\ &= 18\ \text{lb/lbmol} \end{aligned}$$

Therefore:

1. (20 lb H_2O)/(18 lb/lbmol) = 1.11 lb mol H_2O
2. (20 lb H_2O)(454 g/lb)/(18 g/gmol) = (9,080 g H_2O)/(18 g/gmol) = 504.4 gmol H_2O
3. (504.4 gmol H_2O)(6.02 x 10^{23} molecules/gmol) = 3.04 x 10^{26} molecules H_2O

4.9.4 Illustrative Example 4

The heat capacity of a substance is defined as the quantity of heat required to raise the temperature of that substance by 1 degree. If the heat capacity of methanol is 0.61 cal/g-°C at 60°F convert this value to English Engineering units.

Solution. Calculate the heat capacity of methanol in Btu/lb-°F by multiplying the heat capacity of methanol by the corresponding conversion factors to obtain the desired units.

$$(0.61\ \text{cal/g-°C})(454\ \text{g/lb})(1\ \text{Btu/252 cal})(1\text{°C/1.8°F}) = 0.61\ \text{Btu/lb-°F}$$

NOTE: as indicated from these calculations, 1.0 Btu/lb-°F is equivalent to 1.0 cal/g-°C.

4.9.5 Illustrative Example 5

A wastewater contains 250 mg/L aminobenzoic acid (AMB acid), $H_2NC_6H_4CO_2H$. What is the ultimate oxygen demand of this water? The oxygen demand is determined from the stoichiometric equation of the compound combining with O_2 to form CO_2, H_2O, and other fully oxidized end products. The molecular weight of AMB acid is 137.13 g/gmol.

Solution. Calculate the number of moles of AMB acid in 1 L of solution.

$$\text{Number of moles} = (250\ \text{mg/L})/(137.13\ \text{g/gmol}) = 1.82\ \text{millimol/L}$$

Also determine the end products from oxidizing 1 mole of acid and the atoms of oxygen (O) required. Rearranging and simplifying the formula for AMB acid $H_2NC_6H_4CO_2H = C_7H_7NO_2$. Combining this compound with oxygen yields the following balanced, stoichiometric equation:

$$C_7H_7NO_2 + 9.25\ O_2 \rightarrow 7\ CO_2 + 3.5\ H_2O + NO_3$$

The end products are:

CO_2: 7 gmol, O_2:14 atoms of O
H_2O: 3.5 gmol, O_2: 3.5 atoms of O
NO_3: 1 gmol, O_2: 3 atoms of O

The total oxygen needed is the atoms of O or 14 + 3.5 + 3 = 20.5. Since the amount of oxygen already present is 2 atoms, the net or ultimate oxygen demand (UOD) is:

$$\text{UOD} = 20.5 - 2 = 18.5\ \text{atoms of oxygen} = 9.25\ \text{mol}\ O_2$$

The oxygen demand is often described with units of g O_2/g compound, so for AMB acid, this relationship is:

$$\begin{aligned}\text{UOD} &= 9.25\ \text{gmol}\ O_2/\text{gmol AMB acid}\\ &= 9.25(32\ \text{g/gmol})/[\text{gmol}(137.13\ \text{g/gmol})]\end{aligned}$$

$$\text{UOD} = (296\ \text{g}\ O_2)/(137.13\ \text{g AMB acid}) = 2.16\ \text{g}\ O_2/\text{g AMB acid}$$

The UOD for 250 mg/L of AMB acid is then:

$$\text{UOD} = 250\ \text{mg/L}(2.16\ \text{g}\ O_2/\text{g AMB acid}) = 515\ \text{mg/L}$$

4.9.6 Illustrative Example 6

What is the kinematic viscosity of a gas whose specific gravity and absolute viscosity are 0.8 and 0.02 cP, respectively?

Solution. Converting the units of absolute viscosity into English units yields:

$$\mu = (0.02\ \text{cP})\left(\frac{6.720\ \text{x}10^{-4}\ \text{lb/ft-s}}{1\ \text{cP}}\right) = 1.344 \times 10^{-5}\text{lb/ft}^3\text{-s}$$

The density of the gas is then calculated based on its specific gravity as:

$$\rho = (SG)(\rho_{ref}) = (0.80(62.4\ \text{lb/ft}^3) = 49.94\ \text{lb/ft}^3$$

Finally, the kinematic viscosity is calculated as:

$$\nu = \mu / \rho = (1.344\ \text{x}\ 10^{-5}\text{lb/ft}^3\text{-s})/(49.94\ \text{lb/ft}^3) = 2.69\ \text{x}\ 10^{-7}\text{ft}^2/\text{s}$$

4.9.7 Illustrative Example 7

A mixture contains 20 lb of O_2, 2 lb of SO_2, and 3 lb of SO_3. Determine the weight fraction and mole fraction of each component.

Solution. By definition using Equations 4.39, 4.40, and 4.41:

$$\text{Weight fraction} = \text{weight of A/total weight} \tag{4.39}$$

$$\text{Moles of A} = \text{weight of A/molecular weight of} \tag{4.40}$$

$$\text{Mole fraction} = \text{moles of A/total moles} \tag{4.41}$$

First, calculate the weight fraction of each component as shown in Table 4.3:

Calculate the mole fraction of each component as shown in Table 4.4, noting that moles = weight/molecular weight. The molecular weights of O_2, SO_2, and SO_3 are 32, 64, and 80, respectively. Table 4.4 can then be completed.

The reader should note that, in general, weight fraction (or percent) is *not* equal to mole fraction (or percent).

Table 4.3 Weight fraction for each component for illustrative Example 7.

Compound	Weight, lb	Weight fraction
O_2	20	20/25 = 0.8
SO_2	2	0.08
SO_3	3	0.12
Total	25	1.00

Table 4.4 Mole fraction calculations for each component for illustrative Example 7.

Compound	Weight, lb	Molecular weight, lb/lbmol	Moles	Weight fraction
O_2	20	32	$20/32 = 0.625$	$0.625/0.6938 = 0.901$
SO_2	2	64	$2/64 = 0.031$	0.045
SO_3	3	80	$3/80 = 0.0375$	0.054
Total	25		0.6938	1.00

4.10 Problems

1. The composition and combustion properties of a gas mixture are given in Table 4.5. Determine the gross heating value (HVG) of this gas (Table 4.5).
2. Calculate the Reynold's number for a fluid flowing through a 5-in diameter pipe at 10 ft/s with a density of 50 lb/ft^3 and a viscosity of 0.65 cP. Based on the resulting Reynold's number is the flow turbulent or laminar?
3. The height of a liquid column of mercury (density = 848.7 lb/ft^3) open to the atmosphere is 2.493 ft. Determine the pressure at the base of the column in psia, psig, and in inches of water.
4. An analysis of each gas and their mole percent from the analysis of a flue gas from a natural gas boiler is given as:

   N_2 - 79%, O_2 - 5%, CO_2 - 10%, CO - 6%

   Determine the average molecular weight of the flue gas mixture.
5. Express the concentration of 72 g of HCl in 128 cm^3 of water in terms of fraction and percent by weight, parts per million, and molarity.

Table 4.5 Gas composition and component combustion property data for Problem 1.

Component	Mole fraction, x_i	Gross heating value, btu/scf
N_2	0.0515	0
CH_4	0.8111	1,013
C_2H_6	0.0967	1,792
C_3H_8	0.0351	2,590
C_4H_{10}	0.0056	3,370
Total	1.00	

6. Calculate the hydrogen ion and hydroxyl ion concentration of an aqueous solution if the pH of the solution is 1, 3, 5, 7, 8, 10, 12, and 14.
7. Process considerations require pH control in a 50,000 gal storage tank used for waste mixtures (including liquid plus solids) at a desalination plant. Normally, the tank is kept at neutral pH. However, the operation can tolerate pH variation, from 6 to 8. Waste is generated in 5,000-gal increments. Assume that the tank is completely mixed, contains 45,000 gal of pH 7.0 liquid when the waste arrives, and there is negligible buffering capacity in the tank. What is the pH of the most acidic waste that can be handled without neutralization? What is the pH of the most alkaline waste that can be handled without neutralization?

References

Bird, R., Stewart, W., and Lightfoot, E. (2002). *Transport Phenomena*, 2nd Edition. Hoboken, NJ: John Wiley & Sons.

Chopey, N. (1994). *Handbook of Chemical Engineering Calculations*, 2nd Edition. New York, NY: McGraw-Hill.

Green, D.W. and Perry, R.H. (2008). *Perry's Chemical Engineer's Handbook*, 8th Edition. New York, NY: McGraw-Hill.

Santoleri, J., Reynolds, J., and Theodore, L. (2002). *Introduction to Hazardous Waste Incineration*, 2nd Edition. Hoboken, NJ: John Wiley & Sons.

Smith, J.M., Van Ness, H.C., and Abbott, M.M. (2005). *Introduction to Chemical Engineering Thermodynamics*, 7th Edition. New York, NY: McGraw-Hill

Theodore, L. (2014). *Chemical Engineering: The Essential Reference*. New York, NY: McGraw-Hill.

Theodore, L. and Dupont, R.R. (2012). *Environmental Health and Hazard Risk Assessment: Principles and Calculations*. Boca Raton, FL: CRC Press/Taylor Francis Group.

Theodore, L., Hyland, M., McGuinn, Y., Schoen, F., and Taylor, F. (1988). *Principles of Accident and Emergency Management*. Research Triangle Park, NC: USEPA/APTI.

Theodore, L. and Reynolds, J. (1992). *Thermodynamics, a Theodore Tutorial*. East Williston, NY: Theodore Tutorials. Originally published Research Triangle Park, NC: USEPA/APTI.

Theodore, L. and Ricci, F. (2010). *Mass Transfer Operations for the Practicing Engineer*. Hoboken, NJ: John Wiley & Sons.

Theodore, L., Ricci, F., and VanVliet, T. (2009). *Thermodynamics for the Practicing Engineer*. Hoboken, NJ: John Wiley & Sons.

5

Water Properties

5.1 Introduction

There is no question that of all of the compounds on the Earth, water possesses unique properties including:

1. High dielectric constant
2. Low density
3. High melting (fusion) temperature
4. High boiling temperature
5. High heat capacity

In many desalination operations it is often necessary to perform calculations involving the properties of both steam and mixtures of air and water vapor. Such calculations often require knowledge of:

1. The amount of water vapor carried by air under various conditions,
2. The thermal properties of such mixtures, and
3. The changes in enthalpy and moisture content of steam as air containing some moisture is brought into contact with water (or wet solids) and other similar processes.

This chapter provides a brief summary of the hydrologic cycle and then explores the properties of steam (and ice) and mixtures of air and water vapor plus the mechanisms of humidification and drying. Materials on steam tables and psychrometry/psychrometric charts (including some key definitions) are also highlighted in this chapter. An application section provides six illustrative examples related to the general subject of water properties relevant to desalination operations. The reader should note that the bulk of this chapter was drawn from Carbonaro's contribution to the work by Theodore and Theodore (2009).

Although the effect of temperature on the viscosity of air and water is generally not a concern in this field, two tables summarizing this information, Tables 5.1 and 5.2, are provided here.

Introduction to Desalination: Principles, Processes, and Calculations, First Edition. Louis Theodore and R. Ryan Dupont.

Table 5.1 Viscosity of air at 1 atmosphere[a].

T (°C)	Viscosity, (μP)
0	170.8
18	182.7
40	190.4
54	195.8
74	210.2
229	263.8

[a]1 P = 100 cP = 10^6 μP; 1 cP = 6.72 x 10^{-4} lb/ft-s.

Table 5.2 Viscosity of water.

T (°C)	Viscosity, (cP)
0	1.792
5	1.519
10	1.308
15	1.140
20	1.000
25	0.894
30	0.801
35	0.723
40	0.656
50	0.594
60	0.469
70	0.406
80	0.357
90	0.317
100	0.284

5.2 The Hydrologic Cycle

Water is the original renewable resource. Although the total amount of water on the surface of the Earth remains fairly constant over time, individual water molecules carry with them a rich history. The water molecules contained in the fruit one ate yesterday may have fallen as rain last year in a distant place or could have been used decades, centuries, or even millennia ago by one's ancestors.

Water is always in motion, and the hydrologic cycle describes this movement from place to place. The vast majority (96.5%) of water on the surface of the Earth is contained in the oceans. Solar energy heats the water at the ocean surface, and some of it evaporates to form water vapor. Air currents take the vapor up into the atmosphere along with water transpired from plants and evaporated from the soil. The cooler temperatures in the atmosphere cause the vapor to condense into clouds. Clouds move around the world until the moisture capacity of the cloud is exceeded, and the water falls as precipitation. Most precipitation in warm climates falls back into the oceans or onto land where the water flows over the ground as surface runoff. Runoff can enter rivers and streams, which transport the water to the oceans, accumulate and be stored as freshwater in lakes, or soak into the ground as infiltration. Some of this water may infiltrate deep into the ground and replenish aquifers which store huge amounts of freshwater for long periods of time. In cold climates, precipitation falls as snow and can accumulate as ice caps and glaciers, which can store water for thousands of years.

Throughout this cycle, water picks up contaminants originating from both naturally occurring and anthropogenic sources. Depending upon the type and amount of contaminant present, water present in rivers, lakes, and streams or beneath the ground may become unsafe for use.

5.3 The Triple Point of Water

It will be helpful to review certain definitions and concepts associated with phases and the phase rule of J. Willard Gibbs before proceeding to the triple point concept.

A homogeneous system is one for which all parts (on a macro scale) have the same physical and chemical properties. Such a system has a uniform composition and concentration throughout. Air in a cylinder is an example of such a homogeneous system. Water completely occupying a closed container is another. A heterogeneous system has nonuniform composition. Its physical and chemical properties vary from one location to another. A phase is a finite part of a system that is homogeneous throughout and is physically separated from other phases of the system by distinct boundaries called interfaces. When water stands in contact with its own vapor, the surface of the water is the interface between the liquid and vapor phases.

In stating and applying the phase rule, it is essential to determine the number of so-called components. By definition, the number of components is the minimum number of independently variable chemical species needed to express the total composition of the system or any phase present in the system. The phase rule applies to all systems in which a condition of equilibrium exists (Theodore 2014; Theodore et al. 2017, 2009). According to Gibbs:

$$F = C + 2 - P \tag{5.1}$$

where P = the number of phases present, C = the number of components as previously defined, and F = the so-called degrees of freedom or variability of the system. The degrees of freedom, F, may be defined as the number of independent variables, such as temperature, pressure, or concentration, that must be specified in order to completely define the system. A second definition of F is the number of variables, such as temperature and pressure, that may be changed independently without causing the appearance or disappearance of a phase. The number 2 is valid only when there are two variables in addition to concentration. These two are commonly temperature and pressure. If, for example, conditions are such that pressure may be regarded as fixed throughout, then there is only one effective variable in addition to concentration and, in such a system

$$F = C + 1 - P \tag{5.2}$$

Regarding the triple point, the curves DA and DB in Figure 3.1 show the variation with temperature of the vapor pressure of liquid and solid forms, respectively, of a given substance. The curve DA extends only as far as the critical temperature, whereas DB continues to absolute zero. The vapor pressure curve DA of the liquid represents the conditions of equilibrium temperature and pressure, for a system of liquid and vapor, while curve DB similarly indicates the conditions under which solid and vapor phases of a substance are in equilibrium. The two curves meet at D, and hence this point is known as the *triple point* where the three physical states of a substance, i.e. solid, liquid, and vapor, coexist.

That the two vapor pressure curves must cut each other, and in the manner shown, can be deduced qualitatively from the Clapeyron-Clausius equation (Reynolds et al. 2004). Since the density of liquid or solid is large compared with that of the vapor, the slope of the vapor pressure–temperature curve, i.e. dp/dT, is determined essentially by the latent enthalpy (Reynolds et al. 2004). Furthermore, it is evident that the latent enthalpy of sublimation must be greater than that of evaporation so that the slope of the vapor pressure curve of the solid must always be greater than that for the liquid at the same temperature; this conclusion is in agreement with Figure 5.1.

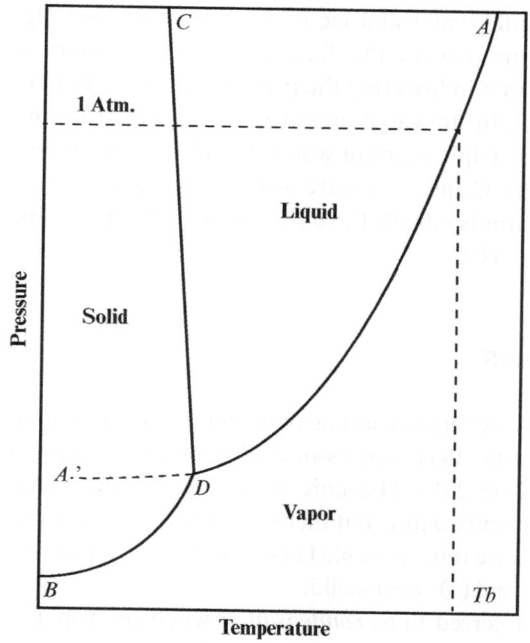

Figure 5.1 Solid-liquid-vapor equilibria diagram for a substance.

Further inspection of Figure 5.1 indicates that the curve DC represents the temperatures and pressures at which the solid and liquid forms can be in equilibrium, i.e. indicating the influence of pressure on the melting point of the substance. Its slope depends on whether an increase of pressure raises or lowers the melting point, i.e. on whether the density of the liquid is greater or less, respectively, than that of the solid. This line in Figure 5.1 slopes to the left, indicating that, as for water, the melting point is lowered by an increase of external pressure. It should be noted that the extent of the slope has been greatly exaggerated for the sake of clarity.

The curve DC must obviously meet the other curves at the triple point D, where the solid, liquid, and vapor phases are in equilibrium. The conditions under which two of the three phases can coexist are given by each of the curves, and all three forms are in equilibrium where the three curves meet. There is consequently only one point where this is possible, and hence there is only one triple point. The normal melting point of a solid is the temperature at which solid and liquid are in equilibrium at atmospheric pressure. At the triple point, however, the pressure is the equilibrium vapor pressure of the system, and the temperature differs slightly from the melting point. For water, the triple point is in the vicinity of 0°C, but the vapor pressure of liquid water and ice is then about 4.6 mm Hg. Since an increase of pressure lowers the freezing point of water by 0.0075°C/atm, it is apparent that in lowering the pressure from 760 mmHg (1 atm) to 4.6 mmHg, the equilibrium vapor pressure will *raise* the freezing point by nearly 0.0075°C. The triple point of water, found by experiment, is at a temperature of + 0.0098°C and a pressure of 4.58 mmHg. The freezing point of *pure* water at 1 atm is 0.0023°C, so the calculated triple point is 0.0023°C + 0.0075°C = 0.0098°C.

5.4 Water Steam Tables

As noted earlier, a pure substance refers to a state of matter that is gas, liquid, or solid. Latent enthalpy (heat) effects are associated with *phase* changes (Theodore et al. 2009, Theodore 2014, Theodore et al. 2017). These phase changes involve no change in temperature, but there is a transfer of energy to and from the substance. There are three possible latent effects, as detailed : (1) vapor–liquid, (2) liquid–solid, and (3) vapor–solid.

Vapor–liquid changes are referred to as *condensation* when the vapor is condensing and *vaporization* when liquid is vaporizing; these changes find application in desalination operations involving evaporation. *Liquid–solid* changes are referred to as *melting* when the solid melts to liquid and *freezing* when a liquid solidifies; these changes find application in desalination operations involving crystallization. *Vapor–solid* changes are referred to as *sublimation*. One should also note that there are enthalpy effects associated with a phase change of a solid to another solid form; however, this enthalpy effect is small compared to the other effects mentioned earlier and find no application in desalination. Specific volume, enthalpy, and entropy data for saturated steam, superheated steam, and steam–ice mixtures are provided in Tables 5.3, Table 5.4, and Table 5.5, respectively (Keenan and Keyes 1930, Theodore et al. 2009). Table 5.3 provides properties for saturated steam/water between 32°F and 700°F. Table 5.4 contains properties for superheated steam between 1.0 psia and 600 psia. Table 5.5 provides properties for saturated steam–ice between 32°F and -40°F. Lowercase notation is employed for specific volume (v), enthalpy (h), and entropy (s) since the values are listed on a mass basis. English units are employed throughout the three tables. Linear interpolation should be employed where necessary. Additional data are available in the two cited references. Other physical and chemical properties were addressed in the previous chapter.

Table 5.3 Specific Volume, Enthalpy, and Entropy Data for Saturated Steam.

Temperature, °F	Absolute pressure, lb_f/in^2	Specific volume, ft^3/lb; v			Enthalpy, btu/lb; h			Entropy, (Btu/lb-°R); s		
		Saturated liquid	Evaporation difference	Saturated vapor	Saturated liquid	Evaporation difference	Saturated vapor	Saturated liquid	Evaporation difference	Saturated vapor
32	0.08854	0.01602	3306	3306	0	1075.8	1075.8	0	2.1877	2.1877
35	0.09995	0.01602	2947	2947	3.02	1074.1	1077.1	0.0061	2.1709	2.177
40	0.1217	0.01602	2444	2444	8.05	1071.3	1079.3	0.0162	2.1435	2.1597
45	0.14752	0.01602	2036.4	2036.4	13.06	1068.4	1081.5	0.0262	2.1167	2.1429
50	0.17811	0.01603	1703.2	1703.2	18.07	1065.6	1083.7	0.0361	2.0903	2.1264
60	0.2563	0.01604	1206.6	1206.7	28.06	1059.9	1088	0.0555	2.0393	2.0948
70	0.3631	0.01606	867.8	867.9	38.04	1054.3	1092.3	0.0745	1.9902	2.0647
80	0.5069	0.01608	633.1	633.1	48.02	1048.6	1096.6	0.0932	1.9428	2.036
90	0.6982	0.0161	468	468	57.99	1042.9	1100.9	0.1115	1.8972	2.0087
100	0.949 2	0.01613	350.3	350.4	67.97	1037.2	1105.2	0.1295	1.8531	0.9826
110	0.2748	0.01617	265.3	265.4	77.94	1031.6	1109.5	0.1471	0.8106	0.9577
120	0.6924	0.0162	203.25	203.27	87.92	1025.8	1113.7	0.1645	0.7694	0.9339
130	0.2225	0.01625	157.32	157.34	97.9	1020	1117.9	0.1816	0.7296	0.9112
140	0.8886	0.01629	122.99	123.01	107.89	1014.1	1122.0	0.1984	0.691	0.8894
150	0.718	0.01634	97.06	97.07	117.89	1008.2	1126.1	0.2149	0.6537	0.8685
160	0.741	0.01639	77.27	77.29	127.89	1002.3	1130.2	0.2311	0.6174	0.8485
170	0.992	0.01645	62.04	62.06	137.9	996.3	1134.2	0.2472	1.5822	0.8293

(*Continued*)

Table 5.3 (Continued)

Temperature, °F	Absolute pressure, lb_f/in^2	Specific volume, ft^3/lb; v			Enthalpy, btu/lb; h			Entropy, (Btu/lb-°R); s		
		Saturated liquid	Evaporation difference	Saturated vapor	Saturated liquid	Evaporation difference	Saturated vapor	Saturated liquid	Evaporation difference	Saturated vapor
180	0.51	0.01651	50.21	50.23	147.92	990.2	1138.1	0.263	1.548	1.8109
190	0.339	0.01657	40.94	40.96	157.95	984.1	1142	0.2785	1.5147	1.7932
200	11.526	0.01663	33.62	33.64	167.99	977.9	1145.9	0.2938	1.4824	1.7762
210	14.123	0.0167	27.8	27.82	178.05	971 .6	1149.7	0.309	1.4508	1.7598
212	14.696	0.01672	26.78	26.8	180.07	970.3	1150.4	0.312	1.4446	1.7566
220	17.186	0.01677	23.13	23.15	188.13	965.2	1153.4	0.3239	1.4201	1.744
230	20.78	0.01684	19.365	19.382.	198.23	958.8	1157	0.3387	1.3901	1.7288
240	24.969	0.01692	16.306	16.323	208.34	952.2	1160.5	0.3531	1.3609	1.714
250	29.825	0.017	13.804	13.82	218.48	945.5	1164	0.3675	1.3323	1.6998
260	35.429	0.01709	11.746	11.763	228.64	938.7	1167.3	0.3817	J.3043	1.686
270	41.858	0.01717	10.044	10.061	238.84	931.8	1170.6	0.3958	1.2769	1.6727
280	49.203	0.01726	8.628	8.645	249.06	924.7	1173.8	0.4096	1.2501	1.6597
290	57.556	0.01735	7.444	7.461	259.31	917.5	1176.8	0.4234	1.2238	1.6472
300	67.013	0.01745	6.449	6.466	269.59	910.1	1179.7	0.4369	1.198	1.635
310	77.68	0.01755	5.609	5.626	279.92	902.6	1182.5	0.4504	1.1727	1.6231
320	89.66	0.01765	4.896	4.914	290.28	894.9	1185.2	0.4637	1.1478	1.6115
330	103.06	0.01776	4.289	4.307	300.68	887	1 187.7	0.4769	1.1233	1.6002
340	118.01	0.01787	3.77	3.788	311.13	879	1190.1	0.49	1.0992	1.5891
350	134.63	0.01799	3.324	3.342	321.63	870.7	1 192.3	0.5029	1.0754	1.5783

Table 5.3 (Continued)

Temperature, °F	Absolute pressure, lb_f/in^2	Specific volume, ft^3/lb; v			Enthalpy, btu/lb; h			Entropy, (Btu/lb-°R); s		
		Saturated liquid	Evaporation difference	Saturated vapor	Saturated liquid	Evaporation difference	Saturated vapor	Saturated liquid	Evaporation difference	Saturated vapor
360	153.04	0.01811	2.939	2.957	332.18	862.2	1194.4	0.5158	1.0519	0.5677
370	173.37	0.01823	2.606	2.625	342.79	853.5	11%.3	0.5286	1.0287	1.5573
380	195.77	0.01836	2.317	2.335	353.45	844.6	1198.1	0.5413	1.0059	1.5471
390	220.37	0.0185	2.0651	2.0836	364.17	835.4	1199.6	0.5539	0.9832	1.5371
400	247.31	0.01864	1.8447	1.8633	374.97	826	1201	0.5664	0.9608	1.5272
410	276.75	0.01878	1.6512	1.67	385.83	816.3	1202.1	0.5788	0.9386	1.5174
420	308.83	0.01894	1.481 1	1.5	396.77	806.3	1203.1	0.5912	0.9166	1.5078
470	514.7	0.0198	0.8811	0.9009	452.8	751.5	1204.3	0.6523	0.8083	1.4606
480	566.1	0.02	0.7972	0.8172	464.4	739.4	1203.7	0.6645	0.7868	1.4513
490	621.4	0.0202	0.7221	0.7423	476	726.8	1202.8	0.6766	0.7653	1.4419
520	812.4	0.0209	0.5385	0.5594	511.9	686.4	1198.2	0.713	0.7006	1.4136
540	962.5	0.0215	0.4434	0.4649	536.6	656.6	1193.2	0.7374	0.6568	1.3942
560	1133.1	0.0221	0.3647	0.3868	562.2	624.2	1186.4	0.7621	0.6121	1.3742
580	1325.8	0.0228	0.2989	0.3217	588.9	588.4	1177.3	0.7872	0.5659	1.3532
600	1542.9	0.0236	0.2432	0.2668	617	548.5	1165.5	0.8131	0.5176	1.3307
620	1786.6	0.0247	0.1955	0.2201	646.7	503.6	1150.3	0.8398	0.4664	1.3062
640	2059.7	0.026	0.1538	0.1798	678.6	452	1130.5	0.8679	0.411	1.2789
660	2365.4	0.0278	0.1165	0.1442	714.2	390.2	1104.4	0.8987	0.3485	1.2472
680	2708.1	0.0305	0.0810	0.1115	757.3	309.9	1067.2	0.9351	0.2719	1.2071
700	3093.7	0.0369	0.0392	0.0761	823.3	172.1	995.4	0.9905	0.1484	1.1389

Table 5.4 Specific volume, enthalpy, and entropy data for superheated steam.

Absolute pressure, lb_f/in^2 (Saturated temperature, °F)	Temperature, °F													
	Phase	200	220	300	350	400	450	500	550	600	700	800	900	1000
1	v	392.6	404.5	452.3	482.2	512.0	541.8	571.6	601.4	631.2	690.8	750.4	809.9	869.5
(101.74)	l	1150.4	1159.5	1195.8	1218.7	1241.7	1264.9	1288.3	1312.0	1335.7	1383.6	1432.7	1482.6	1533.4
	s	2.0512	2.0647	2.1153	2.1444	2.1720	2.1983	2.2233	2.2468	2.2702	2.3137	2.3542	2.3923	2.4283
5	v	78.16	80.59	90.25	96.26	102.26	108.24	114.22	120.19	126.16	138.1	150.03	161.95	173.87
(162.24)	l	1148.8	1158.1	1195.0	1218.1	1241.2	1264.5	1288.0	1311.7	1335.4	1383.6	1432.7	1482.6	1533.4
	s	1.8718	1.8857	1.9370	1.9664	1.9942	2.0205	2.0456	2.0692	2.0927	2.1361	2.1776	2.2148	2.2509
10	v	38.85	40.09	45.00	48.03	51.04	54.05	57.05	60.04	63.03	69.01	74.98	80.95	86.92
(193.21)	l	1146.6	1156.2	1193.9	1217.2	1240.6	1264.0	1264.0	1311.3	1335.1	1383.4	1432.5	1482.4	1533.2
	s	1.7927	1.8071	1.8595	1.8892	1.9172	1.9436	1.9436	1.9924	2.0160	2.0596	2.1002	2.1383	2.1744
	v		27.15	30.53	32.62	34.68	36.73	36.73	40.82	42.86	46.94	51.00	55.07	59.13
14.696	l		1154.4	1192.8	1216.4	1239.9	1263.5	1263.5	1310.9	1334.8	1383.2	1432.3	1482.3	1533.1
(212.00)	s		1.7624	1.8160	1.8460	1.8743	1.9008	1.9008	1.9498	1.9734	2.0170	2.0576	2.0958	2.1319
20	v			22.36	23.91	25.43	26.95	26.95	29.97	31.47	34.47	37.46	40.45	43.44
(227.96)	l			1191.6	1215.6	1239.2	1262.9	1262.9	1310.5	1334.4	1382.9	1432.1	1482.1	1533.0
	s			1.7808	1.8112	1.8396	1.8664	1.8664	1.916	1.9392	1.9829	2.0235	2.0618	2.0978

Table 5.4 (Continued)

Absolute pressure, lb_f/in^2 (Saturated temperature, °F)	Temperature, °F													
	Phase	200	220	300	350	400	450	500	550	600	700	800	900	1000
40	v			11.040	11.843	12.628	13.401	13.401	14.93	15.688	17.198	18.702	20.20	21.70
(267.25)	l			1186.8	1211.9	1236.5	1260.7	1260.7	1308.9	1333.1	1381.9	1431.3	1481.4	1532.4
	s			1.6994	1.7314	1.7608	1.7881	1.7881	1.8384	1.8619	J.9058	1.9467	1.9850	2.0214
60	v			7.259	7.818	8.357	8.884	8.884	9.916	10.427	I1.441	12.449	13.452	14.454
(292.71)	l			1181.6	1208 .2	1233 .6	1258.5	1258.5	1307.4	1331.8	1380.9	1430.5	1480.8	1531.9
	s			1.6492	1.6830	1.7135	1.7416	1.7678	1.7926	1.8162	1.8605	1.9015	1.9400	1.9762
80	v				5.803	6.220	6.624	7.020	7.410	7.797	8.562	9.322	10.077	10.830
(312.03)	l				1204.3	1230.7	1256.J	1281.1	1305.8	1330.5	1379.9	1429.7	1480.1	1531.3
	s				I.6475	1.6791	1.7078	1.7346	1.7598	1.7836	1.8281	1.8694	1.9079	1.9442
100	v				4.592	4.937	5.268	5.589	5.905	6.218	6.835	7.446	8.052	8.656
(327.81)	l				1200.1	1227.6	1253.7	1279.1	1304.2	1329.1	1378.9	1428.9	1479.5	1530.8
	s				1.6188	1.6518	1.6813	1.7085	l.7339	l.7581	1.8029	1.8443	1.8829	1.9193
120	v				3.783	4.081	4.363	4.636	4.902	5.165	5.683	6. 195	6 .702	7.207
(341.25)	l				1195.7	1224.4	1251.3	1277.2	1302.5	1327.7	1377.8	1428.1	1478.8	1530.2
	s				1.5944	1.6287	1.6591	16869	1.7127	1.7370	1.7822	1.8237	1.8625	1.8990

(Continued)

Table 5.4 (Continued)

Absolute pressure, lb_f/in^2 (Saturated temperature, °F)	Temperature, °F													
	Phase	200	220	300	350	400	450	500	550	600	700	800	900	1000
140	v					3.468	3.715	3.954	4.186	4.413	4.861	5.301	5.738	6.172
(353.02)	l					1221.1	1248.7	1275.2	1300.9	1326.4	1376.8	1427.3	1478.2	1529.7
	s					1.6087	1.6399	1.6683	1.6945	1.7190	1.7645	1.8063	1.8451	1.8817
160	v					3.008	3.230	3.443	3.648	3.849	4.244	4.631	5.015	5.396
(363.53)	l					1217.6	1246.1	1273.1	1299.3	1325.0	1375.7	1426.4	1477.5	1529.1
	s					1.5908	1.6230	1.6519	1.6785	1.7033	1.7491	1.7911	1.8301	1.8667
180	v					2.649	2.852	3.044	3.229	3.411	3.764	4.110	4.452	4.792
(373.06)	l					1214.0	1243.5	1271.0	1297.6	1323.5	1374.7	1425.6	1476.8	1528.6
	s					1.5745	1.6077	1.6373	1.6642	1.6894	1.7355	1.7776	1.8167	1.8534
200	v					2.361	2.549	2.726	2.895	3.060	3.380	3.693	4.002	4.309
(381.79)	l					1210.3	1240.7	1268.9	1295.8	1322.1	1373.6	1424.8	1476.2	1528.0
	s					1.5594	1.5937	1.6240	1.6513	1.6767	1.7232	1.7655	1.8048	1.8415
220	v					2.125	2.301	2.465	2.621	2.772	3.066	3.352	3.634	3.913
(389.86)	l					1206.5	1237.9	1266.7	1294.1	1320.7	1372.6	1424.0	1475.5	1527.5
	s					1.5453	1.5808	1.6117	1.6395	1.6652	1.7120	1.7545	1.7939	1.8308

Table 5.4 (Continued)

Absolute pressure, lb_f/in^2 (Saturated temperature, °F)	Temperature, °F													
	Phase	200	220	300	350	400	450	500	550	600	700	800	900	1000
240	v					1.9276	2.094	2.247	2.393	2.533	2.804	3.068	3.327	3.584
(397.37)	l					1202.5	1234.9	1264.5	1292.4	1319.2	1371.5	1423.2	1474.8	1526.9
	s					1.5319	1.5686	1.6003	1.6286	1.6546	1.7017	1.7444	1.7839	1.8209
260	v						1.9183	2.063	2.199	2.330	2.582	2.827	3.067	3.305
(404.42)	l						1232.0	1262.3	1290.5	1317.7	1370.4	1422.3	1474.2	1526.3
	s						1.5573	1.5897	1.6184	1.6447	1.6922	1.7352	1.7748	1.8118
280	v						1.7674	1.9047	2.033	2.156	2.392	2.621	2.845	3.066
(411.05)	l						1228.9	1260.0	1288.7	1316.2	1369.4	1421.5	1473.5	1525.8
	s						1.5464	1.5796	1.6087	1.6354	1.6834	1.7265	1.7662	1.8033
300	v						1.6364	1.7675	1.8891	2.005	2.227	2.442	2.652	2.859
(417.33)	l						1225.8	1257.6	1286.8	1314.7	1368.3	1420.6	1472.8	1525.2
	s						1.5360	1.5701	1.5998	1.6268	1.6751	1.7184	1.7582	1.7954
350	v						1.3734	1.4923	1.6010	1.7036	1.8980	2.084	2.266	2.445
(431.72)	l						1217.7	1251.5	1282.1	1310.9	1365.5	1418.5	1471.1	1523.8
	s						1.5119	1.5481	1.5792	1.6070	1.6563	1.7002	1.7403	1.7777
400	v						1.1744	1.2851	1.3843	1.4770	1.6508	1.8161	1.9767	2.134
(444.59)	l						1208.8	1245.1	1277.2	1306.9	1362.7	1416.4	1469.4	1522.4
	s						1.4892	1.5281	1.5607	1.5894	1.6398	1.6842	1.7247	1.7623

Table 5.5 Specific volume, enthalpy, and entropy data for steam–ice mixtures.

Temperature, °F	Absolute pressure, lb_f/in^2	Saturated ice, v_i	Saturated steam, $v_g \times 10^{-3}$	Saturated ice, h_i	Sublimation difference, h_{sub}	Saturated steam, h_g	Saturated ice, s_i	Sublimation difference, s_{sub}	Saturated steam, s_g
32	0.0885	0.01747	3.306	–143.35	1219.1	1075.8	–0.2916	2.4793	2.1877
30	0.0808	0.01747	3.609	–144.35	1219.3	1074.9	–0.2936	2.4897	2.1961
20	0.0505	0.01745	5.658	–149.31	1219.9	1070.6	–0.3038	2.5425	2.2387
10	0.0309	0.01744	9.05	–154.17	1220.4	1066.2	–0.3141	2.5977	2.2836
0	0.0185	0.Q1742	14.77	–158.93	1220.7	1061.8	–0.3241	2.6546	2.3305
–10	0.0108	0.01741	24.67	–163.59	1221.0	1057.4	–0.3346	2.7143	2.3797
–20	0.0062	0.01739	42.20	–168.16	1221.2	1053.0	–0.3448	2.7764	2.4316
–30	0.0035	0.01738	74.1	–172.63	1221.2	1048.6	–0.3551	2.8411	2.4860

5.5 Psychrometry and the Psychrometric Chart

Some key (and important) terms are introduced before proceeding to the general subject of psychrometry. In the discussion of the physical properties of mixtures of air and water vapor, certain terms need to be defined. The definition of several key terms follow (Theodore and Barden 1996; Theodore and Ricci 2010):

1. The *humidity*, or *absolute humidity*, Y_A, is defined as the mass of water (designated with subscript A) carried by one unit of mass of bone dry air (BDA). The *molal humidity*, Y'_A, can also be defined, but most of the literature on humidification and drying is in terms of Y_A. The two quantities are related by:

$$Y_A = \frac{18}{29} Y'_A \tag{5.3}$$

The humidity at 1.0 atm is then related to the mole fraction, y_A, and the partial pressure, p_A by:

$$Y_A = \frac{18}{29}\left(\frac{y_A}{1.0 - y_A}\right) = \frac{18}{29}\left(\frac{p_A}{1.0 - p_A}\right) \tag{5.4}$$

2. Air is said to be saturated with water vapor at a given temperature and pressure when it contains the maximum amount of water vapor possible at that condition. This is achieved when the air is in equilibrium with liquid water. The *saturation humidity*, Y_{As}, is the value of Y_A corresponding to a partial pressure, p_A, equal to the vapor pressure of water, p'_A, at the given temperature. The *relative humidity*, Y_R, is defined by:

$$Y_R = 100\left(\frac{p_A}{p'_A}\right) \tag{5.5}$$

3. The percent humidity, or percent saturation, *PS*, is given by:

$$PS = 100\left(\frac{Y_A}{Y_{As}}\right) = 100\left(\frac{1.0 - p'_A}{1.0 - p_A}\right) \tag{5.6}$$

where p'_A and p_A must be expressed in atmospheres.

4. The *humid heat capacity*, C_{pH}, is defined as the energy required to raise the temperature of 1 lb of the carrier gas (air, *B*) *and* its accompanying vapor (water, *A*) by 1°F. Thus, if C_{pB} and C_{pA} are the heat capacities of the carrier gas and vapor, respectively, C_{pH} is given by:

$$C_{pH} = C_{pB} + Y_A C_{pA} \tag{5.7}$$

For the air–water system, the earlier equation becomes:

$$C_{pH} = 0.24 + 0.46\ Y_A; \left[\text{Btu/°F-lb BDA}\right] \tag{5.8}$$

Note that heat capacities on a mass basis are designated with a *capital* C in this field. The molal humid heat is based on moles of BDA.

5. The *humid volume*, V_H, is defined as the volume, in ft^3 of "moist" gas per unit mass of bone dry gas where

$$V_H = V_R + Y_A V_A; \left[\text{ft}^3/\text{lb BDA}\right] \tag{5.9}$$

The terms Y_R and V_A represent the volume, in ft^3, of BDA/lb BDA and volume of A/lb A, respectively.

6. The enthalpy of moist air, H_y, at T_y is defined by Equation 5.10 in units consistent with the present discussion:

$$H_y = C_{pH}\left(T_y + T_o\right) + \lambda_o Y_A; \left[\text{Btu/lb BDA}\right] \tag{5.10}$$

The term λ_o is the heat of vaporization at some reference temperature, T_o which is normally assumed to be either 0°F or 32°F. This effectively represents the sensible enthalpy of the air–water mixture relative to T_o plus the enthalpy of vaporization at T_o.

7. The *adiabatic saturation temperature*, T_{as}, of moist air is the temperature that air reaches when it is saturated adiabatically (i.e. at constant enthalpy).
8. The *wet-bulb temperature*, T_{wb}, is the temperature attained by a small reservoir of water in contact with a large amount of air flowing past it. Normally, one can assume that T_{wb} and T_{as} are the same for the water–air system at temperatures low enough to form a dilute gas-phase solution. For other liquid–vapor mixtures, the adiabatic saturation, and wet-bulb temperatures are normally not equal.
9. The *dew point temperature*, T_{dp}, is the temperature at which a given sample of moist air becomes saturated as it is cooled at constant pressure and absolute humidity. Thus, the dew-point pressure is the (total) pressure to which moist air must be compressed at constant temperature and humidity to bring it to saturation. As an example, suppose the air in a room at 70°F is at 50% relative humidity. From the steam tables (see Table 5.3), the vapor pressure of water at 70°F is 0.3631 psi, which means that the air at 50% humidity holds water vapor with a partial pressure of (0.50)(0.3631) or 0.1812 psi. If the temperature is dropped at constant pressure to the point where 0.1812 psi equals the water vapor pressure (around 52°F), the air becomes saturated with water, and any further drop in temperature will cause condensation. The dew point of this air mixture is then 52°F. Obviously, if the air were already saturated at 70°F (i.e. 100% relative humidity), then the dew point would also be 70°F.

A summary of the earlier definitions is provided in Table 5.6.

Table 5.6 Definitions of psychrometric terms.

Term	Definition	Units
1. Absolute humidity	Vapor content of a gas	lb vapor/lb noncondensible gas (BDA)
2. Molal humidity	Vapor content of a gas	moles vapor/ mole noncondensible gas
3. Relative saturation or relative humidity	Ratio of partial pressure of vapor to partial pressure of vapor at saturation	atm/atm, or mole fraction; mole fraction often expressed as a %
4. Percent saturation or percent humidity	Ratio of concentration of vapor to the concentration of vapor at saturation with concentrations expressed as mole ratios	mole ratio/mole ratio, often expressed as a %
5. Humid volume	Volume occupied by 1 lbmol of dry gas plus its associated vapor	ft^3/lbmol of BOA
6. Humid heat capacity	Energy required to raise the temperature of 1 lbmol of bone dry gas plus its associated vapor 1°F	Btu/lbmol of BDA-°F
7. Adiabatic saturation temperature	Temperature that would be attained if the gas were saturated in an adiabatic process	°F
8. Wet-bulb temperature	Steady-state temperature attained by a wet-bulb thermometer under standardized conditions	°F
9. Dew point temperature	Temperature at which vapor begins to condense when the gas phase is cooled at constant pressure	°F

One vapor–liquid phase equilibrium example containing air–water raw data is the psychrometric or humidity chart. A humidity chart is used to determine the properties of moist air and calculate the moisture content in air. The ordinate of the chart is the absolute humidity, Y (with the subscript A dropped for convenience), which was defined earlier as the mass of water vapor per mass of bone dry air. (Note that some charts base the ordinate on moles instead of mass). The previously defined concepts and definitions are normally presented graphically on the aforementioned psychrometric or humidity chart (see Figures 5.2 and Figure 5.3). These charts were originally generated from thermodynamic data. There are also charts that apply to other single noncondensable gases and single condensable components at a fixed pressure (usually 1 atm).

Curves showing the relative humidity (ratio of the partial pressure of the water vapor in the air to the vapor pressure of water at the system temperature) of humid air also appear on the charts. The curve for 100% relative humidity is also referred to as the *saturation curve*. The abscissa of the humidity chart is the air temperature, also known as the *dry-bulb temperature*, T_{db}. The *wet-bulb temperature*, which has also already been defined (see Table 5.6) is another measure of the humidity. As described earlier, it is the temperature at which a thermometer stabilizes when it has a wet wick wrapped around the bulb. No evaporation occurs if the air is saturated with water, hence, T_{wb} and T_{db} are the same and the lower the humidity, the greater the difference between these two temperatures. On the psychrometric chart, constant wet-bulb temperature lines are straight with negative slopes. The value of T_{wb} corresponds to the value of the abscissa at the point of intersection of this line with the saturation curve.

The *humid volume*, also defined earlier, is the volume of wet air per mass of BDA and is linearly related to the humidity. This quantity is used as an alternate ordinate in Figure 5.3. Note the straight parallel lines labeled with units of

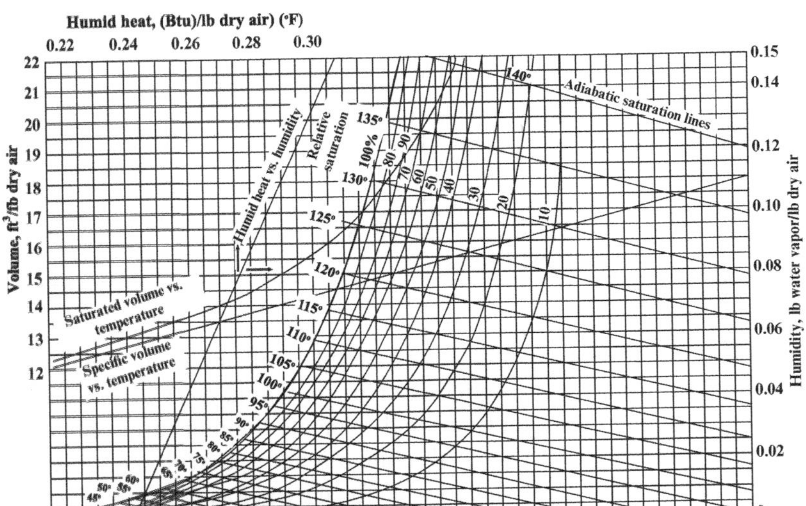

Figure 5.2 Psychrometric chart, low temperatures. Barometric pressure of 1 atm.

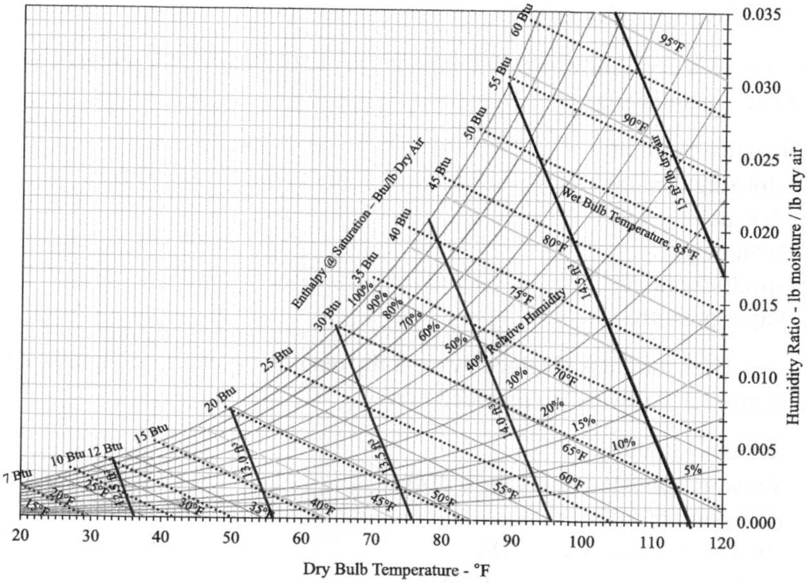

Figure 5.3 Psychrometric chart, high temperatures. barometric pressure of 1 atm.

cubic feet. The *humid enthalpy* is the enthalpy of the moist air on a bone-dry basis. Since enthalpy is a measure of the energy content of the mixture, the enthalpy for saturated air can be read from the chart by extending the approximate wet-bulb temperature line upward to the diagonal scale labeled *enthalpy of saturation*.

Finally, it should be noted that the evaporation of water into air for the purpose of increasing the air humidity is known as *humidification*. Closely allied to this is the evaporation of water into air for the purpose of cooling the water. *Dehumidification* consists of condensing water from air to decrease the air humidity. All these processes are of considerable industrial importance in desalination applications and involve the contacting of air and water, accompanied by heat and mass transfer.

5.6 Illustrative Examples

Eight illustrative examples complete the presentation of materials highlighted in this chapter.

5.6.1 Illustrative Example 1

For air with a dry- and wet-bulb temperature of 80°F and 65°F, respectively, calculate the following:

1. Relative humidity
2. Moisture content
3. Dew point
4. Total heat
5. Specific volume
6. Vapor pressure

Solution. The solution is obtained by carefully reading the charts in Figures 5.2 and Figure 5.3. Answers are provided here.

1. Relative humidity = 45%
2. Moisture content = 0.151 lb/lb BDA
3. Dew point = 56°F
4. Total heat = 30.0 Btu/lb BDA
5. Specific volume = 13.81 ft^3/lb BDA
6. Vapor pressure = 0.228 psi

5.6.2 Illustrative Example 2

A stream of moist air is cooled and humidified adiabatically from T_{db} of 100°F and T_{wb} of 70°F to a T_{db} of 80°F. How much more moisture is added per pound of BDA?

Solution. Refer to the psychrometric chart in Figure 5.2. Adiabatic cooling follows the wet-bulb temperature line upwards (toward the saturation curve). The difference in the final and initial humidities is the moisture added:

State	Y
Initial	0.0090
Final	0.0133

Therefore

$$\Delta Y = 0.0133 - 0.0090 = 0.0043 = 4.3 \times 10^{-3} \text{lb } H_2O/\text{lb BD}$$

5.6.3 Illustrative Example 3

Generate a physical properties table of H_2O that includes the heat capacity (*Cp*), boiling point (*BP*), melting point (*MP*), enthalpy of vaporization (*EV*), and enthalpy of melting (*EM*) at pressures of 1 atm and 0.005 atm.

Solution. These physical properties of water are listed in Table 5.7 employing SI units.

5.6.4 Illustrative Example 4

Employing English and SI units, determine the enthalpy (h_g), specific volume (v_g), entropy (s_g), and temperature (t) of saturated steam at 151 psi (1041 kPa).

Solution. Study of the saturation-pressure table shows that there is no pressure value listed for 151 psi (1041 kPa). Therefore, it will be necessary to interpolate between the next higher and next lower tabulated pressure values. In this instance, these values are 152 and 150 psia (1048.0 and 1034.3 kPa), respectively. The pressure for which properties are being found (151 psia or 1041.1 kPa) is called the "intermediate pressure." At 152 psia (1048.0 kPa); $h_g = 1194.3$ Btu/lb (2777.5 kJ/kg); v_g = 2.977 ft^3/lb (0.19 m^3/kg); s_g = 1.5683 Btu/lb-°F (6.57 kJ/kg-°C); and t = 359.46°F (181.9°C). At 150 psia (1034.3 kPa); $h_g = 1194.1$ Btu/lb (2777.5 kJ/kg); v_g = 3.015 ft^3/lb (0.19 m^3/kg); s_g = 1.5694 Btu/lb-°F (6.57 kJ/kg-°C); and t = 358.42°F (181.3°C).

For the enthalpy, note that as the pressure increases, so does h_g. Therefore, the enthalpy at 151 psia (1041.1 kPa) (the intermediate pressure) will equal the enthalpy at 150 psi a (1034.3 kPa) (the lower pressure used in the interpolation) plus the proportional change (difference between the intermediate pressure and the lower pressure) for a 1 psi (6.9 kPa) pressure increase. Or, at any higher pressure, $h_{gi} = h_{gl} + \left[(p_i - p_l)/(p_h - p_l)\right](h_h - h_l)$, where h_{gi} is enthalpy at the intermediate pressure, h_{gl} is enthalpy at the lower pressure used in the interpolation; h_h is enthalpy at the higher pressure used in the interpolation, p_i is intermediate pressure, and p_h and p_l are higher and lower pressures used in the interpolation, respectively. Thus, using the enthalpy values obtained from the steam table for 150 and 152 psia (1034.3 and 1048.0 kPa), h_{gi} = 1194.1 + [(151 - 150)/(152 - 150)](1194.3 - 1194.1) = 1194.2 Btu/lb (2777.7 kJ/kg) at 151 psia (1041.1 kPa) saturated.

Table 5.7 Physical properties of water at P = 1 atm and 0.005 atm.

Property	P = 1 atm	P = 0.005 atm
Cp, J/mol	33 (solid) 75 (liquid) 35 (vapor)	33 (solid) 75 (liquid) 35 (vapor)
BP, °C	100.0	-2.0
MP, °C	0.0	-2.0
EV, J/mol	4.1 x 10^4	4.5 x 10^4
EM, J/mol	6.0 x 10^3	6.8 x 10^3

Next, study the steam table to determine the direction of change of specific volume, v_g, between the lower and higher pressures. This should show that the specific volume decreases as the pressure increases. Therefore, the specific volume at 151 psia (1041.1 kPa) (the intermediate pressure) will equal the specific volume at 150 psia (1034.3 kPa) (the lower pressure used in the interpolation) minus the proportional change (difference between the intermediate pressure and the lower interpolating pressure) for a 1 psi (6.9 kPa) pressure increase. Or, at any pressure, $v_{gi} = v_{gl} - [(p_i - p_l)/(p_h - p_l)](v_l - v_h)$, where the subscripts are the same as the earlier ones and v is specific volume at the respective pressure. With the volume values obtained from steam tables for 150 and 152 psia (1034.3 and 1048.0 kPa), v_{gi} = 3.015 - [(151 - 150)/(152 - 150)](3.015 - 2.977) = 2.996 ft^3/lb (0.19 m^3/kg) at 151 psia (1041.1 kPa) saturated.

Once again study of the steam table for the direction of temperature change shows that the saturation temperature, like enthalpy, increases as the pressure increases. Therefore, the temperature at 151 psia (1041.1 kPa) (the intermediate pressure) will equal the temperature at 150 psia (1034.3 kPa) (the lower pressure used in the interpolation) plus the proportional change (difference between the intermediate pressure and the lower interpolating pressure) for a 1 psi (6.9 kPa) increase. Or, at any higher pressure, $t_{gi} = t_{gl} + [(p_i - p_l)/(p_h - p_l)](t_h - t_l)$ = 358.42 + [(151 - 150)/(152 - 150)](359.46 - 358.42) = 358.94°F (181.6° C) at 151 psia (1041.1 kPa) saturated.

5.6.5 Illustrative Example 5

Employing English and SI units, determine the enthalpy (h_g), specific volume (v_g), entropy (s_g), and pressure (p_g) of saturated steam at 261°F (127.2°C).

Solution. Study of the saturation-temperature table shows that there is no temperature value listed for 261°F (127.2°C). Therefore, it will be necessary to interpolate between the next higher and next lower tabulated temperature values. In this instance, these values are 262°F and 260°F (127.8°C and 126.7°C), respectively. The temperature for which properties are being found (261°F or 127.2°C) is called the "intermediate temperature."

At 262°F (127.8°C), h_g = 1168.0 Btu/lb (2716.8 kJ/kg); v_g = 11.396 ft^3/lb (0.71 m^3/kg); s_g = 1.6833 Btu/lb-°F) (7.05 kJ/kg-°C); and p_g = 36.646 psia (252.7 kPa). At 260°F (126.7°C), h_g = 1167.3 Btu/lb (2715.1 kJ/kg); v_g = 11.763 ft^3/lb (0.73 m^3/kg); s_g = 1.6860 Btu/lb-°F (7.06 kJ/kg-°C); and p_g = 35.429 psia (244.3 kPa).

For enthalpy, note that as the temperature increases, so does h_g. Therefore, the enthalpy at 261°F (127.2°C) (the intermediate temperature) will equal the enthalpy at 260°F (126.7°C) (the lower temperature used in the interpolation) plus the proportional change (difference between the intermediate temperature and the lower temperature) for a 1°F (0.6°C) temperature increase. Or, at any higher temperature, $h_{gi} = h_{gl} + \left[\left(t_i - t_l\right)/\left(t_h - t_l\right)\right]\left(h_h - h_l\right)$, where h_{gi} is enthalpy at the intermediate temperature, h_{gl} is enthalpy at the lower temperature used in the interpolation; h_h is enthalpy at the higher temperature used in the interpolation, t_i is intermediate temperature, and t_h and t_l are the higher and lower temperatures used in the interpolation, respectively. Thus, using the enthalpy values obtained from the steam table for 260°F and 262°F (126.7°C and 127.8°C), h_{gi} = 1167.3 + [(261 - 260)/(262 - 260)](1168.0 - 1167.3) = 1167.65 Btu/lb (2716.0 kJ/kg) at 261°F (127.2°C) saturated.

Next, study the steam table to determine the direction of change of specific volume, v_g, between the lower and higher temperatures. This should show that the specific volume decreases as the temperature increases. Therefore, the specific volume at 261°F (127.2°C) (the intermediate temperature) will equal the specific volume at 260°F (126.7°C) (the lower temperature used in the interpolation) minus the proportional change (difference between the intermediate temperature and the lower interpolating temperature) for a 1°F (0.6°C)) temperature increase. Or, at any higher temperature,- $s_{gi} = s_{gl} - \left[\left(t_i - t_l\right)/\left(t_h - t_l\right)\right]\left(s_l - s_h\right)$ = 1.6860 – [(261 - 260)/(262 - 260)](1.6860 – 1.6833) = 1.68465 Btu/lb-°F) [7.1 kJ/kg-°C] at 261°F (127.2°C).

Once again study of the steam tables for the direction of pressure change shows that the saturation pressure, like enthalpy, increases as the temperature increases. Therefore, the pressure at 261°F (127.2°C) (the intermediate temperature) will equal the pressure at 260°F (126.7°C) (the lower temperature used in the interpolation) plus the proportional change (difference between the intermediate temperature and the lower interpolating temperature) for a 1°F (0.6°C)) temperature increase. Or, at any higher temperature, $p_{gi} = p_{gl} + \left[\left(t_i - t_l\right)/\left(t_h - t_l\right)\right]\left(p_h - p_l\right)$ = 35.429 + [(261 - 260)/(262 - 260)] (36.646 - 35.429) = 36.0375 psia (248.5 kPa) at 261°F (127.2°C) saturated.

5.6.6 Illustrative Example 6

Employing English and SI units, determine the pressure of steam (p_g) at 1000°F (537.8°C) if its specific volume (v_g), is 2.6150 ft^3/lb (0.16 m^3/kg).

Solution. Choose the superheated-steam table for steam at 1000°F (537.9°C) and 2.6150 ft^3/lb (0.16 m$\hat{}$/kg) because the highest temperature at which saturated steam can exist is 705.4°F (374.1°C). This is also the highest temperature tabulated in some saturated temperature tables. Therefore, the steam is superheated when at a temperature of 1000°F (537.9°C).

Look down the 1000°F (537.9°C) columns in the superheat table until a specificvolume value of 2.6150 is found. This occurs between 325 psia (2240.9

kPa, $v = 2.636$ or 0.16) and 330 psia (2275.4 kPa, $v = 2.596$ or 0.16). Since there is no volume value exactly equal to 2.6150 tabulated, it will be necessary to interpolate. List the values from the steam table thus, $p = 325$ psia (2240.9 kPa); $t = 1000°$ F (537.9° 0); $v = 2.636$ ft^3/lb (0.16 m^3/kg); and $p = 330$ psia (2275.4 kPa); $t = 1000°$F (537.9°C); $v = 2.596$ ft^3/lb (0.16 m^3/kg).

Note that as the pressure rises, at constant temperature, the volume decreases. Therefore, the intermediate (or unknown) pressure is found by subtracting from the higher interpolating pressure (330 psia or 2275.4 kPa in this instance) the product of the proportional change in the specific volume and the difference in the pressures used for the interpolation, or, $p_{gi} = p_{gl} + \left[\left(t_i - t_l\right) / \left(t_h - t_l\right)\right]\left(p_h - p_l\right)$ where the subscripts h, l, and i refer to the high, low, and intermediate (or unknown) pressures, respectively. In this instance, $p_{gi} = 330 - [(2.615 - 2.596)/(2.636 - 2.596)](330 - 325) = 327.62$ psia (2258.9 kPa) at 1000°F (537.9°C) and a specific volume of 2.6150 ft^3/lb (0.16 m^3/kg).

5.6.7 Illustrative Example 7

Employing English and SI units, determine the enthalpy (h_g), specific volume (v_g), and entropy (s_g) of 300 psia (2068.5 kPa) steam at 567.2°F (297.3°C).

Solution. When given a steam pressure and temperature, determine, before performing any interpolation, the state of the steam. Do this by entering the saturation-pressure table at the given pressure and noting the saturation temperature. If the given temperature exceeds the saturation temperature, the steam is superheated. In this instance, the saturation pressure table shows that at 300 psia (2068.5 kPa), the saturation temperature is 417.33°F (214.1°C). Since the given temperature of the steam is 567.2°F (297.3°C), the steam is superheated because its actual temperature is greater than the saturation temperature.

Enter the superheated-steam table at 300 psia (2068.5 kPa) and find the next temperature lower than 567.22°F (297.3°C); this is 560°F (293.3°C). Also find the next higher temperature; this is 580°F (304.4°C). Tabulate the enthalpy, specific volume, and entropy for each of these temperatures: $t =$ 560°F (293.3°C); $h = 1292.5$ Btu/lb (3006.4 kJ/kg); $v = 1.9128$ ft^3/lb (0.12 m^3/kg); $s = 1.6054$ Btu/lb-°F (6.72 kJ/kg-°C)]; and $t = 580$°F (304.4°C); $h = 1303.7$ Btu/lb (3032.4 kJ/kg); $v = 1.9\,594$ ft^3/lb (0.12 m^3/kg); $s = 1.6163$ Btu/lb-°F (6.77 kJ/kg-°C).

Use the same procedures for each property – enthalpy, specific volume, and entropy – as given previously, but change the sign between the lower volume and entropy and the proportional factor (temperature in this instance), because for superheated steam, the volume and entropy increase as the steam temperature increases. Thus,

$$h_{gi} = 1292.5 + \frac{567.22 - 560}{580 - 560}(1303.7 - 1292.50)$$
$$= 1269.6 \text{ Btu/lb} (3015.9 \text{ kJ/kg})$$

$$v_{gi} = 1.9128 + \frac{567.22 - 560}{580 - 560}(1.9594 - 1.9128)$$
$$= 1.9296 \text{ ft}^3\text{/lb} (0.12 \text{ m}^3\text{/kg})$$

$$s_{gi} = 1.6054 + \frac{567.22 - 560}{580 - 560}(1.6163 - 1.6054)$$
$$= 1.6093 \text{ Btu/lb} - {}^\circ\text{F} (6.7 \text{ kJ/kg} - {}^\circ\text{C})$$

Note: Also observe the direction of change of a property *before* interpolating. The reader should use a *plus* or *minus* sign between the higher interpolating value and the proportional change depending on whether the tabulated value increases (+) or decreases (-).

5.6.8 Illustrative Example 8

Describe the difference between percent humidity and relative humidity.

Solution. Do not confuse percent humidity with relative humidity. "Relative humidity" is the ratio of the partial pressure of the water vapor to the vapor pressure of water at the temperature of the air, this ratio usually being expressed as a percent. "Percent humidity" is the ratio of the actual humidity to the saturation humidity that corresponds to the gas temperature, which is also usually expressed as a percent. At all humidities other than 0 or 100%, the percent humidity is less than the relative humidity.

5.7 Problems

1. Convert the vapor pressure of water at 20°C in mmHg to kPa and °F.
2. Convert the result of Problem 1 to units of lb_f/in^2.
3. Estimate the enthalpy and entropy of saturated steam (vapor) at 225°F.
4. Refer to the steam tables. Determine the pressure, enthalpy of liquid water, enthalpy of liquid vapor, and enthalpy of enthalpy of evaporation of saturated steam at 212°F.
5. Refer to the steam–ice tables. Determine the pressure, enthalpy of saturated ice, enthalpy of saturated steam, and enthalpy of sublimation of saturated steam–ice.

6. Employing English and SI units, determine the enthalpy (h_g), specific volume (v_g), and entropy (s_g) of 400 psia (2757.9 kPa) steam at 575°F (301.7°C).
7. Employing English and SI units, determine the enthalpy (h_g), specific volume (v_g), entropy (s_g), and pressure (p_g) of saturated steam at 182°F (83.3°C).

References

Keenan, J. and Keyes, F. (1930). *Thermodynamic Properties of Steam*. Hoboken, NJ: John Wiley & Sons.

Reynolds, J., Jeris, J., and Theodore, L. (2004). *Handbook of Chemical and Environmental Engineering Calculations*. Hoboken, NJ: John Wiley & Sons.

Theodore, L. (2014). *Chemical Engineering: The Essential Reference*. New York, NY: McGraw-Hill.

Theodore, L. and Barden, J. (1996). *Mass Transfer Operations, A Theodore Tutorial*. East Williston, NY: Theodore Tutorials.

Theodore, L., Dupont, R., and Ganesan, K. (2017). *Unit Operations in Environmental Engineering*. Salem, MA: Scrivener-Wiley.

Theodore, L. and Ricci, F. (2010). *Mass Transfer Operations for the Practicing Engineer*. Hoboken, NJ: John Wiley & Sons.

Theodore, L., Ricci, F., and VanVliet, T. (2009). *Thermodynamics for the Practicing Engineer*. Hoboken, NJ: John Wiley & Sons.

6

Water Chemistry

6.1 Introduction

Water chemistry deals with the fundamental chemical properties of water itself, the chemical properties of other constituents that dissolve in water, and the countless chemical reactions that take place in water. The field of natural water chemistry is concerned principally with reactions that occur in relatively dilute solutions (low concentrations), although some natural waters have rather high solute concentrations (Hem 1970). During a chemical reaction, tiny subatomic particles (e.g. electrons) and atoms (e.g. hydrogen) are transferred, shared, and exchanged. When a chemical reaction occurs in water, these changes require transport through the water medium. Water is not passive in these chemical reactions. Instead, it plays an active role, constantly making and breaking chemical bonds, thereby facilitating chemical change.

This chapter introduces the reader to these water chemistry principles including the physical properties of water, the chemical properties of water, the chemical composition of natural waters, chemical reactions and an application section that provides six illustrative examples related to the general subject of water chemistry. The bulk of the material for this chapter was drawn from a contributed chapter by Carbonaro (2021).

6.2 Physical Properties of Water

A water (H_2O) molecule consists of three atoms: two hydrogen atoms each of which are bonded to a central oxygen atom. Water can exist in three states: solid, liquid, and gas. At room temperature and atmospheric pressure water is a liquid, but below 0°C (32°F) it freezes and turns into ice. Water is present in the gaseous state above 100°C (212°F) and 1.0 atmospheric pressure. This is the boiling point of water, at which water will evaporate.

Introduction to Desalination: Principles, Processes, and Calculations, First Edition. Louis Theodore and R. Ryan Dupont.

Water has a number of unusual physical properties that are a consequence of its hydrogen bonding among neighboring water molecules. The hydrogen bond is a weak bond that is the result of the dipolar nature of all water molecules. The hydrogen bonds between water molecules impart water with a relatively large heat capacity, heat of vaporization, and heat of fusion than that expected for a molecule of its size. Hydrogen bonds are also important biologically. Bonding between adjacent base pairs holds double-stranded DNA together. Many proteins also utilize hydrogen bonding to hold their three-dimensional shape and assist in enzymes binding to their substrate.

Hydrogen bonding also causes water to expand upon freezing, which results in solid water being less dense than liquid water. As a result, water collecting in the cracks will expand upon freezing which is an important mechanism for mechanically breaking surfaces apart, i.e. mechanical weathering. This mechanical weathering breaks surfaces into smaller fragments, which increases their surface area. This in turn increases the breakdown of the surface by chemical reactions, a process known as chemical weathering.

6.3 Chemical Properties of Water

Water is called the "universal solvent" because it is capable of dissolving many substances. This chemical properly of water arises from the dipolar nature of water molecules. Water molecules effectively surround positively charged ions (cations) and negatively charged ions (anions) which serve to prevent them from precipitating as solid. This means that wherever water goes, either through the ground or through one's body, it carries with it various solutes such as dissolved minerals, nutrients, organics, and heavy metals.

Even pure water contains some amount of hydrogen ion (H^+) and hydroxide ion (OH^-) as a result of a chemical reaction known as the autoionization of water. The concentration of these ions change as acids and bases are added to water; however, the product of their concentrations is always constant. The relative presence of H^+ and OH^- is measured by the pH, which is defined as the negative logarithm of the H^+ concentration in units of mol/L (see Chapter 4). A decrease in pH indicates an increase in the H^+ concentration and a corresponding decrease in the OH^- concentration.

Pure water has a neutral pH of exactly 7.0. Values of pH less than this are considered acidic, while pH values greater than this are termed alkaline. The typical pH range of water is from 0 to 14, although values outside of this range can be attained under extreme conditions. The pH of water is an important chemical property which controls the distribution of chemical species among various forms. For example, the pH is buffered at precise values in animal cells to maintain functionality of specific enzymes and proteins. Likewise, in the environment, pH controls the distribution of chemicals amongst their various forms and also controls the rate at which many chemical reactions occur. It is therefore often necessary to precisely measure the pH of water to understand its water chemistry.

6.4 Chemical Composition of Natural Waters

The composition of natural waters is often described according to its physical qualities, chemical constituents, and/or its biological inhabitants. The focus to follow is primarily concerned with chemical constituents. Water sampling programs are used to obtain information on the chemical characteristics of potential and existing water sources and the performance of water and wastewater treatment plants (MWH 2005). Typically, prior knowledge of the type of chemical constituent (i.e. organic versus inorganic, dissolved versus suspended) is required to design and implement effective sampling programs. Preservatives are often added to prevent degradation of certain constituents and holding times have been recommended by the EPA and other agencies so as to maintain proper quality control (Clesceri et al. 1998).

6.4.1 Dissolved Minerals

Soil water in the ground reacts with common rock-forming minerals to release ions and form new minerals. Table 6.1 lists the most commonly occurring chemical elements present in the Earth's crust. The most abundant group of

Table 6.1 Mineral composition of the Earth's crust (Adapted from Lutgens and Tarbuck 2000; Nicholls 1975).

Element	%	Element	%
O (oxygen)	46.6	Cl (chlorine)	0.1
Si (silicon)	27.7	Cr (chromium)	0.04
Al (aluminum)	8.1	C (carbon)	0.03
Fe (iron)	5.0	V (vanadium)	0.02
Ca (calcium)	3.6	Ni (nickel)	0.008
Na (sodium)	2.8	Cu (copper)	0.007
K (potassium)	2.6	Co (cobalt)	0.002
Mg (magnesium)	2.1	Pb (lead)	0.001
TI (titanium)	0.6	Sc (scandium)	0.0005
Mn (manganese)	0.1	Zn (zinc)	0.0001

minerals in crustal rock is a family called the feldspars. These minerals are comprised of sodium, potassium, and calcium aluminum silicates. They react with water, thereby producing Na^+, K^+, Ca^{2+}, Al^{3+}, and H_4SiO_4. Magnesium (Mg^{2+}) and iron (Fe^{2+}) are released from other silicate minerals. Carbonate rocks, limestone ($CaCO_3(s)$), and dolomite ($(Ca,Mg)CO_3$ (s)) weather to release Ca^{2+}, Mg^{2+}, and HCO_3^-. Phosphorous (PO_4^{3-}) is released by the chemical weathering of apatite, a calcium phosphate mineral. The soluble constituents described earlier, Na^+, K^+, Ca^{2+}, Mg^{2+}, HCO_3^-, and H_4SiO_4, along with Cl^- and SO_4^{2-} find their way to river and streams and eventually reach the ocean. Over time, the concentrations of these constituents have increased to the levels found in the oceans today.

Aluminum (Al) and iron (Fe), mobilized from chemical weathering processes, have low solubility and are not transported over large distances. Under aerobic conditions, iron either stays behind as a hydroxide or oxide coating on the surface of weathered rocks, or attaches to small particles that remain suspended and are carried with the flowing water. Aluminum precipitates as $Al(OH)_3(s)$, or reacts with H_4SiO_4 to form the mineral kaolinite and other clay minerals.

Hardness is a bulk chemical property that measures the presence of specific dissolved mineral ions. Calcium and magnesium dissolved in water are the two most common minerals that make water "hard." Hard water interferes with washing clothes, dishwashing, and bathing. Clothes laundered in hard water may look dingy and feel harsh or scratchy. Hard water causes a film on surfaces when it evaporates due to the presence of a thin layer of salt that remains. Water flow may be reduced by mineral deposits in pipes. Synthetic detergents are usually less effective in hard water because the active ingredient is partially inactivated by the high levels of calcium and magnesium.

Another bulk measurement of the dissolved ion content of water is total dissolved solids (TDS). TDS is a measure of all of the dissolved ions in solution, and is determined by filtering out any suspended material in the water, evaporating off the water, and weighing the dry residue that remains. TDS levels indicate the potential uses for a water body since the TDS is basically an indicator of the salt content. Freshwater has a TDS of less than 1,000 mg/L. Surface waters with significantly high TDS may not be usable for potable water or irrigation purposes. As noted earlier, the TDS of sea water is approximately 35,000 mg/L. This means that for every kilogram of seawater there are approximately 35 g of dissolved salt.

Alkalinity is a measure of the buffering capacity of water. It does not refer to pH, but instead refers to the ability of water to resist change in pH upon addition of acid or base. Waters with low alkalinity are very susceptible to changes in pH, while waters with high alkalinity are able to resist major shifts in pH. The buffering chemicals in most natural waters are bicarbonate (HCO_3^-), and carbonate (CO_3^-), although borates, silicates, phosphates, ammonium, sulfides, and organic acids can also contribute to a small degree. Water having a pH below 4.5 contains virtually no alkalinity, because all of the bicarbonate and carbonate have been converted to dissolved carbon dioxide. The amount of alkalinity therefore determines the ability of a water body to neutralize acidic pollution from rainfall or wastewater. Most surface waters typically have alkalinity ranging from 10 to 200 mg/L as $CaCO_3$ (Hem 1970).

6.4.2 Dissolved Gases

The Earth's atmosphere is comprised of a layer of gases that are retained by gravity. It contains roughly 78% nitrogen (N_2), 21% oxygen (O_2), 0.93% argon (Ar), 0.038% carbon dioxide (CO_2), trace amounts of other gases, and a variable amount (about 1% on average) of water vapor. The atmosphere protects life on Earth by filtering out harmful ultraviolet solar radiation, and by trapping infrared radiation (heat) from escaping which regulates surface temperatures at habitable levels.

All gases present in the Earth's atmosphere dissolve to some extent into water that is in contact with it. Thus, all surface water has small amounts of N_2, O_2, Ar, CO_2, and other gases dissolved in it. For aquatic life forms, the presence of this small amount of oxygen is essential for survival. Most species of fish, for example, require at least 5 mg/L of dissolved oxygen. CO_2 acts as a weak acid when dissolved in water thereby imparting rain water with its characteristic slightly acidic pH. N_2 and Ar, although present in all waters, do not engage in chemical reactions to any significant extent and their presence is usually ignored.

6.4.3 Heavy Metals

The term "heavy metals" is an ambiguous one, and not necessarily associated with any specific set of elements, and therefore does not imply any common set of properties such as high toxicity, high atomic weight, etc. (Duffus 2002). Nonetheless, the term has been used more and more in the literature. A simple, but useful, definition of a heavy metal is any metallic element on the periodic table with an atomic number larger than that of calcium (atomic number = 20). Examples of heavy metals include titanium (Ti), manganese (Mn), chromium (Cr), vanadium (V), copper (Cu), cobalt (Co), lead (Pb), scandium (Sc). and zinc (Zn).

Heavy metals are natural components of the Earth's crust (Table 6.1). Unlike organic chemicals, they cannot be degraded or destroyed. To a small extent they enter one's body via food, drinking water, and air. Some heavy metals (e.g. copper, selenium, zinc, etc.) are micronutrients, and are essential to maintain the metabolism of the human body. However, at higher concentrations they can cause adverse health effects. Some metals that are in low abundance can have a large environmental impact. Mercury (Hg), arsenic (As), selenium (Se), and silver (Ag), for example, are all widely considered to be environmental stressors. Small amounts of these compounds can be harmful to both human health and aquatic life.

Heavy metals can enter a water supply by industrial and consumer waste, from drinking water contamination (e.g. lead and copper pipes), or from acidic rain breaking down soils and leaching heavy metals into streams, lakes, rivers, and groundwater. Additional routes of exposure include inhalation of high ambient air concentrations near emission sources or ingestion of metals via the food chain (Duffus 2002).

6.4.4 Organic Constituents

Organic chemistry is the study of the properties of chemical compounds containing carbon. Nearly all chemical compounds that have carbon atoms are considered organic. The exceptions, termed inorganic carbon compounds, include carbon dioxide, bicarbonate and carbonate ions, cyanide, metal carbides, and a handful of other compounds. On the other hand, the number of organic carbon compounds is impossibly large to count, a quality that arises from the fact that carbon readily bonds with other carbon atoms in many configurations. This creates countless ways in which carbon atoms can be arranged relative to one another. In many organic molecules, carbon is bound to hydrogen. In addition, carbon atoms are also able to bind with other elements such as oxygen, nitrogen, phosphorous, sulfur, and chlorine, thereby further increasing the possible number of distinct and different organic carbon constituents that may be present in water.

Organic chemicals of environmental interest are usually classified into various groups or categories based upon similar chemical properties or common origins. The most basic distinction is made based upon whether an organic chemical is naturally occurring or synthetic. Naturally occurring organic molecules include fossil fuels such as methane (CH_4) gas and the complex mixture of compounds present in petroleum, sugars, and starches, and biomolecules such as proteins and enzymes. The G, T, A, and C base pairs of DNA are organic, thereby making DNA an organic molecule.

When plant, animal, and microbial material in soil and water undergo decomposition, a variety of complex organic molecules are produced that are called natural organic matter (NOM). Although NOM is ubiquitous in the environment, the structure and properties of the molecules themselves are not well understood. NOM plays an important role in aquatic toxicology because it interacts with metal ions and minerals to form complexes of a widely differing chemical and biological nature (Tipping 2002). When NOM binds with metal ions, the metal ions become less bioavailable, which lowers their potential toxicity to aquatic life. However, NOM creates problems for the water supply industry, requiring removal to minimize water color and odor, and giving rise to potentially harmful chemical byproducts after chlorination. Through a process called "biofouling," NOM can also degrade the performance of membrane filtration systems used for water purification and desalination by supporting the growth of microorganisms on these treatment membranes.

Synthetic organic chemistry is the science of the design, analysis, and/or construction of organic chemicals for practical purposes. As such, synthetic organic chemicals (SOCs) are chemicals that are produced on a large scale for use by humans. SOCs include several subclasses of chemicals such as pesticides, industrial solvents, chelating agents and disinfection byproducts. Many of these compounds are highly toxic and tested routinely in public water supply systems to ensure finished water is safe for human consumption.

A pesticide is any substance or mixture of substances designed to prevent, destroy, repel, or mitigate any pest. Insecticides, herbicides, and fungicides all fall under the pesticide umbrella. Organophosphate pesticides such as malathion and parathion, and carbamate pesticides such as aldicarb and methomyl affect the nervous system of insects by disrupting enzymes that regulate neurotransmitters. Organochlorine insecticides such a DDT and chlordane were common in the past, but many have been banned from use in the United States and many other countries due to their health and environmental effects as well as their persistence. Pyrethroids are a class of pesticides developed as a synthetic version of the naturally occurring pesticide pyrethrin.

Organic solvents are a chemical class of compounds that are used routinely in commercial industries for dissolving other organic compounds. They have found extensive use in dry cleaning (e.g. tetrachloroethylene or PCE), as paint thinner (e.g. toluene, turpentine), as nail polish remover and glue solvents (e.g. acetone), in spot removers (e.g. hexane), in detergents (e.g. limonene), and in perfumes (e.g. ethanol). They are also particularly useful in the synthesis of other organic chemicals. Many organic solvents are recognized by the EPA as known or suspected human carcinogens. Volatile organic compounds (VOCs) are another class of organic chemical compounds that are characterized as having high enough vapor pressures to significantly vaporize and enter the atmosphere. Many of the organic solvents discussed earlier are also VOCs.

Chelating agents, or chelators for short, are a class of SOCs that are used in chemical analysis as water softeners, and are ingredients in many commercial products such as shampoos and food preservatives. The most commonly used synthetic chelating agents are NTA and EDTA (Grundler et al. 2004). Due to its inability to be broken down in many wastewater treatment plants, significant concentrations of EDTA have been found in the environment. Long-term accumulation of chelating agents is not a concern, however, because they are eventually broken down by bacteria (Grundler et al. 2004). Their presence in surface waters is more of a concern due to their ability to solubilize heavy metals thereby making them mobile.

Disinfection byproducts (DBPs) are a class of chemical compounds that are formed when water is disinfected. Disinfection of drinking water with chlorine has been applied since the 1900s, and has prevented the spread of waterborne diseases such as cholera and typhoid. However, during the 1970s, scientists discovered that chlorination of drinking water containing moderate to high levels of NOM produced a new class of compounds, DBPs, which were later shown to be harmful to human health. Alternative disinfectants such as ozone or chlorine dioxide produce their own characteristic DBPs. Thus, switching from chlorine to ozone or chlorine dioxide is not a final remedy.

Pharmaceuticals and personal care product (PPCPs) refer to any product used by individuals for personal health or cosmetic reasons or used by agriculture to enhance growth or health of livestock. PPCPs comprise a diverse collection of thousands of individual chemicals, including prescription and over-the-counter therapeutic drugs, veterinary drugs, fragrances, cosmetics, sunscreen products, bug repellants, diagnostic agents, and vitamins. Individuals contribute PPCPs to the environment through excretion, bathing, and disposal of medications and personal care products to sewers and trash. The various sources of PPCPs include human activity (e.g. bathing, shaving, swimming, etc.), illicit drug use, veterinary drug use – especially antibiotics and steroids, agriculture, and residue from pharmaceutical manufacturing and hospitals. Studies have shown that PPCPs are present in the nation's water bodies. To date, scientists have found no evidence of adverse human health effects from PPCPs in the environment. However, there is strong evidence of ecological harm. PPCPs that can affect the endocrine system in animals, which controls important function through communication of glands, hormones, and cellular receptors, are known as endocrine-disrupting compounds (EDCs). Many EDCs are associated with developmental, reproductive, and other health problems in fish and wildlife, both in the field and the laboratory.

6.4.5 Nutrients

Nutrients are chemical elements critical to the growth of plant and animal life. In healthy rivers and lakes, nutrients are needed for the growth of algae that form the base of a complex food web that supports the entire aquatic ecosystem. The nutrients that receive the most attention in lakes and streams are nitrogen (as nitrate and ammonia) and phosphorus (as orthophosphate or total phosphate).

If provided with an abundance of nutrients, algae and aquatic plants will continue to grow well beyond the amount needed to support the food web. The excess algae and plants will die and consume dissolved oxygen as microorganisms break down their cellular material. As a result, other aquatic organisms may suffer from lack of oxygen. Other problems associated with excessive algal and plant growth include scum and foam formation, and taste and odor problems if the water is used for drinking.

Eutrophication is the natural process of enrichment of lakes and streams with nutrients, and the associated biological and physical changes that result. Human activity has dramatically increased the rate of eutrophication in many water bodies. Lake and pond are particularly vulnerable to eutrophication because the nutrients carried into them continue to build up; in contrast, the nutrients present in rivers and streams can be carried away in moving water.

Phosphorus in the form of phosphates is contributed to surface and groundwater naturally from soil and dissolution of rocks, while natural sources of nitrogen include leaves and other organic debris from riparian vegetation. The primary anthropogenic sources of these nutrients are wastewater treatment plants, septic systems, storm water runoff, suspended sediment resulting from excessive erosion (only P), acid rain (only N), animal manure, and commercial fertilizers. In the past, household detergents brought high loads of phosphorus to treatment plants, which then were discharged with the effluent. In the United States, however, laws restricting the phosphorus content of detergents have produced markedly reduced phosphate levels in many surface water bodies.

6.5 Chemical Reactions

Chemistry is the science of making new substances out of old substances via chemical reactions. All of the chemical constituents described earlier undergo chemical reactions that result in their being degraded, as is the case for many of the organic chemicals, or transformed to another form, as is the case for many of the inorganic chemicals. When studying a chemical reaction, chemists often pose some basic questions. What is the driving force that make this reaction occur? What is the equilibrium state of this chemical system? These questions are answered with the help of a subject known as chemical thermodynamics. Why is this reaction so fast? Why is that reaction so slow? These questions are answered with the help of a subject known a chemical kinetics.

In the same way that water will always find its own level, chemical reactions proceed in a way that minimizes the useful energy that is available. Chemical thermodynamic calculations quantify the change in this energy (known as the

Gibbs free energy) as a reaction proceeds. This allows for one to determine the equilibrium state of this chemical system. The calculations are often relatively simple, and there are many commercially available software programs that automate the task. The results from equilibrium calculations are often a reasonable approximation for many systems, and even if the system is not at equilibrium they provide information about the direction and extent to which reactions will proceed (Benjamin 2002; Theodore et al. 2009).

The subject of chemical kinetics allows one to quantify how fast chemical reactions occur and answer why certain reactions are faster than others (Reynolds et al. 2004; Stumm and Morgan 1996; Theodore 2012). Chemical kinetics is often quantified through the measurement of the rate of change in concentrations of reactant and/or products. The most important factors that influence rates of chemical reactions are the nature and concentration of the reactant(s).

Increasing the temperature of a system imparts more kinetic energy to molecules, thereby serving to increase rates of chemical reactions. The detailed explanation of how a reaction proceeds at a molecular level is called a reaction mechanism. Determination of reaction mechanisms requires a broad and detailed understanding of the properties of reactants and products, and the changes that occur before, during, and after a chemical reaction, and is often difficult if not impossible to confirm unequivocally.

6.6 Illustrative Examples

6.6.1 Illustrative Example 1

Describe the role of nutrients in natural waters.

Solution. Nutrients are chemical elements critical to the growth of plant and animal life. In healthy rivers and lakes nutrients are needed for the growth of algae which form the base of a complex food web that supports the aquatic ecosystem.

6.6.2 Illustrative Example 2

Describe the neutrality aspect of water in terms of pH.

Solution. Pure water has a neutral pH of exactly 7.0. Values of pH less than this are considered acidic, while pH values greater than this are termed alkaline. The typical pH range of water is from 0 to 14, although values outside of this range can be attained under extreme conditions. The pH of water is an important chemical property which controls the distribution of chemical species among various forms.

6.6.3 Illustrative Example 3

An aqueous, industrial wastewater has an alkalinity of 60 mg/L as $CaCO_3$ and a dissolved CO_2 concentration of 7.0 mg/L. Determine the pH of this wastewater.

Solution. Since the wastewater has dissolved CO_2 reported, alkalinity is provided by the equilibrium of carbonic acid (H_2CO_3, dissolved CO_2) and the bicarbonate ion. The first ionization constant of H_2CO_3 is:

$$K_{i,1} = \frac{[H^+][HCO_3^-]}{[H_2CO_3]} = 3.98 \times 10^{-7} \tag{6.1}$$

"Alkalinity" is measured as equivalents of $CaCO_3$ which has a molecular weight of 100 and an equivalent weight of 50. For example, a water solution of 17 mg/L of OH^- (equivalent weight = 17 g/equivalent) contains 10^{-3} equivalent weights of OH^-/L, which is equal to 10^{-3} equivalents of $CaCO_3$/L. The alkalinity of that solution is therefore $(50)(10^{-3})$ g/L or 50 mg/L as $CaCO_3$.

For an alkalinity of 60 mg/L as $CaCO_3$, the concentration of HCO_3^- is:

$$[HCO_3^-] = \frac{60 \text{ mg/L}}{50 \text{ meq/L}} = 1.2 \text{meq} / \text{L} = 1.2 \text{mmol} / L$$

The unit "meq" represents 10^{-3} equivalents. For the dissolved CO_2,

$$[CO_2] = \frac{7 \text{ mg/L}}{44 \text{ mg/mmol}} = 0.159 \text{mmol} / \text{L}$$

Note that CO_2 dissolved in aqueous solution is in the form of H_2CO_3. Substitution in Equation 6.1 gives

$$3.98 \text{ x } 10-7 = \frac{[H^+](1.2 \text{ x } 10^{-3})}{(0.159 \text{ x } 10^{-3})};\ \left[H^+\right] = 5.27 \times 10^{-8} \text{g} - \text{ion} / L$$

The pH is therefore given by:

$$\text{pH} = -\log\left[H^+\right] = -\log\left(5.27 \times 10^{-8}\right) = 7.28$$

6.6.4 Illustrative Example 4

An elementary school science teacher found an old bottle containing 2 L of concentrated sulfuric acid in the science room storage closet. Not having any use for the acid, the teacher considered disposing of it by pouring it down the drain with the faucet running. The principal happened to stop by and informed the teacher that Federal regulations require that any material put down the drain must have a pH greater than 2.0 and less than 12.5 in order to protect the aquatic environment and the bacteria in the local wastewater treatment plant, not to mention the pipe in the sewerage system. The teacher considered three methods of bringing the acid to a pH of 2.1 to comply with the regulations and to dispose of the acid safely and quickly. These three methods included: dilution, neutralization to Na_2SO_4 and H_2O with NaOH, and neutralization to $CaSO_4$ and H_2O with slaked lime, $Ca(OH)_2$. Answer the following four questions: (1) Find the pH of the concentrated H_2SO_4, (2) Find the volume to which the 2 L of the 18 M H_2SO_4 would have to be diluted in order to raise the mixture pH to 2.1, (3) Calculate the quantity of NaOH needed to neutralize the 2 L of 18 M H_2SO_4, and (4) Calculate the quantity of $Ca(OH)_2$ needed to neutralize the 2 L of 18 M H_2SO_4.

Solution. With the strong acid ionizing 100%, the acid dissociates completely as follows:

$$H_2SO_4 \rightarrow 2H^+ + SO_4^- \tag{6.2}$$

1) Therefore, 18 gmol/L H_2SO_4 would ionize to yield 36 g-ion/L H^+ to yield a pH of:

$$pH = -\log\left[H^+\right] = -\log(36) = -1.6$$

Note: a negative pH is possible at very high acid concentrations.

2) Next, the H^+ concentration in a sulfuric acid solution of pH = 2.1 can be calculated as:

$$\left[H^+\right] = 10^{-pH} = 10^{-2.1} = 7.94 \times 10^{-3} \text{g-ion / L}$$

The molarity of this solution is 7.94 x 10^{-3}/2 = 3.97 x 10^{-3} M. Thus, in order to raise the pH of the 2 L of concentrated sulfuric acid from −1.6 to 2.1, the molarity of the solution must be lowered from 18 M to approximately 0.004 M. The volume of the diluted acid can be calculated from a mass balance for hydrogen ion, i.e. $M_1V_1 = M_2V_2$:

$$18\,M(2\,L) = 0.004\,M(V_2);\; V_2 = 9{,}000\,L$$

Note: Since 9,000 L equals approximately 2,250 gallons of solution this is clearly not a practical method of disposal.

3) The number of moles of NaOH required to neutralize the sulfuric acid is based on the number of moles of sulfuric acid in the 18 M solution and the

stoichiometric relationship describing this neutralization reaction, or 2 L (18 gmol/L) = 36 gmol to be neutralized according to the stoichiometry:

$$H_2SO_4 + 2NaOH \rightarrow Na_2SO_4 + 2H_2O \tag{6.3}$$

This requires 2 gmol of NaOH for each gmol of sulfuric acid neutralized for a total of 72 gmol of NaOH for complete neutralization. Converting gmol to pounds of NaOH required yields:

$$\frac{(72\text{ g mol NaOH})(40\text{ g/gmol})}{454\text{ g/lb}} = 6.3\,\text{lb NaOH}$$

Note: As a laboratory scale reaction, this would be a very large reaction, generating a significant amount of heat with considerable chance of splattering and injury.

4) The number of moles of $Ca(OH)_2$ required to neutralize the sulfuric acid is again based on the number of moles of sulfuric acid in the 18 M solution and the stoichiometric relationship describing this neutralization reaction, or 2 L (18 gmol/L) = 36 gmol to be neutralized according to the stoichiometry:

$$H_2SO_4 + Ca(OH)_2 \rightarrow CaSO_4 + 2H_2O \tag{6.4}$$

This requires 1 gmol of $Ca(OH)_2$ for each gmol of sulfuric acid neutralized for a total of 36 gmol of $Ca(OH)_2$ for complete neutralization. Converting gmol to pounds of $Ca(OH)_2$ required yields:

$$\frac{(36\text{ g mol Ca(OH)2})(74\text{ g/gmol})}{454\text{ g/lb}}) = 5.9\,\text{lb}\,Ca(OH)_2$$

Note: Again, as a laboratory scale reaction, this would still be a very large reaction, generating a significant amount of heat with considerable chance of splattering and injury. The teacher and principle should look for alternative labs within their school or their district that could make use of the sulfuric acid as a way to safely manage this potentially hazardous material.

6.6.5 Illustrative Example 5

A biologically contaminated hospital liquid waste requires disinfection with chlorine prior to discharge into a nearby lake. Given a contact time of 30 minutes, a required chlorine dose for a 99.99% kill of the pathogenic organisms = 3.0 mg/L, and a 0.5 mg/L chlorine requirement to carry out oxidation–reduction reactions with compounds in the liquid waste, how many pounds of pure chlorine (in the gaseous form) would be required daily to disinfect a 100,000 gal/d waste flow?

Solution. The total chlorine concentration required to meet the oxidation–reduction reaction requirement in the water is 0.5 mg/L, while 3.0 mg/L chlorine are required for the disinfection reaction. The total chlorine dose necessary then is then 3.5 mg/L. The required chlorine dose (D) with units of lb/d is found by employing appropriate conversion factors as follows:

$$Q, \mathrm{MGD} = (100{,}000\,\mathrm{gal/d})(1\,\mathrm{MG}/10^6\,\mathrm{gal}) = 0.10\,\mathrm{MGD}$$

$$\mathrm{Dose} = (3.5\,\mathrm{mg/L})(0.10\,\mathrm{MG})(8.34\,\mathrm{lb/MG/mg/L}) = 2.94\,\mathrm{lb/d}$$

6.7 Problems

1) Sea water contains approximately 3.5% salt which is predominantly NaCl. Determine the approximate weight% of Na and Cl in sea water.
2) Calculate the composition in units of weight%, of Na and Cl, in NaCl.
3) A waste stream contains 0.013% by weight of aminobenzoic acid that has a molecular formula of $C_7H_7NO_2$. Determine the following characteristics of this aqueous waste stream: (1) the weight% of nitrogen in the acid, (2) the weight% of N in the waste stream, (3) the equivalent concentration of the acid in the waste stream in units of mg/L, and (4) the equivalent concentration of nitrogen in the waste stream in units of mg/L.
4) Calculate the g-ion/L concentration of hydrogen and hydroxide ion in a basic solution with a pH of 8.0.
5) Looking at Illustrative Example 4, calculate the quantity of baking soda, NaHCO3, or quick lime, CaO, that would be required for the neutralization of the concentrated sulfuric acid bottle, and comment on their use as alternative neutralizing agents for this sulfuric acid disposal problem.

References

Benjamin, M.M. (2002). *Water Chemistry*. New York, NY: McGraw-Hill.

Carbonaro, R. (2021). Water chemistry. In: *Introduction to Environmental Management*, 2nd Edition (ed. M.K. Theodore and L. Theodore). Boca Raton, FL: CRC Press/Taylor Francis Group.

Clesceri, L.S., Greenberg, A.E., and Eaton, A.D. (1998). *Standard Methods for the Examination of Water and Wastewater*, 20th Edition. Washington, D.C.: American Public Health Association Publications.

Duffus, J. (2002). "Heavy metals" – A meaningless term? *Pure and Applied Chemistry* 74: 793–807.

Grundler, O., Hans-Ulrich, J., and Witteler, H. (2004). Environmental impact of amniocarboxylate chelating agents. In: *Handbook of Detergents, Part B: Environmental Impact (Surfactant Science)* (ed. U. Zoller). Boca Raton, FL: CRC Press.

Hem, J.D. (1970). *Study and Interpretation of the Chemical Characteristics of Natural Water*. Water Supply Paper 2254. Washington, D.C.: U.S. Geological Survey.

Lutgens, F.K. and Tarbuck, E.J. (2000). *Essentials of Geology*, 7th Edition. Upper Saddle River, NJ: Prentice Hall.

MWH. (2005). *Water Treatment: Principles and Design*. Hoboken, NJ: John Wiley & Sons.

Nicholls, D. (1975). *Complexes and First-Row Transition Elements*. New York, NY: Elsevier.

Reynolds, J., Jeris, J., and Theodore, L. (2004). *Handbook of Chemical and Environmental Engineering Calculations*. Hoboken, NJ: John Wiley & Sons.

Stumm, W. and Morgan, J.J. (1996). *Aquatic Chemistry: Chemical Equilibria and Rates in Natural Waters*. Hoboken, NJ: John Wiley & Sons.

Theodore, L. (2012). *Chemical Reactor Analysis and Applications for the Practicing Engineer*. Hoboken, NJ: John Wiley & Sons.

Theodore, L., Ricci, F., and VanVliet, T. (2009). *Thermodynamics for the Practicing Engineer*. Hoboken, NJ: John Wiley & Sons.

Tipping, E. (2002). *Cation Binding by Humic Substances*. Cambridge, UK: Cambridge University Press.

7

The Conservation Laws, Stoichiometry, and Thermodynamics

7.1 Introduction

In order to better understand the design, as well as the operation and performance of equipment and processes in the environmental industry, it is necessary for engineers (as well as applied scientists) to understand the fundamentals and principles underlying the conservation laws and thermodynamics. How can one predict what products will be emitted from effluent streams? At what temperature must a unit be operated to ensure the desired performance? How much energy in the form of heat is given off during a combustion process? Is it economically feasible to recover this heat? Is the feed high enough in heating value, or must additional fuel be added to assist in the thermal process? If so, how much fuel must be added? How much energy is required to operate a desalination plant? What is the minimum amount of energy that must be expended? One of the keys necessary to answering these questions is often obtained via the application of one or more of the conservation laws or are rooted in the various theories of thermodynamics including thermochemistry, phase equilibrium, and chemical reaction equilibrium.

This chapter provides a brief summary of the conservation laws and then explores in more detail the conservation laws for momentum, mass, and energy. Materials on stoichiometry and thermodynamics are also highlighted in this chapter. An application section provides seven illustrative examples related to the general subject of conservation laws, stoichiometry, and thermodynamics relevant to desalination operations.

7.2 The General Conservation Law

Four important terms are defined below before proceeding to the conservation laws:

1. A *system* is any portion of the universe that is set aside for study.
2. Once a system has been chosen, the rest of the universe is referred to as the *surroundings*.

Introduction to Desalination: Principles, Processes, and Calculations, First Edition. Louis Theodore and R. Ryan Dupont.

3. A system is described by specifying that it is in a certain *state*.
4. The *path*, or series of values certain variables assume in passing from one state to another defines a *process*.

Momentum, energy, and mass are all conserved. As such, each quantity obeys the general conservation law below, as applied within a system.

$$\left\{\begin{matrix}quantity\\ into\\ system\end{matrix}\right\} - \left\{\begin{matrix}quantity\\ out\,of\\ system\end{matrix}\right\} + \left\{\begin{matrix}quantity\\ generated\,in\\ system\end{matrix}\right\} = \left\{\begin{matrix}quantity\\ accumulated\\ in\,system\end{matrix}\right\} \tag{7.1}$$

Equation 7.1 may also be written on a time rate basis:

$$\left\{\begin{matrix}rate\\ into\\ system\end{matrix}\right\} - \left\{\begin{matrix}rate\\ out\,of\\ system\end{matrix}\right\} + \left\{\begin{matrix}rate\\ generated\,in\\ system\end{matrix}\right\} = \left\{\begin{matrix}rate\\ accumulated\\ in\,system\end{matrix}\right\} \tag{7.2}$$

The conservation law may be applied by the practitioner at the macroscopic, microscopic, or molecular level. The differences in these methods can be best illustrated with an example. Consider a system in which fluid is flowing through a cylindrical tube (Figure 7.1). Define the system as the fluid contained within the tube between Points 1 and 2 at any time. If one is interested in determining changes occurring at the inlet and outlet of a system, the conservation law is applied on a "macroscopic" level to the entire system without regard for internal variations *within* the system. It is this approach that is usually applied by the practicing engineer.

The microscopic approach is employed when detailed information concerning the behavior *within* a system is required; this is often requested of and by the environmental engineer or scientist. The conservation law is then applied to a *differential* element within the system that is large compared to an individual molecule, but small compared to the entire system. The resulting differential equation is then expanded via an integration in order to describe the

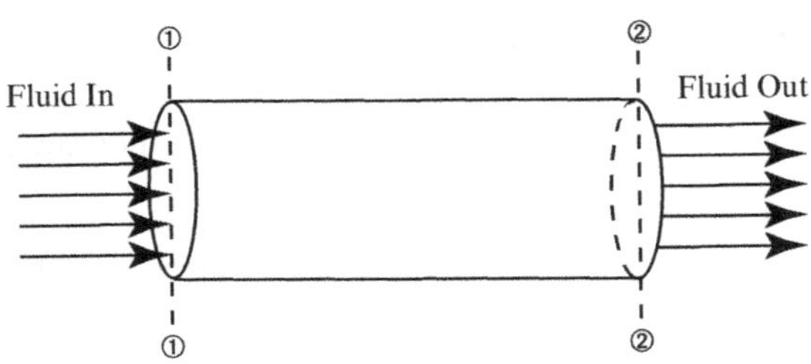

Figure 7.1 Fluid Flow through a Cylinder Tube.

behavior of the entire system. This is defined by some as the *transport phenomena approach* (Bird et al. 2002; Theodore 1970).

The molecular approach involves the application of the conservation laws to individual molecules. This leads to a study of statistical and quantum mechanics – both of which are beyond the scope of this text. In any case, the description at the molecular level is of little value to the practicing engineer. However, the statistical averaging of molecular quantities in either a differential or finite element within a system can lead to a more meaningful description of the behavior of a system.

Both the microscopic and molecular approaches shed light on the physical reasons for the observed macroscopic phenomena. Ultimately, however, for the practicing engineer, these approaches may be valid but are akin to attempting to kill a fly with a machine gun. Developing and solving these differential equations (in spite of the advent of computer software packages) is typically not worth the trouble.

Traditionally, the applied mathematician has developed differential equations describing the detailed behavior of systems by applying the appropriate conservation law to a differential element or shell within the system. Equations were derived with each new application. The engineer later removed the need for these tedious and error-prone derivations by developing a general set of equations that could be used to describe systems. These have come to be referred to by many as the *transport equations*. In recent years, the trend toward expressing these equations in vector form has gained momentum (no pun intended). However, the shell-balance approach has been retained in most texts where the equations are presented in componential form, i.e. in three particular coordinate systems – rectangular, cylindrical, and spherical. The componential terms can be "lumped" together to produce a more concise equation in vector form. The vector equation can be, in turn, re-expanded into other coordinate systems. This information is available in the literature (Bird et al. 2002; Theodore 1970).

It should be noted that the macroscopic approach has been primarily adapted by undergraduate environmental engineering educators. Any attempt to include the microscopic approach has been essentially reserved solely for graduate studies in this area. Only the macroscopic approach is employed in this text. This chapter's aim, then, is to express the laws of conservation for momentum, energy, and mass in algebraic or finite difference form.

7.3 The Conservation Law for Momentum

The general conservation law for momentum on a rate basis when applied to a volume element is provided in Equation 7.3. The law numerates the rate of momentum and forces acting on a moving fluid in a volume element of concern at any time t. Each rate of momentum or force term in the equation can be expressed in terms of lb_f in order to maintain dimensional consistency. It is suggested that the reader refers to the literature for more information (Bird et al. 2002; Theodore 1970). The application of this conservation law finds extensive use in the field of fluid mechanics or fluid flow (Abulencia and Theodore 2009). An example application of this law is provided in Illustrative Example 2, and additional details are proved in Chapter 8 on unit operations.

7.4 The Conservation Law for Mass

The conservation law states that without nuclear reaction, mass can neither be created nor destroyed. From an applications perspective it can be applied to any process, equipment, or system. The general form of this law is given by Equations 7.4 and 7.5.

Nuclear reactions, in which interchanges between mass and energy are known to occur provide a notable exception to this law. Even in chemical reactions, a certain amount of mass-energy interchange takes place. However, in normal environmental engineering applications, nuclear reactions do not occur and the mass–energy exchange in chemical reactions is so minuscule that it is not worth taking into account.

$$\underset{(1)}{\begin{Bmatrix} \text{rate of} \\ \text{momentum} \\ \text{in by} \\ \text{convection} \end{Bmatrix}} - \underset{(2)}{\begin{Bmatrix} \text{rate of} \\ \text{momentum} \\ \text{out by} \\ \text{convection} \end{Bmatrix}} + \underset{(3)}{\begin{Bmatrix} \text{rate of} \\ \text{momentum in} \\ \text{by molecular} \\ \text{diffusion} \end{Bmatrix}} - \underset{(4)}{\begin{Bmatrix} \text{rate of} \\ \text{momentum out} \\ \text{by molecular} \\ \text{diffusion} \end{Bmatrix}} +$$

bulk flow velocity gradients

$$\underset{(5)}{\begin{Bmatrix} \text{external} \\ \text{forces} \\ \text{exerted} \\ \text{on fluid} \end{Bmatrix}} = \underset{(6)}{\begin{Bmatrix} \text{rate of momentum} \\ \text{accumulation} \end{Bmatrix}} \tag{7.3}$$

surface forces inventory
body forces

$$\begin{Bmatrix} \text{mass} \\ \text{in} \end{Bmatrix} - \begin{Bmatrix} \text{mass} \\ \text{out} \end{Bmatrix} + \begin{Bmatrix} \text{mass} \\ \text{generated} \end{Bmatrix} - \begin{Bmatrix} \text{mass} \\ \text{consumed} \end{Bmatrix} = \begin{Bmatrix} \text{mass} \\ \text{accumulated} \end{Bmatrix} \tag{7.4}$$

$$I \quad - \quad O \quad + \quad G \quad - \quad C \quad = \quad A$$

or on a time rate basis by

$$\begin{Bmatrix} \text{rate of} \\ \text{mass} \\ \text{in} \end{Bmatrix} - \begin{Bmatrix} \text{rate of} \\ \text{mass} \\ \text{out} \end{Bmatrix} + \begin{Bmatrix} \text{rate of} \\ \text{mass} \\ \text{generated} \end{Bmatrix} - \begin{Bmatrix} \text{rate of} \\ \text{mass} \\ \text{consumed} \end{Bmatrix} = \begin{Bmatrix} \text{rate of} \\ \text{mass} \\ \text{accumulated} \end{Bmatrix} \tag{7.5}$$

$$I \quad - \quad O \quad + \quad G \quad - \quad C \quad = \quad A$$

In environmental related processes, including desalination operations, it is often necessary to obtain quantitative relationships by writing mass balances on the various elements in the system. This equation may be applied either to the total mass involved or to a particular species on either a mole or mass basis. This law can be applied to steady-state or unsteady-state (transient) processes and to batch or continuous systems. As noted earlier, in order to isolate a system for study, it is separated from the surroundings by a boundary or envelope. This boundary may be real (e.g. the walls of a thermal device) or imaginary. Mass crossing the boundary and entering the system is part of the *mass in* term in Equation 7.4, while that crossing the boundary and leaving the system is part of the *mass out* term. Equation 7.4 may be written for any compound whose quantity is not changed by chemical reaction and for any chemical element whether or not it has participated in a chemical reaction. It may be written for one piece of equipment, around several pieces of equipment, or around an entire process. It may be used to calculate an unknown quantity directly, to check the validity of experimental data, or to express one or more of the independent relationships among the unknown quantities in a particular problem situation.

A steady state process is one in which there is no change in conditions (pressure, temperature, composition, etc.) or rates of flow with time at any given point in the system. The accumulation term in Equation 7.4 is then zero. (If there is no chemical or nuclear reaction, the generation term is also zero.) All other processes are *unsteady state*.

In a *batch process*, a given quantity of reactants is placed in a container, and by chemical and/or physical means, a change is made to occur. At the end of the process, the container (or containers) to which material may have been transferred holds the product or products.

In a *continuous process*, reactants are continuously fed to a piece of equipment or to several pieces in series, and products are continuously removed from one or more points. A continuous process may or may not be steady-state. A coal-fired power plant, for example, operates continuously. However, because of the wide variation in power demand between peak and slack periods, there is an equally wide variation in the rate at which the coal is fired. For this reason, power plant problems may require the use of average data over long periods of time. However, most thermal operations are assumed to be steady-state and continuous.

As indicated previously, Equation 7.4 may be applied to the total mass of each stream (referred to as an *overall* or *total material balance*) or to the individual component(s) of the stream (referred to as a *componential* or *component material balance*). The primary task in preparing a material balance in engineering calculations is often to develop the quantitative relationships among the streams. The primary factors, therefore, are those that *tie* the streams together. An element, compound, or unreactive mass (e.g. ash) that enters or exits in a single stream or passes through a process unchanged is so convenient for his purpose that it may be considered a *key* to the calculations. If sufficient data are given about this component, it can be used in a component balance to determine the total masses of the entering and exiting streams. Such a component is sometimes referred to as the *key component.* Since a key component does not react in a process, it must retain its identity as it passes through the process. Obviously, except for nuclear reactions, elements may always be used as key components because they do not change identity even though they may undergo a chemical reaction. Thus, CO (carbon monoxide) may be used as a key component only when it does not react, but C (carbon) may always be used as a key component. A component that enters the system in only one stream and leaves in only one stream is usually the most convenient choice for a key component. Additional details are provided in the stoichiometric discussion provided later.

Four important processing concepts are *bypass*, *recycle*, *purge*, and *makeup*. With bypass, part of the inlet stream is diverted around the equipment to rejoin the (main) stream after the unit (see Figure 7.2). This stream effectively moves in parallel with the stream passing through the equipment. In recycle, part of the product stream is sent back to mix with the feed. If a small quantity of nonreactive material is present in the feed to a process that includes recycle, it may be necessary to remove the nonreactive material in a purge stream to prevent its building up above a maximum tolerable value. This can also occur in a process without recycle; if a nonreactive material is added in the feed and not totally removed in the products, it will accumulate until purged. The purging process is sometimes referred to as blowdown. Makeup as its name implies, involves adding or making up part of a stream that has been removed from a process. Makeup may be thought of as the opposite of purge and/or blowdown.

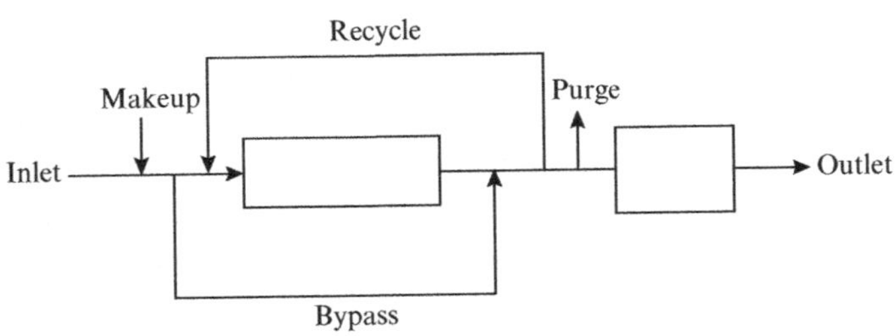

Figure 7.2 Bypass, Recycle, Purge, and Makeup.

7.5 The Conservation Law for Energy

From the early recognition of energy, man has studied its effects upon objects, its transfer from object to object, and its transformation from one form to another. This field of study is an integral part of engineering. Before proceeding to the first law of thermodynamics, often referred to as the conservation law for energy, certain important terms must be defined.

1. *Isothermal* means constant temperature.
2. *Isobaric* means constant pressure.
3. *Isochoric* means constant volume.
4. *Adiabatic* specifies no transfer of heat to or from systems.

The first law of thermodynamics specifies that energy is conserved. In effect, this law states that energy is neither created nor destroyed. Thus, the change in energy of a system is exactly equal to the negative of the change in the surroundings. For a system of constant mass (a closed system), the only way the system and surroundings may interchange energy is by work and heat. Work and heat are defined as "energy in transit." They are not properties and cannot be stored in a system. Two common forms of work are expansion and electrical. Heat is energy in transit because it is transferred through temperature differences. This heat transfer may take place by conduction, convection, or radiation (Theodore 2014).

An energy balance makes use of the conservation law to account for all the energy in a chemical process, or in any other processes for that matter. After a system is defined, the energy balance considers the energy entering the system across a boundary, the energy leaving the system across the boundary, and the accumulation of energy within the system. This may be written in a simplified equation form as:

$$\text{energy in} - \text{energy out} = \text{energy accumulated} \tag{7.6}$$

This expression has the same form as the general law of conservation of mass as well as the conservation law for momentum. It may also be written on a rate basis.

All forms of energy must be included in an energy balance. In many processes, certain energy forms remain constant therefore any changes in them may be neglected. However, these forms should be recognized and understood before their magnitude and constancy can be determined. Some forms of energy are easily recognized in everyday life: the energy of a moving object, the energy given off by a fire, and the energy content of a container of hot water. Other forms of energy are less easily recognized. However, the five key energy terms are kinetic, potential, internal, heat, and work. These are briefly described here.

1. *Kinetic energy*. The energy of a moving object is called "kinetic energy." A baseball thrown by a pitcher possesses a definite energy as it travels toward the catcher. A pound of flowing fluid possesses kinetic energy as it travels through a duct.
2. *Potential energy*. The energy possessed by a mass by virtue of its position in the Earth's gravitational field is called "potential energy." A boulder lying at the top of a cliff possesses potential energy with reference to the bottom of the cliff. If the boulder is pushed off the cliff its potential energy is transformed into kinetic energy as it falls. Similarly, a mass of fluid in a flowing system possesses a potential energy because of its height above a reference level.
3. *Internal energy*. The component molecules of a substance are constantly moving within the substance. This motion imparts internal energy to the material. The molecules may rotate, vibrate, or migrate within the substance. The addition of heat to a material increases its molecular activity and hence, its internal energy. The temperature of a material is a measure of its internal energy.
4. *Heat*. As noted earlier, when energy is transferred between a system and its surroundings, it is transferred either as work or as heat. Thus, heat is energy in transit. This type of energy transfer occurs whenever a hot body is brought into contact with a cold body. Energy flows as heat from the hot body flows to the cold body until the temperature difference is dissipated, i.e. until thermal equilibrium is established. For this reason, heat may be considered as energy being transferred due to a temperature difference.
5. *Work*. Work is also energy in transit. Work is done whenever a force acts through a distance.

The first law of thermodynamics may be stated formally, as opposed to equation form, in many ways. One of these is as follows: although energy assumes many forms, the total quantity of energy is constant, and when energy disappears in one form, it must appear simultaneously in other forms. This law, in steady-state equation form for batch and flow processes is:

$$\Delta U = Q + W, \text{ batch processes} \tag{7.7}$$

$$\Delta H = Q + W_s, \text{ continuous processes} \tag{7.8}$$

where potential, kinetic, and other energy effects have been neglected and Q = energy in the form of heat transferred across the boundaries of the system; W = energy in the form of work transferred across the boundaries of the system; W_s = energy in the form of mechanical work transferred across the boundaries of the system; U = internal energy of the system; H = enthalpy of the system; and ΔU, ΔH = changes in the internal energy and enthalpy, respectively, during the process. Both Q and W (or W_s) are considered positive if added/transferred to the system. Also note that for a continuous process:

$$\Delta H = Q \tag{7.9}$$

if $W_s = 0$.

The internal energy and enthalpy in Equations 7.7 and 7.8, as well as the other equations in this chapter, may be on a *mass* basis (i.e. for 1 gal or 1 lb of material), on a *mole* basis (i.e. for 1 gmol or 1 lbmol of material), or represent the total internal energy and enthalpy of the entire system. As long as these equations are dimensionally consistent, it makes no difference. For the sake of clarity, uppercase letters (e.g. H, U, C_p) represent properties on a mole basis, while lowercase letters (e.g. h, u, c_p) represent properties on a mass basis. Properties for the entire system will rarely be used and therefore require no special nomenclature.

Perhaps the most important thermodynamic function the engineer works with is the aforementioned enthalpy. The enthalpy is defined by:

$$H = U + PV \tag{7.10}$$

where P = pressure of the system, and V = volume of the system.

Summarizing, the law of conservation of energy states that energy can neither be created nor destroyed. As with the law of conservation of mass, this law indicates that one should be able to account for all the energy at any point in time. However, the form of energy can change. This mathematical accounting system is referred to as an *energy balance*.

Finally, Einstein in his theory of equivalence between matter and energy implied that there was a combined law of conservation of matter and energy. This law states essentially that the amount of energy and matter is constant. However, the exchange between mass and energy does not arise in desalination applications. Thus one may treat mass and energy balances separately.

7.6 Stoichiometry

When chemicals react, they do so according to a strict proportion. When oxygen and hydrogen combine to form water, the ratio of the amount of oxygen to the amount of hydrogen consumed is always 7.94 by mass and 0.500 by moles. The term stoichiometry refers to this phenomenon, which is sometimes called the chemical law of combining weights. The reaction equation for the combining of hydrogen and oxygen is:

$$2H_2 + O_2 \rightarrow 2H_2O \tag{7.11}$$

In chemical reactions, atoms are neither generated nor consumed, merely rearranged with different bonding partners. The manipulation of the coefficients of a reaction equation so that the number of atoms of each element on the left of the equation is equal to that on the right is referred to as *balancing* the equation. Once an equation is balanced, the whole number molar ratio that must exist between any two components of the reaction can be determined simply by observation; these are known as *stoichiometric ratios*. There are three such ratios (not counting the reciprocals) in the earlier reaction. These are:

2 mol H_2 consumed/mol O_2 consumed

1 mol H_2O generated/mol H_2 consumed

2 mol H_2O generated/mol O_2 consumed

The unit mole represents either the gmol or the lbmol. Using molecular weights, these stoichiometric ratios (which are molar ratios) may easily be converted to mass ratios. For example, the first ratio shown earlier may be converted to a mass ratio by using the molecular weights of H_2 (2.016) and O_2 (31.999) as follows:

(2 gmol H_2 consumed) (2.016 g/gmol) = 4.032 g H_2 consumed

(1 gmol O_2 consumed) (31.999 g/gmol) = 31.999 g O_2 consumed

The mass ratio between the hydrogen and oxygen consumed is therefore:

4.032/31.999 = 0.126 g H_2 consumed/g O_2 consumed

These molar and mass ratios are used in material balances to determine the amounts or flow rates of components involved in chemical reactions. Multiplying a balanced reaction equation through by a constant does nothing to alter its meaning. The reaction used as an example earlier is often written:

$$H_2 + \tfrac{1}{2}O_2 \rightarrow H_2O \tag{7.12}$$

In effect, the stoichiometric coefficients of Equation 7.11 have been multiplied by 0.5. There are times, however, when care must be exercised because the solution to a problem may depend on the manner or form in which the reaction is written. This is the case with chemical equilibrium problems and problems involving thermochemical reaction equations.

There are two different types of material balances that may be written when a chemical reaction is involved: the *molecular balance* and the *atomic balance*. It is a matter of convenience which of the two types is used. The molecular balance is the same as that described earlier. Assuming a steady-state continuous reaction, the accumulation term, A, is zero for all components involved in the reaction, Equation 7.5 then becomes:

$$I + G = O + C \tag{7.13}$$

If a total material balance is performed, the earlier form of the balance equation must be used if the amounts or flow rates are expressed in terms of moles, e.g. lbmol or gmol/h, since the total number of moles can change during a chemical reaction. If, however, the amounts or flow rates are given in terms of mass, e.g. kg or lb/h, the *G* and *C* terms may be dropped since mass cannot be gained or lost in a chemical reaction. Thus,

$$I = O \tag{7.14}$$

In general, however, when a chemical reaction is involved, it is usually more convenient to express amounts and flow rates using moles rather than mass.

A material balance that is not based on the chemicals (or molecules), but rather on the atoms that make up the molecules, is referred to as an atomic balance. Since atoms are neither created nor destroyed in a chemical reaction, the *G* and *C* terms equal zero and the balance once again becomes Equation 7.14.

As an example, consider once again the combination of hydrogen and oxygen to form water:

$$2H_2 + O_2 \rightarrow 2H_2O \tag{7.11}$$

As the reaction progresses, O and H, molecules (or moles) are consumed while H_2O molecules (or moles) are generated. On the other hand, the number of oxygen atoms (or moles of oxygen atoms) and the number of hydrogen atoms (or moles of hydrogen atoms) do not change. Care must also be taken to distinguish between molecular oxygen and atomic oxygen. If, in the previous reaction, one starts out with 1,000 lbmol of O_2 (oxygen molecules), one may replace this with 2,000 lbmol of O (oxygen atoms).

Thus, a chemical equation provides a variety of qualitative and quantitative information essential for the calculation of the quantity of reactants reacted and products formed in a chemical process. As noted, a balanced chemical equation must have the same number of atoms of each type in the reactants on the left hand side of the equation and in the products on the right hand side of the equation. Thus, a balanced equation for butane combustion (reaction with oxygen to form oxidized end products CO_2 and H_2O) is:

$$C_4H_{10} + (13/2)\,O_2 \rightarrow 4\,CO_2 + 5\,H_2O \tag{7.15}$$

Note that:

Number of carbons in reactants = number of carbons in products = 4

Number of oxygens in reactants = number of oxygens in products = 13

Number of hydrogens in reactants = number of hydrogens in products = 10

Number of moles of reactants is 1 mol C_4H_{10} + 6.5 mol O_2 = 7.5 mol total

Number of moles of products is 4 mol CO_2 + 5 mol H_2O = 9 mol total

The reader should note that although the number of moles on both sides of the equation do not balance, the masses of reactants and products (in line with the conservation law for mass) must balance.

When handling a reaction that does not involve stoichiometric ratios of reactants, readers must carefully define the terms used to describe said reaction to avoid confusion. If the reactants are not present in formula or stoichiometric ratio, one reactant is *limiting* while the others are in *excess*. Limiting and excess reactants involve an extension of the stoichiometric calculations provided earlier. Consider the following example. When methane is combusted completely, the stoichiometric equation for the reaction is:

$$CH_4 + 2O_2 \rightarrow CO_2 + 2H_2O \tag{7.16}$$

The stoichiometric ratio of the oxygen to the methane is: 0.5 mol methane consumed/mol oxygen consumed. If one starts out with 1 mol of methane and 3 mol of oxygen in a reaction vessel, only 2 mol of oxygen would be used up, leaving an excess of 1 mol of oxygen in the vessel. In this case, the oxygen is called the *excess* reactant and methane is the *limiting* reactant. The limiting reactant is defined as the reactant that would be completely consumed if the reaction went to completion. All other reactants are excess reactants. The amount by which a reactant is present in excess of stoichiometric requirements (i.e. the exact number of moles needed to react completely with the limiting reactant is defined as the percent excess, and is given by Equation 7.17:

$$\%excess = \left(\frac{n - n_s}{n_s}\right) x100 \tag{7.17}$$

where n = number of moles of the excess reactant at the start of the reaction; and n_s = the stoichiometric number of moles of the excess reactant.

In the earlier example, the stoichiometric amount of oxygen is 2 mol, since that is the amount that would react with the 1 mol of methane. The excess amount of oxygen is 1 mol, which is a percentage excess of 50% or a fractional excess of 0.50.

7.7 Thermodynamics

Thermodynamics was once defined as "the science that deals with the intertransformation of heat and work." The fundamental principles of thermodynamics are contained in the first, second, and third laws of thermodynamics. These principles have been defined as "pure" or "theoretical" thermodynamics. These laws were developed and extensively tested in the latter half of the nineteenth century and are essentially based on experience. (The third law was developed later in the twentieth century).

Practically all thermodynamics, in the ordinary meaning of the term, is "applied thermodynamics" in that it is essentially the application of these three laws, coupled with certain facts and principles of mathematics, physics, and chemistry, to problems in engineering and science. The fundamental laws are of such generality that it is not surprising that these laws find application in other disciplines, including physics, chemistry, plus environmental, chemical and mechanical engineering.

The first law of thermodynamics is a conservation law for energy transformations. Regardless of the types of energy involved in processes – thermal, mechanical, electrical, elastic, magnetic, etc. – the change in energy of a system is equal to the difference between energy input and energy output. The first law also allows free "convertibility" from one form of energy to another, as long as the overall energy quantity is conserved. Thus, this law places no restriction on the conversion of work into heat, or on its counterpart – the conversion of heat into work.

The "limiting" law to energy conservation is called the "second law of thermodynamics." In principle, the second law states that water cannot flow uphill and heat cannot flow from a cold to hot body of its own accord. In addition, it is impossible to convert heat taken into a system completely to work, and no apparatus, equipment, or process can operate in such a way that its only effect on the system and surroundings is to convert heat taken in completely to work.

The unrestricted conversion of work into heat is well known to most engineers. Frictional effects are frequently associated with mechanical forms of work that result in a temperature rise of the bodies in contact. However, the transformation of heat into work is of greater concern to the practicing environmental engineer. In nations with a partially developed or developing technological society, the ability to produce energy in the form of work takes on more significant importance. Work transformations are necessary to transport people and goods, drive machinery, pump liquids, compress gases, and provide energy input to so many other processes that are taken for granted in highly developed societies such as those in the United States. Much of the work input in these societies is available in the form of electrical energy, which can then be converted to rotational mechanical work. Although some of this electrical energy (work) is produced by hydroelectric power plants or renewable energy sources, by far the greatest part of it is still obtained from the combustion of fossil or nuclear fuels. These fuels allow the engineer to produce a relatively high-temperature gas or liquid stream that acts as a thermal (energy) source for the production of work. Hence, the study of the conversion of heat to work is extremely important, especially in the light of developing shortages of fossil and nuclear fuels, along with the accompanying environmental problems, particularly with concerns regarding global warming.

The brief discussion of energy conversion devices leads to an important second-law consideration: that energy has a *quality* as well as a quantity. Theodore et al. (2009) provide a quantitative review of the second law.

7.8 Enthalpy Effects

There are many different types of enthalpy effects that are included in thermodynamic calculations. These include:

1. Sensible (temperature)
2. Latent (phase)
3. Dilution (with water), e.g. HCl with H_2O
4. Solution (nonaqueous), e.g. HC1 with a solvent other than H_2O
5. Reaction (chemical)

This section is primarily concerned with Topics 1 to 4. Topic 5 is only treated superficially.

7.8.1 Sensible Enthalpy Effects

Sensible enthalpy effects are associated with temperature. There are methods that can be employed to calculate sensible enthalpy changes. These methods include the use of:

1. Enthalpy values.
2. Average heat capacity values.
3. Heat capacity as a function of temperature.

Detailed calculations on these methods are provided by Theodore, Ricci, and Van Vleit (2009). Two expressions for heat capacity are considered in Topic 3 employing a, b, c constants and α, β, and γ constants. If enthalpy values are available, the enthalpy change is given by:

$$\Delta h = h_2 - h_1; \text{ mass basis} \tag{7.18}$$

$$\Delta H = H_2 - H_1; \text{ mole basis} \tag{7.19}$$

If average molar heat capacity data are available,

$$\Delta H = \bar{C}_p \Delta T \tag{7.20}$$

where $\bar{C}_p$ = average molar value of C_p in the temperature range ΔT. Average molar heat capacity data are provided in the literature (Theodore 2007, 2014; Theodore et al. 2009).

A more rigorous approach to enthalpy calculations can be provided if the heat capacity variation with temperature is available. If the heat capacity is a function of the temperature, the enthalpy change is written in differential form:

$$dH = C_p dT \tag{7.21}$$

If the temperature variation of the heat capacity is given by

$$C_p = \alpha + \beta T + \gamma T^2 \tag{7.22}$$

Equation 7.21 may be integrated directly between some reference or standard temperature (T_o) and the final temperature (T_1) employing Equation 7.19:

$$\Delta H = H_1 - H_o \tag{7.23}$$

$$\Delta H = \alpha(T_1 - T_o) + (\beta/2)(T_1^2 - T_o^2) + (\gamma/3)(T_1^3 - T_o^3) \tag{7.24}$$

Equation 7.21 may also be integrated if the heat capacity is a function of temperature of the form:

$$C_p = a + bT + cT^{-2} \tag{7.25}$$

The enthalpy change is then given by:

$$\Delta H = a(T_1 - T_o) + (b/2)(T_1^2 - T_o^2) + c(T_1^{-1} - T_o^{-1}) \tag{7.26}$$

Tabulated values of α, β, γ, and a, b, c for a host of compounds (including some chlorinated organics) are available in the literature (Theodore et al. 2009).

7.8.2 Latent Enthalpy Effects

The term *phase*, when used to describe a pure substance, refers to a state of matter that is a gas, liquid, or solid. *Latent* enthalpy effects are associated with phase changes. These phase changes involve no change in temperature, but there is a transfer of energy to and from the substance. There are three possible latent effects: (1) vapor–liquid, (2) liquid–solid, and (3) vapor–solid.

Vapor–liquid changes are referred to as *condensation* when the vapor is condensing and *vaporization* when liquid is vaporizing. *Liquid–solid* changes are referred to as *melting* when the solid melts to liquid and *freezing* when a liquid solidifies. *Vapor–solid* changes are referred to as *sublimation*. Also, there are enthalpy effects associated with a phase change of a solid to a solid of another form; however, this enthalpy effect is small compared to the other effects mentioned earlier.

Since there is a relatively small difference in volume between a solid and the liquid to which it melts, the heat of fusion represents mainly an increase in internal energy. If the molar heat of fusion is not known, it may be approximated from the absolute temperature of fusion, T_f, using the following equation:

$$\Delta H_f / T_f = 2-3 \text{ for solid elements; } 5-7 \text{ for inorganic solids; } 9-11 \text{ for organic solids} \tag{7.27}$$

The ratio may be expressed in any consistent molar units, such as cal/(gmol-K) or Btu/(lbmol-°R).

Since there is a very substantial increase in volume in passing from the liquid to the vapor state, usually from a few hundred to over a thousand-fold, the heat of vaporization is large and important in many environmental engineering applications involving water. Information on properties of water are available in steam tables in Chapter 5.

7.8.3 Enthalpy of Mixing

When two or more pure substances are mixed to form a solution, a heat effect invariably results. The heat (or enthalpy) of mixing is defined as the enthalpy change that occurs when two or more pure substances at a reference state are mixed at constant temperature and pressure to form a solution. When the mixing occurs in the presence of water, the process is normally defined as a *dilution* effect. If mixing occurs in the presence of a solvent other than water, it is defined as a *solution* effect. For example, the mixing of HCI with H_2O is considered a dilution process while the mixing of HCI with H_2SO_4 is a solution process.

The reader should note that there is also an enthalpy effect when gases or solids are dissolved in liquids. This effect is normally small but can be significant in some applications, particularly with acid gases in water. This type of calculation should be included in some absorber systems when temperature changes through the unit are significant. Specific details of this enthalpy effect are provided by Theodore et al. (2009).

7.8.4 Chemical Reaction Enthalpy Effects

The standard enthalpy (heat) of reduction can be calculated from standard enthalpy of formation data. To simplify the presentation that follows, examine the authors' favorite equation:

$$\mathrm{aA + bB = cC + dD} \tag{7.28}$$

If the given reaction is assumed to occur at a standard (or reference) state, the standard enthalpy of reaction, ΔH^o, is given by:

$\Delta H^o = c(\Delta H^o_f)\mathrm{C} + d(\Delta H^o_f)\mathrm{D} - a(\Delta H^o_r)\mathrm{A} - b(\Delta H^o_f)\mathrm{B}$

$$\Delta H^\circ = c\left(\Delta H^\circ{}_f\right)_C + d\left(\Delta H^\circ{}_f\right)_D - a\left(\Delta H^\circ{}_r\right)_A - b\left(\Delta H^\circ{}_f\right)_B \tag{7.29}$$

where $(\Delta H_f^\circ)_\mathrm{i}$ = standard enthalpy of formation of species i.

Thus, the (standard) enthalpy of a reaction is obtained by taking the difference between the (standard) enthalpy of formation of products and reactants multiplied by their respective stoichiometric coefficients. If the (standard) enthalpy of reaction or formation is negative (exothermic), as is the case with most combustion reactions, then energy is liberated due to the chemical reaction. Energy is absorbed and ΔH^o is positive (endothermic).

Tables of enthalpies of formation and reaction are available in the literature (particularly thermodynamics text/reference books) for a wide variety of compounds. It is important to note that these are valueless unless the stoichiometric equation and the state of the reactants and products are included.

Theodore and Reynolds (1991) and Theodore et al. (2009) provide equations to describe the effect of temperature on the enthalpy of reaction. For heat capacity data in α, β, γ form:

$$\Delta H_T^o = \Delta H_{298}^o + \Delta\alpha\left(T - 298\right) + \frac{1}{2}\Delta\beta\left(T^2 - 298^2\right) + \frac{1}{3}\Delta\gamma\left(T^3 - 298^3\right) \tag{7.30}$$

For the reaction presented in Equation 7.28:

$$\Delta\alpha = c\alpha_c + d\alpha_D - a\alpha_A - b\alpha_B \tag{7.31}$$

$$\Delta\beta = c\beta_c + d\beta_D - a\beta_A - b\beta_B \tag{7.32}$$

$$\Delta\gamma = c\gamma_c + d\gamma_D - a\gamma_A - b\gamma_B \tag{7.33}$$

For heat capacity in a, b, c form,

$$\Delta H_T = \Delta H^o_{298} + \Delta a(T - 298) + \frac{1}{2}\Delta b\left(T^2 - 298^2\right) + \Delta c\left(T^{-1} - 298^{-1}\right) \tag{7.34}$$

7.9 Phase Equilibrium

Relationships governing the equilibrium distribution of a substance between two phases, particularly gas and liquid phases, are the principal subject matter of phase equilibrium thermodynamics. These relationships form the basis of calculational procedures that are employed in the design and prediction of the performance of several unit operations equipment and processes involving mass transfer.

Engineering calculations rarely involve single (pure) components. Phase equilibria for multicomponent systems are considerably more complex, mainly because of the addition of composition variables. The most important equilibrium phase relationship is that between a liquid and a vapor. Raoult's and Henry's Laws theoretically describe liquid–vapor behavior and, under certain conditions, are applicable in practice. Raoult's Law is sometimes useful for mixtures of components of similar structure. It states that the partial pressure of any component in the vapor phase is equal to the product of the vapor pressure of the pure component and the mole fraction of that component in the liquid:

$$p_i = p_i' x_i \tag{7.35}$$

where p_i = partial pressure of component i in the vapor, p_i' = vapor pressure of pure component i at the same temperature, and x_i = mole fraction of component i in the liquid. This expression may be applied to all components. If the gas phase is ideal, this equation becomes:

$$y_i = (p_i'/P)x_i \tag{7.36}$$

where y_i = mole fraction of component i in the vapor, and P = total system pressure.

Unfortunately, relatively few mixtures follow Raoult's law, but it can be applied to estimating the vapor pressure of individual compounds above a hydrocarbon mixture, i.e. benzene above gasoline, in underground storage tank and soil and groundwater remediation applications.

Henry's law is a more empirical relation used for representing data for many environmental engineering systems where the liquid phase is water and the solutions are dilute. The law finds widespread use in absorber calculations since the concentration of a solute in many process gas streams is often dilute. Henry's law is expressed as:

$$p_i = H_i x_i \tag{7.37}$$

where H_i = Henry's law constant for component i (in units of pressure). If the gas behaves ideally, the earlier equation may be written as:

$$y_i = K_H x_i \tag{7.38}$$

where K_H = dimensionless Henry's law constant.

In some environmental engineering applications, mixtures of condensable vapors and non-condensable gases must be handled. A common example is water vapor and air; a mixture of organic vapors and air is another such example that often appears in air pollution applications. Condensers can be used to control organic emissions to the atmosphere by lowering the temperature of the gaseous stream, although an increase in pressure will produce the same result. The calculation for this is often accomplished using the *phase equilibrium constant*, K. This constant has also been referred to in the chemical industry as a *phase componential split factor* since it provides the ratio of the mole fractions of a component in the two equilibrium phases. The defining equation is

$$K_i = y_i \,/\, x_i \tag{7.39}$$

where K_i = phase equilibrium constant for component i (dimensionless).

As a first approximation, K_i is generally treated as a function only of the temperature and pressure. For ideal gas conditions, K_i may be approximated by

$$K_i = p_i' \,/\, P \tag{7.40}$$

where p_i' is the vapor pressure of component i, and P is the total system pressure

7.10 Illustrative Examples

Seven illustrative examples complete the presentation of materials highlighted in this chapter.

10.1 Illustrative Example 1

Briefly discuss the impulse-momentum principle.

Solution. The momentum balance discussed in Section 7.3 is also termed the *impulse-momentum principle* and is important in flow problems where forces need to be determined. This analysis is inherently more complicated than those previously presented (i.e. forces possess both magnitude and direction), because the force, F, and momentum, M, are vectors. In order to describe force and momentum vectors, both direction and magnitude must be specified; for mass and energy, only the magnitude is required.

10.2 Illustrative Example 2

Consider a 10-cm-diameter horizontal line carrying saturated steam at a velocity of 420 m/s. Water is entrained by the steam at a rate of 0.15 kg/s. The line has a 90° bend. Employing momentum principles, calculate the force components in the horizontal and vertical directions (and the magnitude and direction of the resultant force) required to hold the bend in place due to the entrained water.

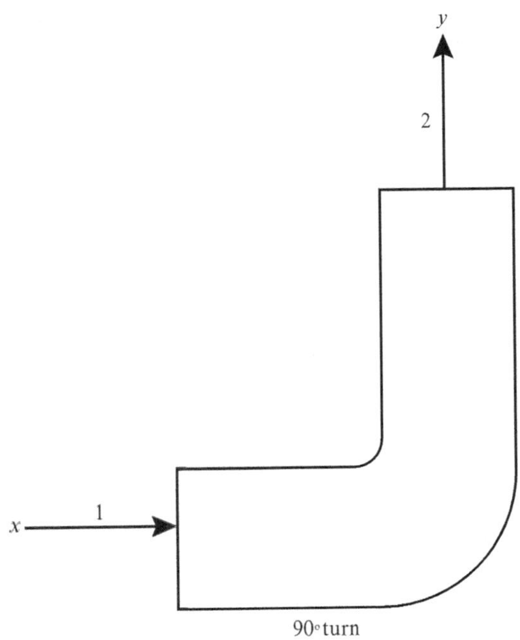

Figure 7.3 Diagram to Calculate the Magnitude and Direction of a Resultant Force for Illustrative Example 2.

Solution. A line diagram of the system is provided in Figure 7.3. Select the fluid in the bend as the system and apply the conservation law for mass:

$$\dot{m}_1 = \dot{m}_2 \tag{7.41}$$

Since the density and cross sectional area of the line are constant,

$$v_1 = v_2 \tag{7.42}$$

where $\dot{m}_1, \dot{m}_2 =$ mass flow rate at Location 1 and 2, respectively; and $v_1, v_2 =$ velocity at Location 1 and 2, respectively.

A linear momentum $(\dot{M})$ balance in the horizontal direction provides the force applied by the channel wall on the fluid in the x-direction, F_x:

$$F_x g_c = \dot{M}_{x,out} - \dot{M}_{x,in} = \frac{d}{dt}(mv)_{x,out} - \frac{d}{dt}(mv)_{x,in} \tag{7.43}$$

Note that the equation assumes that the pressure drop across the bend is negligible. Since $v_{x,out} = 0$ and dm/dt $= \dot{m}$,

$$F_x g_c = 0 - \dot{m}_{x,in} = -\frac{(0.15)(420)}{1} = -63\text{ N} = -14.1\text{ lb}_f$$

The *x*-direction supporting force acting on the 90° elbow is 14.1 lbf acting toward the left. A linear momentum balance in the vertical direction results in:

$$F_y g_c = \dot{M}_{y,out} - \dot{M}_{y,in} = \dot{m}v_{y,out} - \dot{m}v_{y,in} = \dot{m}v_2 - 0 = (0.15)(420)$$
$$= 63N = 14.1\text{ lb}_f$$

The *y*-direction supporting free force on the 90° elbow is 14.1 lbf acting upward. The resultant supporting force is given by:

$$F_{res} = \sqrt{F_x^2 + F_y^2} = \sqrt{(-63)^2 + (63)^2} = 89.1\text{ N} = 19.1\text{ lb}_f \tag{7.44}$$

The direction is given by:

$$tan\theta = \frac{F_y}{F_x} = \frac{63}{-63} = -1; \theta = 135° \tag{7.45}$$

where θ is the angle between the positive x-axis and the direction of the force. The counter-clockwise rotation of the direction from the x-axis is defined as positive. The supporting force is therefore 19 .1 lb_f acting in the "northwest" direction.

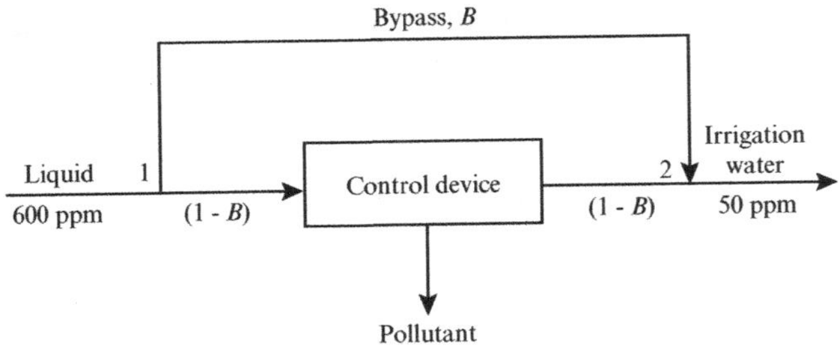

Figure 7.4 Flow Diagram for Illustrative Example 3.

10.3 Illustrative Example 3

A water stream contaminated with a pollutant is being cleansed with a stripping control device (Figure 7.4). If the water has 600 ppm of pollutant and it is permissible to have 50 ppm of this pollutant in the discharge stream, what fraction of the water can bypass the control device?

Solution. Using a basis of 1 lb of water fed to the control device, assuming the contaminant concentration out of the Control device = 0, and using the flow diagram in Figure 7.4, note that B = fraction of liquid by passed, 1-B = fraction of liquid treated. Performing a pollutant balance around Point 2 yields:

$$(1-B)(0)+600\,B=(50)(1.0);B=0.0833 \tag{7.46}$$

Note that in some operations, a process does a more complete job than is required. For example, if moist air is passed through a fresh silica gel dryer, the air will leave the system almost bone dry. If it were desirable to have air containing some moisture, one would have to reintroduce water vapor into the air. This would be a wasteful process compared to bypassing the proper amount of original moist air. In general, a finished product is made only as good as it has to be to meet competition and/or to satisfy the user. "Product quality giveaway" is costly and is often minimized by bypassing.

10.4 Illustrative Example 4

Consider the flow diagram in Figure 7.5 for a wastewater treatment system in a desalination plant. The following flow rates are given. Are the flow rates provided here consistent?

$$\dot{m}_1 = 1{,}000\,\text{lb/min}$$

$$\dot{m}_2 = 1{,}000\,\text{lb/min}$$

$$\dot{m}_4 = 200\ \text{lb/min}$$

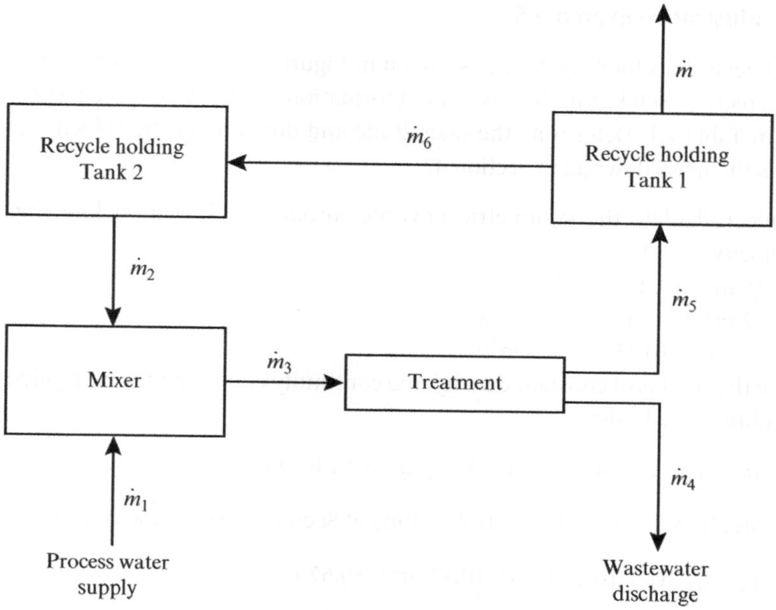

Figure 7.5 Flow Diagram for Illustrative Example 4.

Solution. This problem requires several mass balance calculations to be performed. For the Mixer,

$$\dot{m}_3 = \dot{m}_1 + \dot{m}_2 = 1{,}000 + 1{,}000 = 2{,}000\,\text{lb}/\text{min}$$

For the Treatment Unit,

$$\dot{m}_3 = \dot{m}_5 + \dot{m}_4;\ \dot{m}_5 = \dot{m}_3 - \dot{m}_4 = 2{,}000 - 200 = 1{,}800\,\text{lb}/\text{min}$$

For Tank 2,

$$\dot{m}_6 = \dot{m}_2 = 1{,}000\,\text{lb}/\text{min}$$

For Tank 1,

$$\dot{m}_5 = \dot{m}_6 + \dot{m}; \dot{m} = \dot{m}_5 - \dot{m}_6 = 1{,}800 - 1{,}000 = 800\,\text{lb/min}$$

And for the entire process,

$$\dot{m}_1 = \dot{m} + \dot{m}_4 = 800 + 200 = 1{,}000\,\text{lb/min}$$

Yes, based on the results of the mass balance calculations, these flow rates are consistent.

10.5 Illustrative Example 5

A fluid device has four openings, as shown in Figure 7.6. The fluid has a constant density of 800 kg/m^3. Steady-state information on the system is also provided in Table 7.1. Determine the magnitude and direction of the velocity v_4. What is the mass flowrate at Section 4?

Solution. Calculate the volumetric flowrate through each section knowing q = velocity (area):

$q_1 = (5 \text{ m/s})(0.2 \text{ m}^2) = 1 \text{ m}^3\text{/s}$
$q_2 = (7 \text{ m/s})(0.3 \text{ m}^2) = 2.1 \text{ m}^3\text{/s}$
$q_3 = (12 \text{ m/s})(0.25 \text{ m}^2) = 3 \text{ m}^3\text{/s}$

Since the fluid is of constant density, the continuity equation may be applied on a volume rate basis:

$$q_1 + q_2 = q_3 + q_4; 1 + 2.1 = 3 + q_4; q_4 = 0.1 \ m^3/s$$

Therefore, flow is out of the control volume at Section 4. The velocity v_4 is:

$$v_4 = q_{4/}A_4 = \left(0.1 \ m^3/s\right)/\left(0.15 \ m^2\right) = 0.67 \ m/s$$

Calculate the mass flowrate across Section 4, $\dot{m}_4$ as:

$$\dot{m}_4 = (\text{Density})(\text{Flow rate}) = \left(800 \text{kg}/m^3\right)\left(0.1 \ m^3/s\right) = 80 \text{ kg}/s \quad (7.47)$$

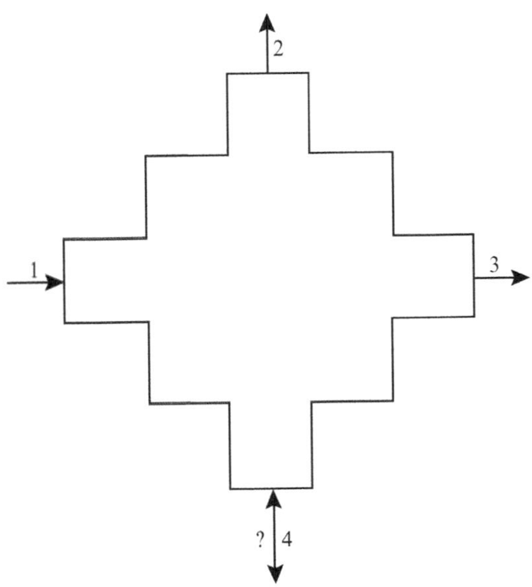

Figure 7.6 Diagram for Illustrative Example 5.

Table 7.1 Data for Illustrative Example 5.

Section	Flow area, m^2	Velocity, m/s	Direction, relative to device
1	0.2	5	In
2	0.3	7	In
3	0.25	12	Out
4	0.15	?	?

10.6 Illustrative Example 6

Consider the following scenario. A lake is located at the top of a mountain. A desalination plant has been constructed at the bottom of the mountain. The potential energy of the water traveling downhill can be used to spin turbines and generate electricity. This is the operating mode in the daytime during peak electrical demand. At night when demand is reduced, the water is pumped back up the mountain. The operation is shown in Figure 7.7. By applying conservation of energy law principles, determine how much power (Watts) is generated by the lake located at an elevation of 3,000 ft above the desalination plant. The flowrate of water is 500,000 gpm. The turbine efficiency is 30%. Neglect friction effects.

Solution. First, convert height and flowrate to SI units in order to solve for the power in Watts:

$$\left(3{,}000\ ft\right)\left(0.3048\ m\ /\ ft\right) = 914.4\ m$$

$$\left(500{,}000\,\text{gal/min}\right)\left(0.00378\ m^3\ /\ \text{gal}\right) = 1{,}890\ m^3\ /\ \text{min}$$

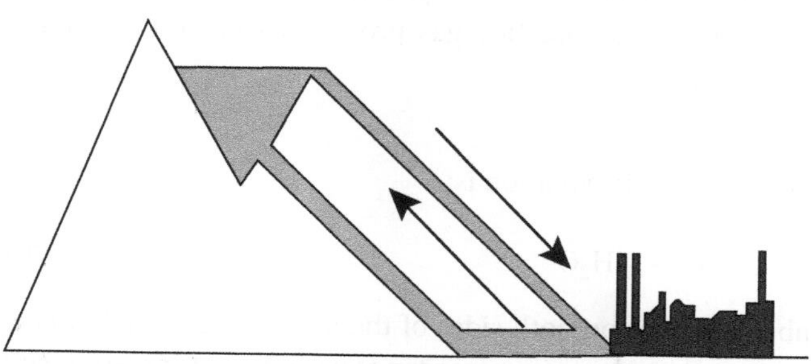

Figure 7.7 Schematic for Lake Pumped Storage Power Plant for Illustrative Example 6.

The mass flow rate of the water in kilograms/second is:

$$\frac{\left(1{,}890\ \mathrm{m}^3/\mathrm{min}\right)\left(1{,}000\ \mathrm{kg}/\mathrm{m}^3\right)}{60\ \mathrm{s}/\mathrm{min}} = 31{,}500\ \mathrm{kg}/\mathrm{s}$$

The loss in potential energy, ME, of the water flow is given by:

$$\Delta PE = \frac{mg\left(\Delta Z\right)}{g_c} \tag{7.48}$$

Substituting yields:

$$\Delta PE = \left(31{,}500\ kg/s\right)\left(9.8\ m/s^2\right)\left(9.14\ m\right) = 2.82\ x10^8 kg - m/s^3$$
$$= 2.82\ N/s = 282\ MW$$

Note that $g_c = 1$ in the SI system of units.

Assuming that the potential energy decrease is entirely converted to energy input to the turbine, the actual power output is:

$$P = \left(0.30\right)\left(282\ MW\right) = 84.6\ MW$$

This is enough power for not only a desalination plant, but also for a small town. No pollutants or/greenhouse gases are generated because no fossil fuel is required. The initial construction expense would be quite high, but the long-term cost of producing electricity in this fashion would be very economical.

10.7 Illustrative Example 7

The reaction equation for the combustion of propane is as follows:

$$C_3H_8 + O_2 \rightarrow CO_2 + H_2O \tag{7.49}$$

1. Balance this equation and determine the ratio of reactants to products.
2. Using the balanced equation, determine the scf of air required for stoichiometric combustion of 1.0 scf of propane (C_3H_8) and calculate both the stoichiometric air requirements and flue gas produced from the complete combustion of propane.

Solution.

1. The balanced stoichiometric equation is:

$$C_3H_8 + 5\ O_2 \rightarrow 3\ CO_2 + 4\ H_2O \tag{7.50}$$

Although the number of moles on both sides of the equation do not balance, the masses of reactants and products (in accordance with the conservation law for mass) must balance.

2. The scf of O_2 required for the complete combustion of 1 scf of propane can now be determined. From the balanced equation, there are 5 moles of O_2 required per mole of propane. Since the moles of all ideal gases are directly proportional to their volume,

scf of O_2 = 5 (1 scf of propane) = 5 scf

From a mass balance, 44 lb of propane has reacted with 160 lb of O_2 to form 132 lb of CO_2 and 72 lb of H_2O, with and initial and final total mass of 204 lb. The amount of N_2 in a quantity of air that contains 5 scf of O_2 is given by the following knowing that the nitrogen to oxygen volume (or mole) ratio in air is 79/21.

scf of N_2 = (79/21) (scf of O_2) = (79/21) 5 scf = 18.81 scf

The stoichiometric amount of air is then

scf of air = (scf of O_2) + (scf of N_2) = (18.81 scf + 5 scf) = 23.81 scf

In order to calculate the amount of flue gas produced, one must include the nitrogen from the air in addition to the products of combustion, i.e. carbon dioxide and water vapor. This calculation follows:

scf of flue gas = (scf of CO_2) + (scf of N_2) + (scf of H_2O) = (18.81 scf + 3 scf + 4 scf) = 25.81 scf

7.11 Problems

1. If 2,000 lb of water has its velocity increased from 8 to 30 ft/s, calculate the change in kinetic energy of the water in units of Btu and ft-lbf along with the minimum energy required to accomplish this change.
2. The heat-generating unit of a natural gas fired power plant may be simply described as a continuous flow reactor into which natural gas (mass flow rate F) and air (mass flow rate A) are fed and from which effluents ("flue gas," mass flow rate E) are discharged.
   a. Draw a flow diagram representing this process. Show all flows into and out of the unit.
   b. Write a mass balance equation for this process.
   c. Suppose the natural gas contains a mass fraction y of noncombustible Component C (for example, ash). Assume that all of the Component C is carried out of the boiler with the flue gas. Write a mass balance equation for Component C.
   d. What is the mass fraction (z) of C in the exit stream E and what is the effect of increasing the combustion air flow upon z?
3. A cooling tower handles 1,000 gal/min (0.063 m^3/s) of circulating water that is cooled from 110°F to 80°F (316 K to 300 K). How much blowdown

and makeup are required if the concentration of dissolved solids is allowed to reach three times the concentration in the makeup?

4. A process plant pumps 2,000 lb of water to an elevation of 1,200 ft above its turbogenerators. Determine the change in potential energy of this water in units of Btu and ft-lb_f.
5. Benzene is incinerated at 2100°F in the presence of 50% excess air (EA). Balance the combustion reaction for this process.
6. The balanced reaction for the combustion of 1.0 lbmol of chlorobenzene (C_6H_5Cl) in stoichiometric oxygen is

$$C_6H_5Cl + 7\ O_2 \rightarrow 6\ CO_2 + 2\ H_2O + HCl$$

Since combustion processes employ air, not oxygen, this reaction with air becomes:

$$C_6H_5Cl + 7\ O_2 + 26.3\ N_2 \rightarrow 6\ CO_2 + 2\ H_2O + HCl + 26.3\ N_2$$

where the nitrogen in the air has been retained on sides of the equation since it does not participate in the combustion reaction. Calculate the moles and masses involved in this reaction, based on the stoichiometric combustion of 1.0 lbmol of C_6H_5C.

References

Abulencia, P. and Theodore, L. (2009). *Fluid Flow for the Practicing Engineer.* Hoboken, NJ: John Wiley & Sons.

Bird, R., Stewart, W., and Lightfoot, E. (2002). *Transport Phenomena*, 2nd Edition. Hoboken, NJ: John Wiley & Sons.

Theodore, L. (1970). *Introduction to Transport Phenomena.* Scranton, PA: International Textbook Co.

Theodore, L. (2007). *Air Pollution Control Calculations.* Hoboken, NJ: John Wiley & Sons.

Theodore, L. (2014). *Chemical Engineering: The Essential Reference.* New York, NY: McGraw-Hill.

Theodore, L. and Reynolds, J. (1991). *Thermodynamics.* East Williston, NY: Theodore Tutorials, originally published by the USEPA/APTI, RTP, NC.

Theodore, L., Ricci, F., and VanVliet, T. (2009). *Thermodynamics for the Practicing Engineer.* Hoboken, NJ: John Wiley & Sons.

8

Unit Operations

8.1 Introduction

In many industrial sectors, including the chemical, food, and biological processing industries, there are many similarities in the manner in which the entering feed materials are modified or processed into final products. These seemingly different physical, chemical, or biological processes can be broken down into a series of separate and distinct steps called *unit operations*. These unit operations are common to all types of diverse process industries. For example, the unit operation *distillation* is used to purify or separate alcohol in the beverage industry and hydrocarbons in the petroleum industry. *Drying* of grain and other foods is similar to drying of lumber, filtered precipitates, and rayon yarn. The unit operation *absorption* occurs in the absorption of oxygen from air in a fermentation process or in a wastewater treatment plant and in absorption of hydrogen gas in a process for liquid hydrogenation of oil. *Evaporation* of salt solutions in the desalination field and chemical industry is similar to evaporation of sugar solutions in the food industry. *Settling* and *sedimentation* of suspended solids are similar in the wastewater and the mining industries. Flow of liquid hydrocarbons in a petroleum refinery and flow of milk in a dairy plant are carried out in a similar fashion. Thus, the most efficient method of organizing the subject matter of unit operation is based on two facts:

1. Although the number of individual processes is great, each one may be separated into a series of steps, called operations, each of which in turn appears in process after process; and
2. Individual operations have common techniques and are based on the same scientific principles.

The unit operation concept, therefore, is this: By systematically studying these operations themselves – operations that clearly cross industry and process lines – the treatment of all processes is unified and simplified.

Introduction to Desalination: Principles, Processes, and Calculations, First Edition. Louis Theodore and R. Ryan Dupont.

Chemical changes, as opposed to physical changes, have been traditionally defined as *unit processes*. These unit processes are important and well-defined reactions of chemistry, a study of which, although beyond the scope of this text, includes determination of the reaction mechanism, and the design of equipment used for carrying out the reactions. Halogenation, in some water processes, is one of these *types of* reactions that includes chlorination, bromination, and iodination. Others have defined a unit process as one that consists of a series of unit operations. To further confuse the issue, the term unit process in the field of water management has been classified as physical (e.g. screening, filtration, sedimentation, etc.), chemical (e.g. chlorination), and biological (e.g. any biological transformation).

This chapter will focus on a description of physical means for the transfer and change of energy and materials through a review of the quantitative approach to addressing unit operations, the classification of unit operations, the basics and fundamentals of the principles of mass and heat transfer, and a description of fluid flow. An application section provides seven illustrative examples of concepts and implementation of analyses of unit operations.

8.2 Quantitative Approaches

It is necessary to understand the basic function of each unit operation in a process in order to properly apply mass and energy balances to it. The conservation laws discussed in the previous chapter may be applied at the macroscopic, microscopic, or molecular level. One can best illustrate the differences in these methods with an example. Consider a system (Figure 8.1) in which a fluid is flowing through a cylindrical tube. The system is defined as the fluid contained within the tube between the Points 1 and 2 at any time, t. If one is interested in determining changes occurring at the inlet and outlet of the system, the conservation law is applied on a macroscopic level to the entire system. The resultant equation (usually algebraic) describes the overall changes occurring to the system without regard for internal variations to the system.

This macroscopic approach is the conventional approach for the analysis of unit operations, has received the bulk of attention in the literature, and is the

Figure 8.1 Hypothetical Flow System.

approach employed in this text. The microscopic approach is employed when detailed information concerning the behavior within the system is required. In the microscopic approach, the conservation laws are then applied to a differential element within the system, which is large compared to an individual molecule but small compared to the entire system. The resultant equations (usually differential) are then expanded, via an integration, to describe the behavior of the entire system. This is defined as the transport phenomena approach. Details of transport phenomenon principles and applications are also available in the literature (Bird et al. 2002).

The molecular approach involves the application of the conservation laws to individual molecules. This leads to a study of statistical and quantum mechanics, both of which are beyond the scope of this text. In any case, the description of individual elements at the molecular level is or little value to the engineer. However, the statistical averaging or molecular quantities in either a differential or finite element within a system (i.e. the microscopic approach) leads to a more meaningful description of the behavior of a system. Using the macroscopic approach, the transfer process, whether it be mass, energy, or momentum, can be simply described as shown in Equation 8.1:

$$\text{Rate of Transfer} = (\text{Driving Force})\,(\text{Transfer Area})/(\text{Transfer Resistance}) \quad (8.1)$$

For mass transfer, the equation becomes:

$$\text{Mass Transfer Rate} = (\Delta\text{Concentration})\,(\text{Mass Transfer Area})/(\text{Mass Transfer Resistance}) \quad (8.2)$$

8.3 Unit Operation Classification

This section introduces a variety of classes of unit operations, particularly those employed in environmental activities. The basic principles underlying these operations are reviewed. Application to specific unit operations (and description of the accompanying equipment) is presented in Part II of this book.

Unit operations deal mainly with the transfer and change of energy and materials by physical means. Some but not all of the important unit operations that can he combined in various sequences in a process are classified and described in the following categories. It should be noted that Categories 3 through 10 are classified as mass transfer operations and are elaborated on later in this chapter.

1. *Fluid flow* principles govern the flow or transportation of any fluid from one point to another. The flow of fluids is an important factor in many unit

operations. Fluids are understood to include liquids, gases, and vapors. The handling of fluids is usually simpler, cheaper, and less troublesome than the handling of solids. Consequently, attempts are made to handle and transport materials in the form of fluids, solutions, or suspensions whenever possible. When materials cannot be handled in this form, one resorts to the handling of solids. In most chemical and physical processes, materials must be moved through process equipment. When fluids are moved, energy must be supplied by a pump, blower, or other means. A quantitative evaluation of the energy balance in flowing fluids requires a knowledge of the fundamental thermodynamics (see Chapter 7) and mechanics of fluids. This information is available in the literature (Theodore 2014).

Farag and Reynolds (1996) and Abulencia and Theodore (2009) provide details regarding pipes, ducts, fittings, valves, fans, pumps, blowers, compressors, etc. More extensive information is available in other literature (Green and Southard 2019).

2. Heat transfer is a unit operation that deals with the principles that govern accumulation and transfer of heat and energy from one position to another. Many chemical reactions progress more rapidly or are more efficient if temperature is elevated above ambient conditions. Furthermore, chemical reactions usually release or absorb heat. Therefore, it is often necessary to heat or cool the reactants and products, making heat transfer an extremely important unit operation in many industrial processes. Heat transfer is also involved in the steam vaporization or condensation of a process stream. It is often possible to heat one process while cooling another in a piece of equipment called a heat exchanger. In this device, the fluids flow past each other and are usually separated by a metal wall through which heat is transferred from the hotter stream to the colder. Recover of heat from product streams is often economically favorable. More discussion of heat transfer equipment and basic heat transfer concepts is provided by Kern (1950), and Flynn et al. (2018).
3. *Evaporation* is a special case of heat transfer that deals with the thermal removal of a volatile solvent, such as water, from a nonvolatile solute, such as a salt or any other material in solution, through the application of thermal energy (Flynn et al. 2018; Theodore and Ricci 2010).
4. *Drying* is an operation in which volatile liquids (usually water) are removed from solid materials not in solution through the application of thermal energy (Flynn et al. 2018; Theodore and Ricci 2010).
5. *Distillation* is an operation whereby components of a liquid mixture are separated by boiling because of their difference in vapor pressure (Theodore and Ricci 2010; Treybal 1955).

6. *Absorption* (Theodore and Ricci 2010; Treybal 1955) is a process whereby a component is removed from a gas stream by dissolution of it in a liquid.
7. *Liquid–liquid extraction* (Theodore and Ricci 2010; Treybal 1951) is a process in which a solute in a liquid solution is removed by dissolution into another liquid solvent that is relatively immiscible upon contact with the solution.
8. *Liquid–solid leaching* (Theodore and Ricci 2010; Treybal 1951, 1955) involves treating a finely divided solid with a liquid that dissolves and removes a solute contained in the solid.
9. *Crystallization* (Theodore and Ricci 2010; Treybal 1955) is the removal of a solute, such as a salt, from a solution by precipitating the solute from the solution.
10. *Mechanical–physical separations* involve separation of solids, liquids, or gases by mechanical means such as filtration, settling, and size reduction which are often classified as separate unit operations.

A more detailed discussion of mass transfer (Section 8.4), heat transfer (Section 8.5), and fluid flow (Section 8.6) follows.

8.4 Mass Transfer

Unit operations are applicable to processes that are essentially physical in nature. A substantial number of the unit operations are concerned with the problems of changing the composition of solutions and mixtures through methods that do not involve chemical reactions. Most frequently, these composition changes are analytical; that is, it is often desirable to separate the original substance into its component parts. Such separations may be entirely mechanical, such as the separation of solid from liquid during filtration or the classification of a granular solid into fractions of different particle size by screening. On the other hand, if the operations involve changes in composition of solutions, they are known as diffusional, or mass transfer operations, and it is these with which engineers are most often concerned. The substance undergoing the change in composition is most frequently itself a solution, but it may be a physical mixture or one that exhibits the properties of both.

Diffusional operations are characterized by a transfer of one substance through another, usually on a molecular scale. For example, when water evaporates from a pool into an air stream flowing over the water surface, molecules of water vapor diffuse through those of the gas at the surface into the main portion of the air stream, where they are carried away. On other occasions, one molecular species may diffuse through another that is itself diffusing in the opposite direction. In every case, mass transfer is a result of a concentration difference, the diffusing substance moving from a location of high concentration to one of low concentration. This concentration difference is the driving force for the transfer of mass in the same way that a temperature difference is the driving force for the transfer of heat between bodies.

It is useful to classify diffusional operations and to cite examples of each. Diffusional separations may be brought about by the following:

1. Contact of two immiscible phases, with mass transfer or diffusion through the surface between the phases.
2. Contact of miscible phases separated by a permeable or semipermeable membrane, with diffusion through the membrane.
3. Direct contact of miscible phases.

These separations are discussed in the next three sections. Equipment descriptions and design information for all the mass transfer equipment associated with these processes are available in the literature (Theodore 2014; Theodore and Ricci 2010).

8.4.1 Contact of Immiscible Phases

The three states of aggregation permit six possible combinations of states: gas–gas, gas–liquid, gas–solid, liquid–liquid, liquid–solid, and solid–solid, each of which is discussed in the following sections.

8.4.1.1 Gas–Gas

Since all gases are completely soluble in each other, this category of operation cannot generally be practically realized.

8.4.1.2 Gas–Liquid

If all components of the system are present in appreciable amounts in both gas and liquid phases, the operation, known as distillation, may be employed. In this instance, the gas phase is created from the liquid phase by application of heat. For example, if a liquid solution of acetic acid and water is partially vaporized by heating, it is found that the newly created vapor phase and the residual liquid both contain acetic acid and water, but in proportions that are different for the two phases and different from those in the original solution. If the vapor and residual liquid are separated physically from one another and the vapor is condensed, two solutions, one richer and the other poorer in acetic acid, are obtained. In this way, a certain degree of separation of the original components is accomplished. Conversely, should a vapor mixture of the two substances be partially condensed, the newly formed liquid phase and the residual vapor will differ in composition. In both instances, an interdiffusion of both components between the phases eventually establishes their final composition.

All the components of the solutions involved may not be present in appreciable amounts in both gas and liquid phases, however. If the liquid phase is a pure liquid containing but one component whereas the gas contains two or more, the operations are known as *humidification* or *dehumidification*, depending upon the direction of the diffusion. For example, contact of dry air with liquid water results in the evaporation of some of the water into the air (humidification of the air). Conversely, contact of very moist air with pure liquid water may result in condensation of part of the moisture in the air (dehumidification). In both cases, relatively little air dissolves in the water, and for most practical purposes, it is considered that only water vapor diffuses from one phase to the other.

Both phases may be solutions, each containing only one common component that distributes between phases. For example, if a mixture of ammonia and air is brought into contact with liquid water, a large portion of the ammonia, but relatively little air, will dissolve in the liquid, and in this way, the air–ammonia mixture may be separated. The operation is known as gas *absorption*. On the other hand, if air is brought into contact with an ammonia–water solution, some of the ammonia leaves the liquid and enters the gas phase. This operation is known as *desorption*, or *stripping*, a common wastewater treatment method.

To complete the classification, consider the case where the gas phase contains but one component and the liquid several, as in evaporation of a saltwater solution by boiling. Here, the gas phase contains only water vapor, since the salt is essentially nonvolatile. Such operations do not depend on concentration gradients, but rather on the rate of heat transfer to the boiling solution and are consequently not considered diffusional separations. Should the salt solution be separated by diffusion of the water into an air stream, however, the operation then becomes one of desorption, or stripping, and is considered a diffusional separation process.

8.4.1.3 Gas–Solid

It is convenient to classify mass transfer operations in the gas–solid category according to the number of components that appear in two phases. If a solid solution were to be partially vaporized without the appearance of a liquid phase, the newly formed vapor phase and the residual solid would each contain all the original components, but in different proportions. The operation is then *fractional sublimation*. As in distillation, the final compositions are established by interdiffusion of the components between the phases. Although such an operation is theoretically possible, it is usually not practical because of the inconvenience of dealing with solid phases in this manner. However, all components may be present in both phases.

If a solid that is moistened with a volatile liquid is exposed to a relatively dry gas, the liquid leaves the solid and diffuses into the gas, an operation generally known as *drying*, or sometimes as *desorption*. An example of this process is the drying of laundry by exposure to air. There are many industrial counterparts such as the drying of lumber or the removal of moisture from a wet filter cake by exposure to dry gas. In these cases, the diffusion is, of course, from the solid to the gas phase. If the diffusion takes place in the opposite direction, the operation is known as *adsorption*. For example, if a mixture of water vapor and air is brought into contact with activated silica gel, the water vapor diffuses to the solid, which retains it strongly, and the air is thus dried.

In other instances, a gas mixture may contain several components, each of which is adsorbed on a solid but to different extents (*fractional adsorption*). For example, if a mixture of propane and propylene gases is brought into contact with a molecular sieve, the two hydrocarbons are both adsorbed, but to different extents, thus leading to a separation of the gas mixture.

In a case in which the gas phase is a pure vapor, such as in the sublimation of a volatile solid from a mixture with one that is nonvolatile, the operation is dependent more on the rate of application of heat than on the concentration difference. This process is essentially a nondiffusional separation. The same is true of the condensation of a vapor to a pure solid, where the rate of condensation depends on the rate of heat removal, not on mass transfer limiting diffusional processes.

8.4.1.4 Liquid–Liquid

Separations involving the contact of two insoluble liquid phases are known as *liquid-extraction* operations. A simple example is a familiar laboratory procedure: if an acetone–water solution is shaken in a separatory funnel with carbon tetrachloride and the liquids are allowed to settle, a large portion of the acetone will be found in the carbon-tetrachloride-rich phase and will thus have been separated from the water. A small amount of the water will also have been dissolved by the carbon tetrachloride, and a small amount of the latter will have entered the water layer, but these effects are relatively minor. As another possibility, a solution of acetic acid and acetone may be separated by adding it to the insoluble mixture of water and carbon tetrachloride. After shaking and settling, both acetone and acetic acid will be found in both liquid phases, but in different proportions. Such an operation is known as *fractional extraction*.

8.4.1.5 Liquid–Solid

Fractional solidification of a liquid, where the solid and liquid phases are both of variable composition containing all components but in different proportions, is theoretically possible but is not ordinarily carried out because of practical difficulties in handling the solid phase and because of the very slow diffusion rates in a solid.

Cases involving distribution of a substance between the solid and liquid phases are common, however. Dissolution of a component from a solid mixture by a liquid solvent is known as *leaching* (sometimes called *solvent extraction*). The leaching of gold from ore by cyanide solutions and the leaching of cottonseed oil from cottonseeds by hexane are two examples of this leaching process. Diffusion is, of course, from the solid to the liquid phase. If the concentration gradient driving diffusion is in the opposite direction, the operation is known as *adsorption*. Thus, colored material that contaminates impure cane sugar solutions may be removed by contacting the liquid solutions with activated carbon, whereupon the colored substances are retained on the surface of the solid carbon and removed from solution.

When the solid phase is a pure substance and the liquid solution is being separated, the operation is called *crystallization*. But as ordinarily carried out, crystallization rates are more dependent on heat transfer rates than on solution concentrations. The reverse operation is *dissolution*. No known operation is included in the category involving a pure liquid phase.

8.4.1.6 Solid–Solid

Because of the extraordinarily slow rates of diffusion within solid phases, there is no industrial separation operation in the solid–solid category.

8.4.2 Miscible Phases Separated by a Membrane

In operations involving miscible phases separated by a membrane, the membrane is necessary to prevent intermingling of the phases. It must be differentially permeable to the components of the solutions, however, if diffusional separations are to be possible.

8.4.2.1 Gas–Gas

The operation in the gas–gas category is known as *gaseous diffusion* or *effusion*. If a gas mixture whose components are of different molecular weights is brought into contact with a porous diaphragm, the various components of the gas will diffuse through the pores at different rates. This leads to different compositions on opposite sides of the diaphragm and, consequently, to separation of the gas mixture. In this manner, large-scale separation of the isotopes of uranium, in the form of uranium hexafluoride, is carried out.

8.4.2.2 Liquid–Liquid

The separation of a crystalline substance from a colloid, by contact of their solution with a liquid solvent with an intervening membrane permeable only to the solvent and the dissolved crystalline substance, is known as *dialysis*. For example, aqueous beet sugar solutions containing undesired colloidal material are freed of the latter by contact with water with an intervening semipermeable membrane. Sugar and water diffuse through the membrane, but the larger colloidal particles cannot. *Fractional dialysis* for separating two crystalline substances in solution makes use of the difference in membrane permeability of the substances. If an electromotive force is applied across the membrane to assist in the diffusion of charged particles, the operation is *electrodialysis*. If a solution is separated from the pure solvent by a membrane that is permeable only to the solvent, the solvent diffuses into the solution – an operation known as *osmosis*. This is not a separation operation, of course, but if the flow of solvent is reversed by superimposing a pressure to oppose the osmotic pressure, the process is labeled *reverse osmosis*.

8.4.2.3 Solid–Solid

The operation in the solid–solid category has found little, if any, practical application in the chemical process or environmental industries.

8.4.3 Direct Contact of Miscible Phases

Operations involving direct contact of miscible phases, because of the difficulty in maintaining concentration gradients without mixing of the fluid, are not generally considered practical industrially except in unusual circumstances.

Thermal diffusion involves the formation of a concentration difference within a single liquid or gaseous phase by imposition of a temperature gradient upon the fluid, thus making possible a separation of the components of the solution. This process has been used, for example, in the separation of uranium isotopes in the form of uranium hexafluoride.

If a condensable vapor, such as steam, is allowed to diffuse through a gas mixture, it will preferentially carry one of the components along with it, thus making a separation by the operation known as *sweep diffusion*. If the two zones within the gas phase where the concentrations are different are separated by a screen containing relatively large-size openings, the operation is called *atmolysis*.

If a gas mixture is subjected to very rapid centrifugation, the components will be separated because of the slightly different forces acting on the various molecules, owing to their different masses. The heavier molecules thus tend to accumulate at the periphery of the centrifuge.

The engineer faced with the problem of separating the components of a solution ordinarily must choose among several possible methods. Although the choice is usually limited because of the peculiar physical characteristics of the materials to be handled, the necessity for making a decision nevertheless almost always exists. Until the fundamentals of the various operations are clearly understood, of course, no basis for such a decision is available, but it is good at least to establish the nature of the alternatives at the beginning.

The engineer may sometimes choose between using a diffusional operation of the sort discussed in this chapter or a purely mechanical separation method. For example, in the separation of a desired mineral from its ore, it may be possible to use either the diffusional operation of leaching with a solvent or the purely mechanical methods of flotation. Vegetable oils may be separated from the seeds in which they occur by extraction or by leaching with a solvent. A vapor may be removed from a mixture with a permanent gas by the mechanical operation of compression or by the diffusional operations of gas absorption or adsorption. Sometimes, both mechanical and diffusional operations are used, especially where the former are incomplete, as in processes for recovering vegetable oils wherein extraction is followed by leaching. A more commonplace example is the wringing of water from wet laundry followed by air drying. It is characteristic that at the end of a solely mechanical operation, the substance removed is pure, whereas if removed by diffusional separation methods, it is associated with another substance.

Frequently a choice may be made between a diffusional operation and a chemical reaction to bring about a separation. For example, water may be removed from an ethanol–water solution either by causing it to react with unslaked lime or by special methods of distillation. Hydrogen sulfide may be separated from other gases either by absorption in a liquid solvent or by chemical reaction with ferric oxide. Chemical methods ordinarily change the substance removed, whereas diffusional methods usually permit its eventual recovery in an unaltered form without great difficulty.

There are also choices to be made within diffusional operations. For example, a gaseous mixture of oxygen and nitrogen may be separated by preferential adsorption of the oxygen on activated carbon, by absorption, by distillation, or by gaseous effusion. A liquid solution of acetic acid may be separated by distillation, by liquid extraction with a suitable solvent, by absorption with a suitable solvent, or by adsorption with a suitable adsorbent.

The principal basis for choice in most cases is economics: the method that costs least is usually the one to be used. Other factors, including legal and/or environmental constraints, also influence the decision, however. The simplest operation, although it may not be the least costly, is sometimes desired because it will be trouble-free. Sometimes, a method will be discarded because imperfect knowledge of design methods or unavailability of data for design will not permit results to be guaranteed. Favorable previous experience with a particular method is often given strong consideration. Cost, however, remains one of the prime factors in process selection.

8.4.4 Mass Transfer Equipment

There are numerous types of mass transfer equipment employed in the manufacturing industry. Some of the more common pieces of equipment are presented here. The reader is referred to the literature (Green and Southard 2019; McCabe et al. 2014; Theodore and Ricci 2010) for additional details on not only the mass transfer equipment listed here but also equipment not covered in this section.

8.4.4.1 Distillation

Distillation is probably the most widely used separation process in the chemical industry. Its applications range from the rectification of alcohol, which has been practiced since antiquity, to the fractionation of crude oil. The separation of liquid mixtures by distillation is based on differences in volatility among components. The greater the relative volatilities, the easier the separation. Vapor flows up a column and liquid flows countercurrently down a column. The vapor and liquid are brought into contact on plates or inert packing material. A portion of the condensate from the condenser is returned to the top of the column to provide liquid flow above the feed point (reflux), and a portion of the liquid from the base of the column is vaporized in the reboiler and returned to provide the vapor flow.

In the stripping section, which lies below the feed, the more volatile components are stripped from the liquid. Above the feed, in the enrichment or rectifying section, the concentration of the more volatile components is increased. Distillation columns produce two product streams, referred to as tops and bottoms, from a single feed. These columns are occasionally used with more than one feed, and with side streams withdrawn at points throughout the column. This does not alter the basic operation, but it does complicate the analysis of the process to some extent. If the process requirement is to strip a volatile component from a relatively nonvolatile solvent, the rectifying section may be omitted, and the column is then called a *stripping column*.

In some operations where the top product is required as a vapor, the liquid condensed is sufficient only to provide the reflux flow to the column, and the condenser is referred to as a *partial condenser*. When the liquid is totally condensed, the liquid returned to the column will have the same composition as the top product. In a partial condenser, the reflux will be in equilibrium with the vapor leaving the condenser. Virtually pure top and bottom products can be achieved by using multiple distillation stages or, sometimes, additional columns.

8.4.4.2 Adsorption

In the *adsorption* process, one or more components in a mixture are preferentially removed from the mixture by a solid (referred to as the adsorbent) (Theodore 2008). Adsorption is influenced by the surface area of the adsorbent, the nature of the compound being adsorbed, the pressure of the operating system (liquid application), and the temperature of operation. These are important parameters to be aware of when designing or evaluating an adsorption process since the possibility for an explosion or fire exists in the adsorption column.

The adsorption process is normally performed in a column. The column is run as either a packed- or fluidized-bed reactor. The adsorbent, after it has reached the end of its useful life, can be either discarded or regenerated. This operation can be applied to either a gas mixture or a liquid mixture. The reader is directed to the literature for further information on the adsorption process (Green and Southard 2019; McCabe et al. 2014; Theodore and Ricci 2010) where additional information is provided.

8.4.4.3 Absorption

The process of *absorption* conventionally refers to the intimate contacting of a mixture of gases with a liquid so that part of one or more of the constituents of the gas will dissolve in the liquid (Theodore 2008). The contact usually takes place in some type of packed column.

Packed columns are used for the continuous contact between liquid and gas. The countercurrent packed column is the most common type of unit encountered in gaseous pollutant control for the removal of the undesirable gases, vapors, or odors. This type of column has found widespread application in the chemical industry. The gas stream moves upward through the packed bed against an absorbing or reacting liquid that is injected at the top of the packing. This results in the highest possible efficiency. Since the concentration in the gas stream decreases as it rises through the column, there is constantly fresher liquid available for contact. This provides a maximum average driving force for the diffusion process throughout the bed. The reader is directed to the literature for further information on the absorption process (Green and Southard 2019; McCabe et al. 2014; Theodore and Ricci 2010) where additional information is provided.

8.4.4.4 Extraction

Extraction (sometimes called *leaching*) encompasses liquid–liquid as well as liquid–solid systems. Liquid–liquid extraction involves the transfer of solutes from one liquid phase into another liquid solvent. It is normally conducted in mixer-settlers, plate and agitated-tower contracting equipment, or packed or spray towers. Liquid–solid extraction, in which a liquid solvent is passed over a solid phase to remove some solute, is carried out in fixed-bed, moving-bed, or agitated-solid columns.

8.4.4.5 Drying

Drying involves the removal of relatively small amounts of water or organic liquids from a solid phase. This can be contrasted to evaporation processes that remove large amounts of water from liquid solutions. In many applications, such as in corn processing, drying equipment follows an evaporation step to provide an ultrahigh-solid-content final product stream.

Drying, in either a batch or continuous process, removes liquid as a vapor by passing warm gas (usually air) over or indirectly heating the solid phase. The drying process is carried out in one of four basic dryer types. The first type is a *continuous tunnel dryer*. In a continuous dryer, trays with wet solids are moved through an enclosed system and warm air is blown over the trays. Similar in concept to the continuous tunnel dryer, *rotary dryers* consist of an inclined rotating hollow cylinder. The wet solids are fed in one side and hot air is usually passed countercurrently over the wet solids. The dried solids then pass out the opposite side of the dryer unit. *Indirect drying* equipment heats either a drum or paddle surface that is in contact with the wet solids. The solids are fed across the outside of the hot transfer surface, dried, and discharged continuously. The final type of dryer is a *spray dryer*. In spray dryers a liquid or slurry is sprayed through a nozzle, and fine droplets are dried by a hot gas passed either concurrently, countercurrently, or concurrently/countercurrently past the falling droplets. Despite the differing methods of heat transfer, the continuous tunnel, indirect, rotary, and spray dryers can all reduce the moisture content of solids to less than 0.01% when designed and operated properly.

8.5 Heat Transfer

Heat transfer is often only one of several unit operations involved in a process, and the interrelations among the different unit operations involved must be understood. Heat transfer requires that there be a driving force present for heat to flow. This driving force is the temperature difference between a warmer and colder body (Theodore 2011).

Practically all of the operations that are carried out in industry involve the production or absorption of energy in the form of heat. The laws governing the transfer of heat and the types of equipment that consider heat flow as one of their main objectives are therefore of great importance. Heat may flow by one or more of the following three basic mechanisms:

1. *Conduction*: When heat flows through a body by the transference of the momentum of individual atoms or molecules without mixing, it is said to flow by conduction. For example, the flow of heat through the wall of a furnace or the metal shell of a boiler takes place by conduction.
2. *Convection*: When heat flows by actual mixing of warmer portions with cooler portions of the same material, the heat transfer mechanism is

known as convection. Convection is restricted to the flow of heat in fluids. Rarely does heat flow through fluids by pure conduction without some convection resulting from the eddies caused by the changes of density of the fluid with changes in temperature. For that reason, the terms *conduction* and *convection* are often used together, although in many cases, heat transfer is primarily by convection. For example, the heating of water flowing by a hot surface is an example of heat transfer due primarily to convection. Heat transfer by convection due to density differences is defined as *natural convection*.

3. *Radiation*: A body emits radiant energy in all directions. If this energy strikes a receiver, a portion of it may be transmitted, a portion may be reflected, and a portion may be absorbed. It is this absorbed portion that represents heat transfer in the form of radiation, i.e. the transfer of energy through space by means of electromagnetic waves. Radiation passing through empty space is not transformed to heat or any other form of energy. If, however, matter appears in its path, the radiation will be transmitted, reflected, or absorbed by that matter. For example, fused quartz transmits practically all the radiation that strikes it. A polished opaque surface or mirror will reflect most of the radiation impinging on it, while a black, dull, or matted surface will absorb most of the radiation it receives. The absorbed energy will be transformed quantitatively into heat. Note that it is only the absorbed energy that appears as heat.

The transfer of heat to and from process fluids is an essential part of most chemical processes. The chemical process industries use four principal types of heat exchangers.

1. *Double-pipe exchanger* – the simplest type, used for cooling and heating
2. *Shell and tube exchangers* – used for all applications
3. *Plate and frame exchangers (plate heat exchangers)* – used for heating and cooling
4. *Direct or contact exchangers* – used for cooling and quenching

The word exchanger applies to all types of equipment in which heat is exchanged but is often used specifically to denote equipment in which heat is transferred between two process streams. An exchanger in which a process fluid is heated or cooled by a plant service stream is referred to as a heater or cooler.

One of the simplest and cheapest types of heat exchangers is the concentric pipe arrangement known as the *double-pipe heat exchanger*. Such equipment can be made from standard fittings and is useful where only a small heat transfer area is required. Several units can be connected in series to extend their capacity.

The *shell and tube exchanger* is by far the most commonly used type of heat transfer equipment in the chemical and allied products industries. The advantages of this type of heat transfer device include the following:

1. Large surface area in a small volume
2. Good mechanical layout, i.e. good shape for pressure operations
3. Reliance on well-established fabrication techniques
4. Wide range of construction materials available
5. Easily cleaned equipment
6. Well-established design procedures

Essentially, shell and tube exchangers consist of a bundle of tubes enclosed in a cylindrical shell. The ends of the tubes are fitted into tube sheets, which separate the shell-side and tube-side fluids. Baffles are provided in the shell to direct the fluid flow and increase heat transfer.

In *direct contact heat exchange*, there is no wall to separate hot and cold streams, and high rates of heat transfer are achieved. Applications include reactor off-gas quenching, vacuum condensers, de-superheating, and humidification. Water-cooling towers are particularly common examples of direct contact heat exchangers. In direct contact cooler condensers, the condensed liquid is frequently used as the coolant.

Use of direct contact heat exchangers should be considered whenever the process stream and coolant are compatible. The equipment is simple and cheap and is suitable for use with heavily fouling fluids and with liquids containing solids. Spray chambers, spray columns, and plate and packed columns are used for direct contact heat exchanger reactors. Heat transfer equipment details and design information is available in the literature (Flynn et al. 2018; Kern 1950; Theodore 2011, 2014; Theodore et al. 2017).

The key heat transfer unit employed in desalination is the evaporator. The processing industry has given the operations involving heat transfer to a boiling liquid the general name *evaporation*. The most common application of evaporation is the removal of water from a processing stream. Evaporation is used in the food, chemical, and petrochemical industries, and it usually results in an increase in the concentration of a solution until it forms a thickened slurry or syrup. Applications of evaporation for material processing include sugar slurry thickening, the concentration of dispersed kaolin clay, and bentonite clay dewatering. The factors that affect the evaporation process are concentration in the liquid, compound solubility, system pressure and temperature, scaling, and materials of construction.

An *evaporator* is a type of heat transfer device designed to induce boiling and evaporation of a liquid. The major types of evaporators include the following:

1. Open kettle or pan evaporators
2. Horizontal-tube natural convection evaporators

3. Vertical-tube natural convection evaporators
4. Forced-convection evaporators

The efficiency of an evaporator can be increased by operating the equipment in a single-effect or multi-effect mode. Triple-effect evaporators are implemented in kaolin clay processing. In this process, the steam utilized for evaporating the solution liquid is passed by the slurry three times before exiting the system. This approach decreases the volume or pressure of steam required to meet the target evaporation rates, thus reducing energy and materials used in the process.

8.6 Fluid Flow

When a fluid is at rest, there is no shear stress and the pressure at any point in the fluid is the same in all directions. The pressure is also the same across any longitudinal section parallel with the Earth's surface; it varies only in the vertical direction, i.e. from height to height. This phenomenon gives rise to the study of hydrostatics.

Consider a differential element of fluid of height, dz, and uniform cross section area, S. The pressure, P, is assumed to increase with height, z. The pressure at the bottom surface of the differential fluid element is P; at the top surface, it is $P + dP$. Thus, the net pressure difference, dP, on the element is acting downward. A force balance on this element in the vertical direction yields:

$$\begin{aligned}&\text{Downward pressure force} - \text{Upward pressure force}\\&+ \text{Gravitational force} = 0\end{aligned} \tag{8.1}$$

Therefore,

$$(P+dP)S - PS + \rho S\frac{g}{g_c}dz = 0;\ (dP)S + \rho\frac{g}{g_c}S(dz) = 0 \tag{8.2}$$

One has a choice as to whether to include g_c in the describing equation(s). The term g_c is a conversion constant with a given magnitude and units, e.g. 32.2 1b/ lb$_f$)(ft/s^2) or dimensionless with a value of unity, e.g. $g_c = 1.0$. In this development, g_c is retained. Rearrangement of Equation 8.2 yields

$$\frac{dP}{dz} = -\rho\frac{g}{g_c} \tag{8.3}$$

where z = vertical displacement (upward is considered positive), ρ = fluid density, and g = acceleration due to gravity.

This equation is the hydrostatic or barometric differential equation. ordinary gradient Equation. The term dP/dz is often referred to as the pressure gradient. Equation 8.3 is a first-order ordinary differential equation. It may be the integrated by separation of variables:

$$\int dP = -\int \rho \frac{g}{g_c} dz = -\frac{g}{g_c}\int \rho dz \tag{8.4}$$

For most engineering applications involving liquids, and many applications involving gases, the density may be considered constant, i.e. the fluid is incompressible. Taking ρ outside the integration sign and integrating between any two limits in the fluid (Station 1 is where the pressure equals P_1 and the elevation is z_1, and Station 2 has a pressure of P_2 and elevation z_2), the pressure–height relationship is:

$$P_2 - P_1 = -\rho \frac{g}{g_c}\left(z_2 - z_1\right) \tag{8.5}$$

Equation 8.5 may also be written as:

$$\frac{P_1}{\rho\left(g / g_c\right)} + z_1 = \frac{P_2}{\rho\left(g / g_c\right)} + z_2 \tag{8.6}$$

The term $P / \rho\left(g / g_c\right)$ is defined as the pressure head of the fluid, with units of m or ft of fluid. Equation 8.6 states that the sum of the pressure head and potential head is constant in hydrostatic "flow." Equation 8.6 is sometimes termed Bernoulli's hydrostatic equation. It is useful in calculating the pressure at any liquid depth.

The *phenomenological* law governing the transfer of momentum by molecular diffusion (Newton's second law) was briefly discussed earlier. In addition to molecular diffusion, momentum (and energy) may also be transferred by bulk motion. Since bulk motion involves transfer of mass from one point in a system to another point in the same system, the conservation law for mass was also discussed earlier. These serve as an excellent warm-up for the discussion of the equation of motion (equation of momentum transfer or the conservation law for momentum).

A momentum balance (also termed the impulse-momentum principle) is important in flow problems where forces need to be determined. This analysis is inherently more complicated than those presented earlier (i.e. forces possess both magnitude and direction), because force F and momentum M are vectors. To describe force and momentum vectors, both direction and magnitude must be specified; for mass and energy, only the magnitude is required (Bird et al. 2002).

Newton's law is applied in order to derive the linear momentum balance equation. Newton's law states that the sum of all forces equals the rate of change of linear momentum or:

$$\sum F = \frac{d}{dt}\left(\frac{mv}{g_c}\right) = \frac{dM}{dt} = \dot{M} \tag{8.7}$$

Here M is the rate (with respect to time) of linear momentum, while m and v represent the mass and velocity, respectively. Newton's law in this form must be applied in a specified direction (e.g. horizontal or vertical). The product mv is defined as linear momentum. When this is applied to a fluid entering or leaving a control volume, the following terms may be defined:

$\dot{M}_{out}$ = momentum rate of fluid leaving the control volume
$\dot{M}_{in}$ = momentum rate of fluid entering the control volume

Equation 8.7 may now be written in finite form as:

$$\sum F = \dot{M}_{out} - \dot{M}_{in} \tag{8.8}$$

This balance essentially means that for steady-state flow, the force on the fluid equals the net rate of outflow of momentum across a control surface. These force rate of momentum terms are the contributing effects that are employed in the development of the conservation law for momentum.

8.7 Key Fluid Flow Equations

8.7.1 Total Energy Equation

The conservation law for energy finds application in many chemical and environmental process units such as heat exchangers, reactors, and distillation columns, where shaft work plus kinetic and potential energy changes are negligible compared with heat flows and either internal energy or enthalpy changes. Energy balances on such units therefore reduce to $Q = \Delta E$ (closed nonflow system) or $Q = \Delta H$ (open flow system).

Another important class of operations is one for which heat flows and internal energy changes are secondary in importance to kinetic and potential energy changes and shaft work. Most of these operations involve the flow of fluids to, from, and between tanks, reservoirs, wells, and process units. Accounting for energy flow in such processes is most conveniently accomplished through mechanical energy balances. Details of this approach follow.

Consider the steady-state flow of a fluid in the process depicted in Figure 8.2. The mass entering at Location 1 carries with it a certain amount of energy, existing in various forms. Thus, because of its elevation z_1 ft above any

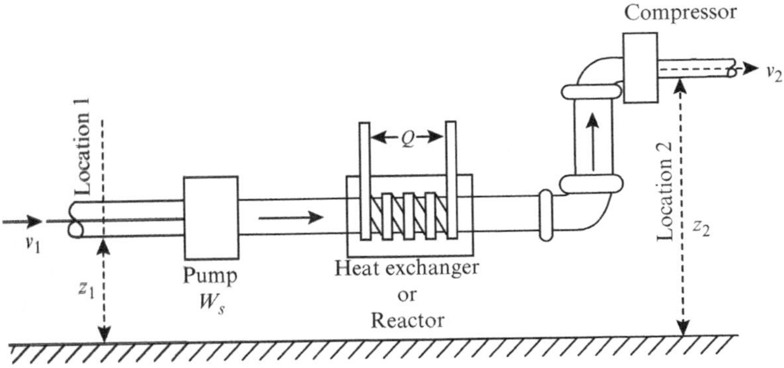

Figure 8.2 Process Flow Diagram.

arbitrarily chosen horizontal reference plane, for example, $z = 0$, it possesses a potential energy $(g/g_c)z_1$ (which can be recovered by allowing the fluid to fall from the height at Location 1 to that of the reference point). Because of its velocity v_1, the mass also possesses and carries with it into Location 1 an amount of kinetic energy $v_1^2/2g_c$. It also possesses so-called internal energy E_1 because of its temperature. Furthermore, the mass of fluid in question entering at Location 1 is forced into the section by the pressure of the fluid behind it and this form of flow energy must also be included. The amount of this energy is given by the force exerted by the flowing fluid times the distance through which it acts; this force is clearly given by the pressure per unit area P_1 times the area S_1 of the cross-section. The distance through which the force acts is the volume V_1 of the fluid divided by the cross-sectional area S_1. Since the work is the force times the distance, i.e. $(P_1S_1)(V_1/S_1) = P_1V_1$, the energy associated with this term is, in turn, the product of the pressure times the volume of the fluid. This was referred to earlier as flow work.

Two additional energy terms need to be included in this analysis. Both involve energy exchange in the form of heat Q and work W between the fluid and the surroundings. In the discussion to follow, it will be assumed (consistent with the notation recently adopted by the scientific community) that any energy in the form of heat or work *added* to the system is treated as a *positive* term. Applying the conservation law of energy under steady-state conditions mandates that all forms of energy entering the system equal that of those leaving. Expressing all terms in consistent units (e.g. energy per unit mass of fluid flowing) results in the total energy balance presented in the following equation:

$$P_1V_1 + \frac{v_1^2}{2g_c} + \frac{g}{g_c}z_1 + E_1 + Q + W_s = P_2V_2 + \frac{v_2^2}{2g_c} + \frac{g}{g_c}z_2 + E_2 \quad (8.9)$$

As written, each term in Equation 8.9 is associated with a mechanical energy effect. For this reason, it has also been defined as a form of the *mechanical energy balance equation* and may be considered a special application of the conservation law for energy. Also note that as written, the volume term, V, by necessity is the specific volume, i.e. the volume per unit mass. In terms of the density of the fluid (mass per unit volume), Equation 8.9 becomes:

$$\frac{P_1}{\rho}+\frac{v_1^2}{2g_c}+\frac{g}{g_c}z_1+E_1+Q+W_s=\frac{P_2}{\rho}+\frac{v_2^2}{2g_c}+\frac{g}{g_c}z_2+E_2 \tag{8.10}$$

and

$$V=\frac{1}{\rho} \tag{8.11}$$

Note once again that Q and W_s can be written on a time-rate basis in this equation by simply dividing by the mass flow rate through the system; the equation then dimensionally reduces to an energy per mass balance, such as (ft-lb$_f$)/lb.

As noted, the solutions to many fluid flow problems are based on the aforementioned mechanical energy balance equation. This equation was derived, in part, from the general (or total) energy equation, Equation 8.9 presented earlier. One may now add some assumptions regarding this equation:

1. Assume adiabatic flow, that is, $Q = 0$.
2. For isothermal, or near-isothermal flow (valid in most applications), the internal energy is constant, so that $E_1 = E_2$.
3. A term $\sum F$, representing the total friction arising as a result of fluid flow, is added to Equation 8.10. This is treated as a positive term in Equation 8.12.
4. An efficiency (fractional) term η is combined with the shaft work term W_s. If work is imparted on the system, the term becomes ηW_s; if work is extracted (with an engine or turbine), the term appears as W_s/η. The efficiency term must be included since part of the work added to or extracted from the system is lost as a result of irreversibilities associated with the mechanical device. The notation h_s will be employed later for this term.

$$\frac{\Delta P}{\rho}+\frac{\Delta v^2}{2g_c}+\frac{g}{g_c}\Delta z-\eta W_s+\sum F=0 \tag{8.12}$$

This equation is also defined as the mechanical energy balance equation.

8.7.2 The Bernoulli Equation

The *Bernoulli equation* is commonly used in fluid flow problems in civil, chemical, and environmental engineering and one form of the equation can be developed from Equation 8.12 by neglecting both work and frictional effects as:

$$\frac{\Delta P}{\rho}+\frac{\Delta v^2}{2g_c}+\frac{g}{g_c}\Delta z=0 \tag{8.13}$$

Equation 8.12 can also be written as:

$$\frac{P_1}{\rho}+\frac{v_1^2}{2g_c}+z_1=\frac{P_2}{\rho}+\frac{v_2^2}{2g_c}+z_2-h_s'\frac{g}{g_c}+h_f'\frac{g}{g_c} \tag{8.14}$$

where h_s' and h_f' effectively replace ηW_s and ΣF, respectively, in Equation 8.12. Note that h_s' is a positive term, as is h_f'. The units of both h_s and h_f are (ft-lb$_f$)/lb. However, each term in this equation, as written, has units of feet of flowing fluid (a pressure term). If the equation is multiplied by g/g_c (keeping in mind once again that g_c is a conversion constant), each term returns to units of energy/mass, e.g. (ft-lb$_f$/lb). The result is as follows:

$$\frac{\Delta P}{\rho}+\frac{\Delta v^2}{2g_c}+\frac{g}{g_c}\Delta z-h_s+h_f=0 \tag{8.15}$$

Note that the prime notation is retained if the units of h_s' and h_f' are given in height of flowing fluid, such as ft H_2O. The choice of units for h_s and h_f is, of course, optional and arbitrary.

8.8 Illustrative Examples

Seven illustrative examples complete the presentation of materials highlighted in this chapter.

8.8.1 Illustrative Example 1

Discuss the difference among the macroscopic, microscopic, and molecular transport approaches.

Solution. The macroscopic approach is concerned with the examination of the overall changes between the inlet and the outlet of a system without regard to internal variations within the system. The microscopic approach is concerned with the examination of changes within and across a system, which ultimately leads to a determination of the net changes across the entire system. Finally, the molecular approach is concerned with the examination of the behavior of individual molecules within a system as a means to determine the change across the entire system.

8.8.2 Illustrative Example 2

List some of the factors influencing the choice of a mass transfer separation process.

Solution. Some of the relevant factors affecting the choice of a mass transfer separation process include the following:

1. Feasibility – can it do the job?
2. Product value generated from the process.
3. Process capacity.
4. Damage to product – can the process damage the resulting product?
5. Classes of available processes.
6. Experience – problems with new processes and experience with traditional processes.

8.8.3 Illustrative Example 3

A heat pump takes in 3,500 gpm of water at a temperature of 38°F and discharges back to a lake at 36.2°F. How many Btu are removed from the water per day (C_p for H_2O = 75.4 J/(gmol-°C), ρ = 62.4 lb/ft^3)?

Solution. To calculate the heat load, the following equation can be employed:

$$\dot{Q} = \dot{m}C_p\left(T_2 - T_1\right) \tag{8.16}$$

$$\dot{m} = \frac{\left(3{,}500 \text{ gal/min}\right)\left(62.4 \text{ lb/ft}^3\right)\left(1440 \text{ min/d}\right)}{7.48 \text{ gal/ft}^3} = 4.2 \times 10^7 \text{ lb/d}$$

The heat capacity given is converted to consistent units and placed on a mass basis as follows:

$$C_p = \frac{\left(75.4 \text{ J/gmol} - {}^\circ\text{C}\right)\left(454 \text{ g/lb}\right)}{\left(1{,}054 \text{ J/Btu}\right)\left(18 \text{ g/gmol H}_2\text{O}\right)\left(1.8^\circ\text{F}/^\circ\text{C}\right)} = 1.00 \text{ Btu/lb} - {}^\circ\text{F}$$

Therefore,

$$\dot{Q} = \left(4.2 \times 10^7 lb/d\right)\left(1.00 \text{Btu/lb} - {}^\circ\text{F}\right)\left(38 - 36.2^\circ\text{F}\right) = 7.56 \times 10^7 \text{Btu/d}$$

8.8.4 Illustrative Example 4

Water enters a boiler at 60°F and exists as two streams: steam at 100°C and liquid water at 100°C. Draw a flow diagram of the system.

Solution: The flow diagram of this system is provided in Figure 8.3.

8.8.5 Illustrative Example 5

Refer to Illustrative Example 4. If H and Q are specified to be 20 lb/min and 10^6 Btu/hr, respectively, outline how to calculate the mass flow rates of the two exit streams, S and W.

Solution: Referring to Figure 8.3, a material balance around the boiler operation gives:

$$H = S + W$$

An energy (enthalpy) balance across the boiler yields:

$$Q + H(h_H) = S(h_S) + W(h_W)$$

If one obtains enthalpy values for the three streams from Chapter 4, the above reduces to two equations and two unknowns and can be solved.

8.8.6 Illustrative Example 6

Refer to Illustrative Examples 4 and 5. Solve the problem for the mass flow rates of the two exit streams, S and W.

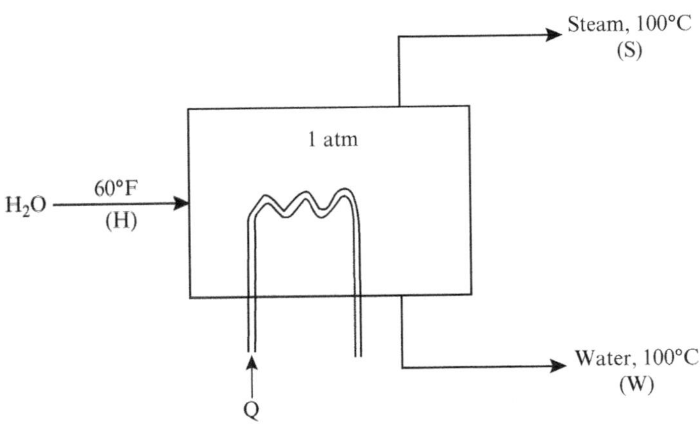

Figure 8.3 Boiler Flow Diagram for Illustrative Example 4.

Solution: A mass balance employing the input data provided gives:

$$H = S + W; 20\ \text{lb/min} = S + W$$
$$S = 20 - W$$

An energy balance with input from Table 5.3 yields the following (h_H for liquid water at 60°F = 28 Btu/lb; h_S for water vapor at 212°F = 1150 Btu/lb; h_W for liquid water at 212°F = 180 Btu/lb):

$$Q + H(h_H) = S(h_S) + W(h_W) \tag{8.17}$$

(10^6 Btu/hr)/(60 min/hr) + (20 lb/min) (28 Btu/lb) = S (1105 Btu/lb) + W (180 Btu/lb)
16,667 Btu/min + 560 Btu/min = (1105 S + 180 W) Btu/min

$$17{,}227\ \text{Btu/min} = 1105S + 180W$$

Substituting Equation 8.17 into Equation 8.18 and solving yields the following:

$$17{,}227\ \text{Btu/min} = 1105\left(20 - W\right) + 180W$$
$$17{,}227\ \text{Btu/min} = 22{,}100 - 1105W + 180W; -4{,}873\ \text{Btu/min} = -925W$$
$$W = 4783/925 = 5.17\ \text{lb/min}$$
$$S = 20 - 5.17 = 14.83\ \text{lb/min}$$

8.8.7 Illustrative Example 7

The height of the liquid (mercury) column is 2.493 ft, the density of mercury is 848.7 lb/ft^3, and atmospheric pressure is 2116 lb/ft^2. Determine the pressure exerted at the bottom of the column and calculate the pressure difference of a column whose top is open to the atmosphere.

Solution: If P_1 is assumed to represent atmospheric pressure, $P_2 - P_1$ reduces to the gauge pressure (the pressure difference) at the bottom of the column. The equation describing the gauge pressure in terms of column height and liquid density is:

$$P_g = \rho \frac{g}{g_c} h;\ h = z \tag{8.18}$$

Calculate the gauge pressure from the input data as follows:

$$P_g = \left(848.7\ \text{lb/ft}^3\right)\left(1.0\right)\left(2.493\ \text{ft}\right) = 2{,}116\ \text{lb/ft}^2$$

The absolute pressure at the bottom of the column is determined from the gauge pressure and atmospheric pressure, P_a, as follows:

$$P = P_g + P_a = 2{,}1162{,}116\,\text{lb}/\text{ft}^2 + 2{,}116\,\text{lb}/\text{ft}^2 = 4{,}232\,\text{lb}/\text{ft}^2$$
$$= 29.4\,\text{lb}/\text{in}^2 \text{ absolute}$$
$$P = 2\,\text{atm absolute}$$

8.9 Problems

1. Determine the depth in the Atlantic Ocean at which the pressure is equal to 10 atm absolute. Assume that seawater is incompressible with a density of 1000 kg/m^3, and the pressure at the ocean surface is 1.0 atm absolute or 0.0 atm gauge.
2. Briefly describe the term critical Reynold's number.
3. Discuss the differences among a fan, a pump, and a compressor.
4. As part of his investigation, Detective Theodore paid a visit to the office of his missing bookmaker, the location where the bookmaker was last seen. In examining the premises, he noticed a half-empty, 7 cm tall coffee mug on the missing bookmaker's desk. Based on his experience and the stain left in the mug, Detective Theodore estimated that the bookmaker had been missing for approximately 2 weeks. Qualitatively explain how this brilliant sleuth reached this conclusion. It may be assumed that the coffee mug was initially full to 0.5 cm from the brim at the time of the bookmaker's disappearance.
5. An air-conditioning duct in a humidifier/dehumidifier desalination application has a rectangular cross-section of 1 m by 0.25 m. If the kinematic viscosity of the air is approximately 1 x 10^{-5} m^2/s, determine the maximum air velocity before the flow becomes turbulent. Assume the critical Reynolds number is 2300. Note: the definition of the hydraulic diameter for input into the Reynolds number equation is $D_h = (2wh)/(w + h)$, where w = width of a channel and h = height of a channel; the equation for the critical Reynolds number is $Re_{crit} = (v\ Dh)/\nu$, where v = velocity and ν = fluid kinematic viscosity.
6. Describe the main differences between gaseous adsorption and gaseous absorption.
7. A 250 psig boiler requires a maximum of 750 gpm of preheated feedwater during peak production. The temperature of the preheated water is kept relatively constant at 120°F due to a heat exchanger that uses steam from the boiler. The feedwater originates from a 46 ft high vented tank that

collects the condensate. Note: The tank is kept at a constant 23 ft level due to a makeup water system and at 90°F due to condensate return and ambient losses. The total pressure loss across the piping, heat exchanger, and fittings is 75 psi. Assuming that the pump efficiency is 75% and the motor efficiency is 92%, determine the pump motor horsepower and available net positive suction head (NPSHA) if the pump is located at the base of the tank and the boiler inlet is 46 ft above the pump. Note: The required pump pressure increase, ΔP, is given by $\Delta P = \rho\,(h_2 - h_1) + (P_2 - P_1) + P_L$, where P_L is the system pressure loss; the required pump hydraulic horsepower (HHP) is given by HHP = (Q) (ΔP); the required pump break horsepower, BHP, is HHP/P_{eff}; the required pump motor horsepower (MHP) is given by MHP = (BHP) (SF)/(M_{eff}),where SF = pump service factor = 115%; and the NPSHA = $H_s + H_a - H_v$, where H_s = static water pressure head, H_a = ambient pressure head, and H_v = water vapor pressure head.

8. A piping system at a desalination facility has an inner diameter (ID) of 6 in. A flow rate of 900 gal/min is desired. Calculate the average velocity, $\bar{v}$, the Reynold's number, *Re*, and determine if the flow is laminar or turbulent.

References

Abulendia, P. and Theodore, L. (2009). *Fluid Flow for the Practicing Chemical Engineering*. Hoboken, NJ: John Wiley & Sons.

Bird, R.B., Stewart, W.E., and Lightfoot, E.N. (2002). *Transport Phenomena*, 2nd Edition. Hoboken, NJ: John Wiley & Sons.

Farag, I. and Reynolds, J. (1996). *Heat Transfer. A Theodore Tutorial*. East Williston, NY: Theodore Tutorials. Originally published by the U.S. EPA/APTI, RTP, NC.

Flynn, A.M., Akashige, T., and Theodore, L. (2018). *Kern's Process Heat Transfer*, 2nd Edition. Salem, MA: Schrivener-Wiley.

Green, D. and Southard, M. (2019). *Perry's Chemical Engineers' Handbook*, 9th Edition. New York, NY: McGraw-Hill.

Kern, D. (1950). *Process Heat Transfer*. New York, NY: McGraw-Hill.

McCabe, W.L., Smith, J.C., and Harriot, P. (2014). *Unit Operations of Chemical Engineering*, 7th Edition. New York: McGraw-Hill.

Theodore, L. (2008). *Air Pollution Control Equipment Calculations*. Hoboken, NJ: John Wiley & Sons.

Theodore, L. (2011). *Heat Transfer for the Practicing Engineer*. Hoboken, NJ: John Wiley & Sons.

Theodore, L. (2014). *Chemical Engineering: The Essential Reference*. New York, NY: McGraw-Hill.

Theodore, L., Dupont, R., and Ganesan, K. (2017). *Unit Operations in Environmental Engineering*. Salem, MA: Scrivener-Wiley.

Theodore, L. and Ricci, F. (2010). *Mass Transfer for the Practicing Engineer*. Hoboken, NJ: John Wiley & Sons.

Treybal, R. (1951). *Liquid Extraction*. New York, NY: McGraw-Hill.

Treybal, R. (1955). *Mass Transfer Operations*. New York, NY: McGraw-Hill.

Part II

Desalination and Water Treatment Processes

Chapters in this Part include the following:

9

The Desalination Process

9.1 Introduction

The world's total water supply is enormous compared with the presently conceivable needs of man, yet there is a growing water concern. Approximately 98% of the 320 million cubic miles of water in the Earth's crust is salty and useful neither for irrigation by present techniques nor for the majority of man's other needs. Precipitation provides many times the world's annual water needs, but fresh water supplies vary widely not only over the Earth's surface but also from time to time in a given region. This accounts for what has come to be referred to as the "water problem."

In the United States, the actual "consumptive use" is only approximately 15% of the potential supply. In addition, water prices are low, and based on statistical analyses there should be no major problem in most areas for many years into the future. Although the United States is relatively well supplied with water, it continues to turn to sources other than rainfall and runoff. Parts of the Southwest have too little water to support rapid population expansion and California has embarked on projects to bring water nearly 1,000 miles to Los Angeles and the southern part of the state from outside the region. Outside of the United States, expensive desalination plants have been in operation for years, including in the Virgin Islands, the Persian Gulf, the West Indies, Libya, the Union of South Africa, Israel, and many other places. An obvious way to increase water availability is to recover fresh potable water through desalination from seawater or from the large underground stores of brackish water which are available in many arid regions. The authors are currently investigating other feed sources for producing potable water. Another alternative is to find ways to use saline waters for agriculture and other purposes which currently require fresh water.

This chapter introduces the reader to the desalination process. Desalination principles and applications are addressed through a review of desalination basics, energy options for use in desalination, separation options, traditional

Introduction to Desalination: Principles, Processes, and Calculations, First Edition. Louis Theodore and R. Ryan Dupont.

desalination processes, and new options for water desalination. An application section provides six illustrative examples of concepts and implementation of the desalination process.

9.2 Basics of Desalination

The overall desalination process is presented in Figure 9.1. Feed water (1) is introduced to the desalination unit (2), where – following the application of some form of energy (3), it is separated into (relatively) pure water (4), and a more concentrated brine solution (5).

The various methods (to be discussed in the next two sections) for separating pure water and the dissolved salts from seawater have two things in common:

1. They all involve some form of energy input; for example, the same theoretical minimum possible energy of approximately 3.67 kWh is required to separate 1,000 gal of fresh water from seawater at 25°C leaving a concentrated brine solution containing the same 1,000 gal of water (50 percent recovery).
2. They all involve some sort of physical separation process.

The earlier two considerations and the accompanying mode of energy delivery serve to differentiate among the various desalination processes, i.e. the type of energy employed and its mode of delivery to the solution plus the subsequent separation technique employed are the key factors. After reviewing the art of

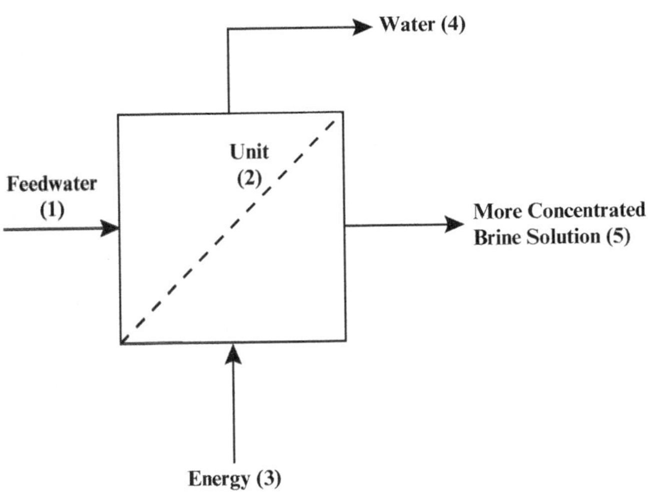

Figure 9.1 Basic Desalination Process.

Table 9.1 Saline water conversion options.

Substance to be separated	State of matter	Technically feasible methods	Commercially feasible methods
Water	Vapor	Evaporation	Evaporation
Water	Solid	Freezing Hydrates	Freezing
Water	Liquid	Extraction Reverse osmosis	Reverse osmosis
Salt	Vapor	Unknown	
Salt	Solid	Ion exchange	
Salt	Liquid	Electrodialysis Osmionic Thermal diffusion Polarization	Electrodialysis

saline water conversion, the authors have identified several possible options. These are summarized in Table 9.1.

In processes in which water crosses the phase boundary, energy is required in proportion to the water recovered; the practical work requirements depend little on the salt concentration of the water fed, whether it is sea water or slightly brackish water. In processes in which the salt passes the phase boundary, the energy required increases roughly in proportion to the salt concentration of the feed. Therefore, processes of the second type are accordingly less expensive to operate with brackish than with sea water supplies.

Of all of the processes, the only ones which are known to currently be economically feasible are: (1) the distillation (evaporation), (2) reverse osmosis, and (3) freezing (crystallization). At present, the economic feasibility of all other processes is considered doubtful with the exception of solvent extraction and hydrate formation which may have some possibility as the art progresses with time.

9.3 Energy Options

Energy is the keystone of life and prosperity. Adequate energy supplies and a satisfactory living environment are goals of overriding importance to every individual. There is no question that as energy is made available, the quality of life improves. At the same time, however, economic and national stability must also be maintained. Recent developments provide evidence that the discourse on all of these aspects will continue and be the prime determinants of domestic and foreign policy for many years to come.

Over the past 40+ years there has arisen among informed leaders of industry, governments, and the environmental movement an acute awareness of energy as an issue of critical importance to everyone's well-being and survival. An energy crisis; or problem, or shortage, or dilemma, as it has been called; is created by the continually increasing demand for energy. This demand has resulted in three issues becoming critical concerns of the entire international community. First is the issue of ensuring an adequate, reliable supply of all forms of energy. Second is the growing public concern with the environmental consequences of producing energy. Finally, and equally important, are the social ramifications associated with the accompanying financial expenditures required to meet the energy demand.

The solutions to the problems that arise from energy demand may simply be conservation and the development of new, less expensive energy source. Energy conservation can sharply reduce the waste of resources that has been at the very heart of many environmental problems. Moreover, an extensive conservation program can be implemented in a very short period of time. Such an effort can play a major role in slowing the growth in the demand for energy and in causing energy to be used more efficiently. At the same time, new sources of energy must be developed to ensure the availability of adequate, inexpensive, long-term energy supplies. The feasibility of solar, wind, tidal, geothermal, fusion, and other less traditional sources of energy must continue to be investigated and developed further (Skipka and Theodore 2014).

Because energy has been relatively cheap and plentiful in the past, many energy-wasting practices were allowed to develop and continue in all sectors of the economy. Industries have wasted energy by discharging hot process water instead of recovering the heat, and by wasting the energy discharged from power plant stacks. Waste hydrocarbons have been discharged or combusted with little consideration for recovering their energy value.

For purposes of desalination, the three major sources of energy are the fossil fuels (particularly natural gas), solar and nuclear, although geothermal does receive some consideration later. However, since most of the water problems mentioned earlier exist in arid regions, it would seem that solar energy would presently be the most attractive option for driving desalination systems. These four energy options are briefly described below.

9.3.1 Natural Gas

Natural gas is a nonrenewable fossil fuel, as natural gas (as with oil) was formed from the remains of tiny sea animals and plants that died 300 to 400 million years ago. When these sea animals and plants died, they sank to the bottom of the oceans where they were buried by layers of sediment that turned into rock. Over the millennia these layers increased in height to several thousands of feet, subjecting the remains to high pressure, which, when combined with the high temperature of the Earth, converting this biological material into not only oil, but also natural gas. Over time, high concentrations of natural gas became trapped in the layers of rock, becoming the natural gas deposits being recovered around the world today.

As with other fossil fuels, natural gas was formed in the Earth's crust, possessing the ability of rising toward the Earth's surface. It built up in pockets when its progress was blocked by an upper layer of solid impervious rock. This process has been described as being analogous to trapping water with a sponge. The gas accumulation can be reached by drilling wells in a manner similar to how oil is recovered from an oil field. It should be mentioned that natural gas is produced on all continents except Antarctica.

Natural gas is often found with oil. The oil fields of the Middle East once contained about one-fifth of the world's natural gas reserves. In the past, large quantities of natural gas recovered along with oil were burned because there was no use for the energy locally. Today, this gas is recovered and liquefied for transport to user nations. The leading producers of natural gas are Russia, the United States, Canada, and Western Europe. Prior to the commercialization of natural gas, gas was manufactured from coal and oil. However, this manufactured gas had only about half the heating value of natural gas.

Natural gas consists of a mixture of different gases. The main component is methane CH_4, which is odorless, colorless, and tasteless. Companies add a chemical odorant called mercaptan (with an odor like rotten eggs) so that any escaping gas can be detected. This process is employed as a safety measure.

9.3.2 Solar Energy

Solar energy is a renewable resource that cannot be depleted (at least for millions of years). Solar energy is radiant energy that is produced by the sun. Every day the sun radiates an enormous amount of energy that maintains the planet Earth at a habitable temperature.

Where does this energy come from? The sun's extremely high pressure and temperature cause nuclear fusion. Four hydrogen nuclei fuse to become one helium atom, a process that is accompanied by a loss in mass. This loss in mass result in the emission of radiant energy. It takes millions of years for the energy in the sun's core to make its way to the solar surface, and then just a little over 8 minutes to travel the 93 million miles to Earth. The solar energy travels to the Earth at a speed of 186,000 miles/sec, which is the speed of light.

Only a small portion of the energy radiated by the sun into space strikes the Earth – approximately one part in two billion. Yet, this amount of energy is enormous. Enough solar energy strikes the United States every day to supply the nation's energy needs for one and a half years. Approximately 15% of the energy that strikes the Earth is reflected back into space. Another 30% evaporates water, and some is absorbed by plants, the land, and the ocean. The remainder is available to supply society's energy needs.

In a very real sense, almost all the radiant energy impacting the surface of the Earth can be converted to solar energy. Solar energy is not a single energy technology, but rather a term that in reality covers a diverse set of renewable energy technologies. Solar energy application can be divided into two groups – heating and cooling, and electricity generation.

In terms of solar energy, it should be noted that, for most points on Earth, the sun does not shine 24 hours a day. It is unavailable the rest of the time when compared to a conventional power plant that is not subjected to such constraints. The angle of incidence of radiation also depends on the motion of the sun and upon the geographic position on the Earth. The angle of radiation incidence becomes more pronounced the farther away a receptor site is located from the position where the sun is directly overhead.

In terms of energy strength, the sun constantly delivers 1.36 kW (1,360 J/s) of power per square meter (430 Btu/hr-ft^2 or 123 W/ft^2) to the Earth. However, some of this power is absorbed by the atmosphere so that even under ideal conditions, a receptor on Earth may receive less than 1,000 W/m^2 or 01 W/cm^2. From an overall global perspective, of the total radiation reaching the Earth's atmosphere (1.5 x 10^{18} kW-hr) its ultimate approximate distribution is as follows:

- 50% is reflected back into space by the clouds;
- 27.8% is absorbed in evaporating water;
- 15% is reflected back by the Earth's surface;
- 5.3% is absorbed by bare soil;
- 1.7% is absorbed by marine vegetation;
- 0.2% is absorbed by land vegetation.

9.3.3 Nuclear Energy

Nuclear energy is energy that is obtained from the nucleus of an atom. The energy is released from an atom through one of two processes: (1) *nuclear fusion* or (2) *nuclear fission.*

Energy is released in nuclear fusion when atoms are combined or fused together. This is how energy is generated in the sun. In nuclear fission, energy is released when atoms are split apart. Nuclear fission is the only method currently used by nuclear plants to generate electricity. However, the terms nuclear fusion and nuclear fission are referred to synonymously as nuclear energy. Both involve changes in (nuclear) mass.

Regarding fission, uranium is the heaviest of the 92 naturally occurring elements. Since it is also one of the few elements that is easily fissioned, it is the fuel of choice used by today's nuclear power plants. This element was formed when the Earth was created and is usually found in rocks. Rocks that contain significant quantities of uranium are referred to as uranium ore, or *pitchblende*. Two forms (isotopes) of uranium are found in these rocks: uranium-235 and uranium-238. These three-digit numbers refer to the number of neutrons and protons in each atom. Uranium-235 is the form normally used for energy production because, unlike uranium-238, the nucleus splits more easily when bombarded by a neutron. During fission, the uranium-235 atom absorbs a bombarding neutron, causing its nucleus to split apart into two atoms of lighter weight. The fission reaction simultaneously releases energy as both heat and radiation. It also releases more neutrons. These released neutrons go on to bombard other uranium atoms, and the process repeats itself. This repetitive process is referred ts as a *chain reaction* (Skipka and Theodore 2014).

When a uranium atom is split apart (fissioned), the mass of the fragments is less than the · mass of the original atom. The energy corresponding to this loss of mass is defined as fission energy. It is represented in equation form by Einstein's equation (Theodore et al. 2007):

$$E = mc^2 \tag{9.1}$$

where E = energy, J; m = mass, kg; and c = the speed of light, 3×10^8 m/sec.

One can show that each kilogram of mass converted liberates approximately 10^{17} J. This thermal energy is enough to run a large turbine generator (a 1-GW unit) for a year at 35% thermal efficiency and is equivalent to the energy possessed by 2.5 million tons of coal.

9.3.4 Geothermal Energy

Geothermal energy refers primarily to the generation of electricity from the stored heat in the Earth's crust. Other uses of geothermal energy for process or space heating are termed *direct heat*. Geothermal power may be natural (i.e. from emissions of steam and/or hot water). In addition, cold water may be injected into deep wells where it is heated and returned to the surface for use.

Geothermal energy is heat from within the Earth that is generated in the Earth's core. This energy (heat) can be recovered and used to heat buildings or generate electricity. Temperatures hotter than the sun's surface are continuously produced inside the Earth by the slow decay of radioactive particles, a process that occurs in all rocks.

Geothermal energy (also referred to by some as *geothermics*) is virtually limitless and is based on the fact that the Earth is hotter the deeper one drills below the surface. Geothermal energy is derived from:

1) Steam trapped deep in the Earth that is brought to the surface and used to drive a turbine to produce electricity;
2) Water heated by pumping it through deep, hot rocks.

The Earth's core lies almost 4,000 miles beneath the Earth's surface. The double-layered core is made up of very hot molten iron surrounding a solid iron center. Estimates of the temperature of the core range from 5,000 to 11,000°F (Skipka and Theodore 2014). Surrounding the Earth's core is the mantle, which is partly rock and partly magma. The mantle is about 1,800 miles thick.

The Earth's geothermal resources are theoretically more than adequate to supply humanity's energy needs, but only a very small fraction may be profitably exploited. Drilling and exploration for deep resources is very expensive. Forecasts for the future of geothermal power depend on assumptions about technology, energy prices, existing subsidies, and the value of money (interest rates).

No discussion of energy options would be complete without a review of the second law of thermodynamics. The law of conservation of energy, as detailed in Chapter 7, is defined by many as the first law of thermodynamics. Its application allows calculations of energy relationships associated with various processes. The second law of thermodynamics is referred to as the *limiting law*. Historically, the basis of the second law was developed by individuals such as Carnot, Clausius, and Kelvin in the middle of the nineteenth century. This development was made purely on a macroscopic scale and is referred to as the "classical approach" to the second law (Theodore et al. 2007).

The terms *energy conservation* and *energy efficiency* have definitions that are many and varied when applied to measuring heat. This mainly occurs because of the many approaches to describing, defining, and measuring the energy efficiency of energy-related processes. The discussion that follows attempts to provide the reader with information on what it really means to conserve energy. This includes a new term – entropy – that is employed in many thermodynamic second-law applications (Theodore et al. 2007).

The aforementioned first law of thermodynamics is a conservation of law concerned with energy transformations. Regardless of the types of energy involved in a process – thermal, mechanical, electrical, elastic, magnetic, etc. – the change in energy of a system is equal to the difference between energy input and energy output. The first law also allows free convertibility from one form of energy to another, as long as the overall quantity is conserved. Thus, this law places no restriction on the conversion of work into heat or its counterpart – the conversion of heat into work. However, the second law of thermodynamics is another matter.

The preceding brief discussion of energy conversion leads to an important second law consideration – energy has *quality* as well as quantity. Because work is 100% percent convertible to heat, but the reverse situation is not true, work is a more valuable form of energy than heat. Although it is not as obvious, it can also be shown through second-law arguments that heat also has quality in terms of the temperature at which it is discharged from a system. The higher the temperature the greater is the potential for the transformation of energy into work. Thus, thermal energy stored at high temperatures is generally more useful to society than that available at lower temperatures. While there is an immense quantity of energy stored in the oceans, for example, its present availability to society for performing useful tasks is quite low. This implies, as noted earlier, that thermal energy loses some of its quality, or is degraded, when it is transferred by means of heat transfer from one temperature to a lower one. Other forms of energy degradation include energy transformations due to frictional effects and electrical resistance. Such effects are undesirable if the use of energy for practical purposes is to be maximized (Skipka and Theodore 2014; Theodore 2014; Theodore et al. 2007).

The second law provides some means of measuring this energy degradation through a thermodynamic term referred to as *entropy*, and it is the second law that serves to define this important property. Entropy is normally designated as S with units of energy per absolute temperature (e.g. Btu/°R or cal/K). Furthermore, entropy calculations can provide quantitative information on the "quality" of energy and energy degradation (Theodore 2014; Theodore et al. 2007).

There are a number of other phenomena that cannot be explained by the law of conservation of energy. It is the second law of thermodynamics that provides an understanding and analysis of these diverse effects. However, among these considerations, it is the second law that can allow the measuring of the aforementioned quality of the energy, including its effect on the design and performance of heat exchangers. Additional details on the second law and its applications are available in the literature (Flynn et al. 2018; Skipka and Theodore 2014; Theodore 2011, 2014; Theodore et al. 2017, 2007).

9.4 Separation Options

There are various mass transfer operations. The three that are most commonly encountered in desalination applications are evaporation, crystallization, and reverse osmosis. As such, they receive the bulk of the treatment in Chapters

10–12. However, there are other mass transfer processes, some of which are briefly described in this chapter. Theodore and Barden (1995) along with Theodore and Ricci (2010) provide more extensive details plus calculations and design procedures.

Mass transfer operations are generally applicable to processes that essentially involve either componential or physical change, or both. A substantial number of these operations are concerned with changing the composition of solutions and mixtures through methods that do not involve chemical reactions; these are defined as componential separation processes. It is also often desirable to separate the original substance into its component parts by phase. Such separations may be entirely mechanical, such as the separation of a solid from a liquid during filtration or the classification of a granular solid into fractions of different particle size by screening. On the other hand, if the operation(s) involve the aforementioned changes in the composition of solutions, they are defined componential separation operations.

For many, mass transfer has come to be defined as the tendency of a component in a mixture to travel from a region of high concentration to one of low concentration. For example, if an open beaker with some alcohol in it is placed in a room in which the air is relatively dry, alcohol vapor will diffuse out through the column of air in the beaker. Thus, there is mass transfer of alcohol from a location where its concentration is high (just above the liquid surface) to a place where its concentration is low (at the top of the beaker). If the gas mixture in the beaker is stagnant (void of any motion), the transfer is said to occur by molecular diffusion. If there is a bulk mixing of the layers of gas in the beaker, mass transfer is said to occur by the mechanism of either forced or natural convection, or both. These two mechanisms are analogous to the transfer of heat by conduction and convection (Flynn et al. 2018). Analogies with fluid flow are not nearly as obvious.

As noted here, the problem of transferring materials from one phase to another is encountered in many engineering operations. There are generally three classes of phases encountered in practice: gas, liquid, and solid. If mass transfer occurs between two phases, with one phase being treated and the other performing the treatment, a total of $(3)^2$ or nine possible combinations of operations with two phases is possible. These mini-scenarios are listed in Table 9.2.

Any of the traditional unit operations are potential candidates for providing the separation technology (Green and Southard 2019; Theodore and Barden 1995; Theodore and Ricci 2010). These can include:

1) Evaporation
2) Reverse osmosis
3) Crystallization
4) Electrodialysis

Table 9.2 Mass Transfer Scenarios.

Scenario	Phase undergoing treatment	Phase performing treatment
1	Gas	Gas
2	Gas	Liquid
3	Gas	Solid
4	Liquid	Gas
5	Liquid	Liquid
6	Liquid	Solid
7	Solid	Gas
8	Solid	Liquid
9	Solid	Solid

5) Solvent extraction
6) Thermal diffusion
7) Polarization
8) Ion exchange
9) Hydration

It was noted earlier that the energy requirement for separation is proportional to the quantity of water recovered in separation processes in which water crosses a phase boundary, and the actual work requirements are effectively independent of the salt concentration of the water processed, whether it is seawater or slightly brackish water. In processes in which the salt crosses the phase boundary, the energy required increases roughly in proportion to the salt concentration of the feed. Therefore, processes of the second type are generally less expensive to operate with brackish water than with seawater. Of all the separation processes mentioned earlier, the only ones which appear to be economically feasible are evaporation, reverse osmosis, and crystallization. As noted earlier, the economic feasibility of all other processes, irrespective of the energy employed, is considered doubtful with the exception of solvent extraction (Treybal 1951) and hydrate formation which may have some possibility in the future.

Any separation process chosen for desalination must be compatible with the scale of operation of a commercial plant. Many separation options have an upper limit on the capacity which can be handled by a single unit, and this in turn limits the production scale for which that operation should be considered. In some cases, the upper limit represents a true limitation imposed by a physical phenomenon, while in other cases, the upper limit is simply the maximum scale at which the commercial equipment is presently being manufactured.

Of all factors influencing the decision to choose one desalination process in preference to another, design reliability is the most important. Regardless of any other consideration, the plant, when constructed, must work properly to produce an acceptable potable water product that can be sold at a profit. The economic consequences of a design failure are too dire to accept an unworkable separation process design. Design reliability is an elusive term to define in quantitative terms because it actually relates to the amount of testing and demonstration that must be done before a suitable commercial scale design is produced. Additional details are provided in Table 9.3. General comments regarding the three key separation processes plus two potential candidates for the future are provided shortly.

9.4.1 Evaporation

Methods for evaporation have been developed over many years of industrial experience and extensive testing of commercial-scale equipment. A competent engineering firm usually can design an evaporation process given a knowledge of the pertinent physical properties of the feed seawater. Occasionally, some small-scale testing of the mixture to be separated is required, but scale-up methods for evaporation are the most reliable of all the separation methods.

9.4.2 Reverse Osmosis

The products from a reverse osmosis separation step consist of a purified solvent stream and a solution somewhat more concentrated than the feed stream.

Table 9.3 Factors Influencing the Choice of Separation Processes.

Factor	Considerations
Reliability	An absolute must
Feasibility	Can the process do the job?
Product value and process capacity	Value of product(s) being isolated
Damage to product	Can process damage product?
Experience	Problems with new process
Generation of process alternatives	Applies to new systems
Classes of other processes	List of advantages and disadvantages
Separation factor and molecular weight	Degree of separation and applicable properties of product(s)
Chemical complexing	Choice of solvent if applicable

Therefore, it is not a complete separation process and must be accompanied by additional processing steps if complete separation or high recovery of solvent is desired. The degree of recovery or extent of concentration of the reject stream is limited by the osmotic pressure of the concentrated solution. The membrane must maintain its integrity and permit reasonable flux of solvent with a pressure difference across the membrane exceeding the osmotic pressure. In general, excessive osmotic pressures are encountered when the solute mole fraction in the concentrated solution exceeds design values, generally 0.03–0.06.

9.4.3 Freeze Crystallization

Freeze crystallization is a phase separation process that is in the process of becoming fully developed or achieving its full potential for significant commercial use. Most of the development has been for use in instant coffee and water desalination. On a laboratory scale, this process has been used in the treatment of feeds containing ammonium nitrate, paper mill bleach solutions, plating liquors, and arsenal redwater. The freeze crystallization process also has potential use in the recovery/removal of 1 to 10% total dissolved solids (TDS) in aqueous streams. Seawater or contaminated wastewater can be subcooled during freeze crystallization in order to form purified ice crystals. The remaining liquid is more concentrated in soluble organic or inorganic materials. The crystals are subsequently removed, washed, and melted to recover the water. The liquid could be freeze crystallized again to further concentrate the feed and make treatment and/or disposal much easier.

9.4.4 Adsorption

Adsorption equipment and operating cycles usually can be designed reliably on the basis of measured adsorption isotherm data and a reasonable number of relatively small-scale experiments to determine the mass transfer characteristics of the stream to be treated in the chosen adsorbent bed. If the stream treated contains several adsorbed components it is generally necessary to run experiments on the actual mixture to be treated since multicomponent isotherm behavior cannot be predicted in general from the individual isotherms.

9.4.5 Ion Exchange

Although ion exchange systems can be designed with the aid of only a few laboratory tests, a pilot plant operation of substantial duration generally is recommended. Ion exchangers are subject to fouling and loss of capacity which cannot be detected readily by laboratory tests. Therefore, it is desirable to operate a pilot plant throughout a number of cycles using the commercial plant regeneration procedures and, if at all possible, the actual stream to be treated in the commercial installation.

9.4.6 Extraction

Except in the rare case in which the desired product form is a solution of the extracted product in the extraction solvent and the solvent does not contaminate the raffinate, complete component separation cannot be accomplished by extraction. In fact, extraction is a selective transfer process rather than a separation process. To obtain the components of the original mixture in the specified purity, it is necessary to separate the extract stream by recovering the extracted component from the extracting solvent. Usually, it is also necessary to recover the small amount of solvent dissolved in the raffinate stream to prevent either excessive solvent make-up costs or contamination of the raffinate product. Frequently, these auxiliary separations are accomplished by distillation. Therefore, an extraction process virtually always requires at least two steps to separate a feed into two acceptable product streams.

9.5 Traditional Desalination Processes

As noted earlier, three prime candidate desalination processes have emerged and been employed in the past. Each are briefly reviewed in more detail in the subsections to follow. The next three chapters (Chapters 10–12) address these processes in even more detail.

9.5.1 Evaporation Processes

The oldest and best developed process for saline water conversion is the evaporation (or what some define as distillation) method. Evaporation remains one of the major methods today for commercial production of fresh water from seawater.

In principle, evaporation is the simplest method. Seawater is boiled in an evaporator by passing hot steam through a steam chest where the steam condenses on the inside of the tubes of the chest and is usually returned to the boiler (Theodore 2011). The vapors rising from the seawater feed are cooled in a condenser and thus converted into pure liquid water which is collected in a storage vessel.

The resulting concentrated brine solution is continuously or intermittently withdrawn from the evaporator. The most advanced commercial evaporation processes are the following:

1. Multiple-effect evaporation
2. Flash evaporation
3. Vapor compression evaporation

In multi-effect units, the vapors from the first evaporator condense in the second effect where their heat of condensation is employed to boil the seawater in the second effect (chest), etc. In the flash evaporation process, seawater is first heated in tubes and then made to evaporate in chambers where the pressure is lower than in the heating tubes. While multiple-effect and flash evaporation units use an external supply of heated steam as the primary heat source, vapor compression evaporation literally uses its own steam as a heat source after it has been compressed. Here, seawater is preheated in a tubular heat exchanger by the outgoing streams of concentrated brine and freshwater, and boil in the tubes of the still. The (steam) vapors are compressed and fed back to the still to condense outside the tubes, thus providing the heat necessary for the boiling process. The two major drawbacks to the evaporation processes are corrosion and scale formation (Flynn et al. 2018).

9.5.2 Reverse Osmosis

Electrodialysis was the membrane separation process employed for desalination a century ago. Here, the ions forming the salt are removed from the saltwater by electric forces and concentrated in separate compartments. As noted earlier, the higher the salinity of the raw water, the more electric power is needed for this process. Hence, this process was applied primarily in the treatment of moderately brackish waters containing 1,000 to 2,000 ppm of dissolved salts.

In recent years, reverse osmosis (Green and Southard 2019; Theodore and Barden 1995; Theodore and Ricci 2010) has displaced electrodialysis as the primary membrane separation desalination process, leaving the latter as the choice for medical kidney applications. Reverse osmosis (RO) consists of an intake, a pre-treatment component, a high-pressure pump, membrane apparatus, remineralization and pH adjustment components, as well as a disinfection step. Generally, a pressure of about 1.7 to 6.9 MPa is required to overcome the osmotic pressure (described shortly) of saltwater.

This advanced separation technique may be used whenever low molecular weight solutes such as inorganic salts or small organic molecules (e.g. glucose) are to be separated from a solvent (usually water). In normal (as opposed to reverse) osmosis, water flows from a less concentrated salt solution to a more concentrated salt solution as a result of an innate concentration driving force (thermodynamically referred to as the chemical potential). As a result of the migration of water, an “osmotic pressure” is created on the side of the membrane to which the water flows. In reverse (as opposed to normal) osmosis, the membrane is permeable to the solvent or water and relatively impermeable to the solute or salt. In order to make water pass through an RO membrane in the desired direction (i.e. away from a concentrated salt solution), a pressure must be applied that is *higher* than the osmotic pressure. The membranes used for RO processes are characterized by a high degree of semi-permeability, high water fluxes, mechanical strength, chemical stability, and relatively low operating and capital costs. Early RO membranes were composed of cellulose acetate, but restrictions on process stream pressure, temperature, and organic solute rejection spurred the development of noncellulosic and composite membrane materials (membrane “sandwiches”). These membranes may be configured into certain geometries for system operation, including: plate and frame, tubular, spiral wound (composite), hollow fiber, etc.

Obviously, the heart of the membrane process is the membrane itself. As noted previously, a membrane is an ultra-thin semi-permeable barrier separating two fluids that permits the transport of certain species through the barrier from one fluid to the other. The membrane is selective since it permits the transport of certain species while rejecting others. The term semi-permeable is frequently used to describe this selective action. New and improved membranes will probably involve nanomaterials (Theodore and Kunz 2005).

9.5.3 Crystallization

Crystallization is an important mass transfer operation that is often employed in the preparation of a pure product. In the process, a crystal usually separates out as a substance of definite composition from a solution of varying composition. Any impurities in the liquid (often referred to as the *mother liquor*) are carried in the crystalline product only to the extent that they adhere to the surface or are occluded (retained) within the crystals that may have grown together during or after the crystallization operation.

The separation of a solid from a solution onto a crystal occurs only if there is a state of imbalance involving a mass driving force; namely, a decrease in chemical potential (or concentration) between the bulk of the liquid solution and the crystal interface exists. This effectively means that the solution must be supersaturated.

Although crystallization is ordinarily thought of as the deposition of a solid crystalline phase from a liquid phase by cooling, evaporation, or both, the same principles apply to crystal formation by precipitation caused by the addition of a third substance. This other component may either react to form a precipitate or simply decrease the solubility of the precipitated material. It should also be noted that when crystallization begins, heat is evolved (liberated). The amount of heat evolved is defined as the heat (or enthalpy) of crystallization.

There are several different ways that crystallization can occur. The four most often encountered in practice are:

1. cooling,
2. evaporation,
3. cooling and evaporation (also denoted adiabatic evaporation), and
4. a salting out process.

9.5.4 New Options for Water Desalination

Seven new or modified processes are proposed by GADUTH (GAniaris, DUpont, and THeodore) Consulting Engineers that may be employed for desalination purposes. These include one which utilizes solar evaporation, one which uses crystallization, one which utilizes a greenhouse solar system, one which is specific for domestic applications, one which involves a humidification/dehumidification unit, one involving a heat conduit evaporator, and one involving a geothermal evaporator. New systems involving reverse osmosis apparently have already been studied and exploited, and all that appears to remain is developing improved membranes, perhaps involving nanomaterials (Theodore and Kunz 2005). The aforementioned new options are discussed in more detail in Chapter 13.

9.6 Illustrative Examples

Six illustrative examples complete the presentation of materials highlighted in this chapter.

9.6.1 Illustrative Example 1

Briefly describe the role shale oil and tar sands can play in the desalination process.

Solution. Shales are geologic formations composed of thinly layered sedimentary rocks that are widely dispersed on the Earth. In certain shale deposits, organic material has been laid down and intermixed with the inorganic material. It has partially decomposed and metamorphosed over the ages until it now exists as a mixture of organic substances locked within the shale. The organic material contained in shale is called *kerogen*. The shale itself contains no oil as it is known today. However, if the shale is pretreated and fractionated at elevated temperatures, the kerogen decomposes and part of it is discharged as an oily vapor which can be condensed into a viscous liquid referred to as shale oil. Unlike typical crude oil, this shale oil contains large quantities of compounds of nitrogen and large amounts of sulfur. These must be removed before a product similar to crude oil results.

"Tar sands" is a unique term. The word *tar* has been defined as oily, dark-colored bituminous products obtained by the distillation of various petroleum feedstocks, including crude oil. Tars are typically rich in organic compounds that are related to benzene. High molecular tars (or pitches) are used in the manufacture of roofing papers and for making coal briquettes. Alternately, the term sand has been defined as a loose, incoherent mass of mineral materials in a finely granular condition, usually consisting of quartz (silica), with small quantities of mica, feldspar, magnetite, and other resistant minerals possibly present. It is the product of the chemical and mechanical disintegration of rocks under the condition of weathering and abrasion. When formed, the particles are usually angular and sharply pointed, becoming smaller and more rounded by attrition. It is also a constituent of most soils and is extremely abundant as a surface deposit along the courses of rivers, on the shores of lakes and oceans, and in arid regions. Tar sands (also alternately referred to as bituminous or oil sands) are located in certain beds. They contain a somewhat dense, viscous form of petroleum, with the assigned technical term of bitumen. The interstices between the grains of sand are largely filled with a thick tarry hydrocarbon mixture that can be extracted. This extract can be processed to yield 65 to 75 percent synthetic crude oil.

In reality the potential use of either of these two fossil fuels as a source of energy for desalination processes is quite low. The likelihood of their use in the future is also quite low due to their environmental impact during the extraction, processing and use stages of their lifecycle.

9.6.2 Illustrative Example 2

Qualitatively discuss the overall problem associated with most of the desalination processes.

Solution. Seawater desalination is one of the most difficult problems ever proposed for chemical engineering process analysis. This is mainly due to process complexities, the concentrated brine reject solution that is difficult and costly to dispose of properly, and the unusual economic requirements of the processes. However, the authors believe that the market for producing drinking water from seawater is not only near unlimited but also astronomical. Domestic clean water ministills in arid regions short on pure water supplies will soon be as common as filtered water systems in many domestic homes in the United States.

9.6.3 Illustrative Example 3

It has been reported that seawater contains 1.23% Na and 1.9% Cl by mass. Suggest the chemical formula of the Na and Cl.

Solution. Assume a mass of 100 g of the chemical in question.

$$\text{Moles of Na} = \frac{(0.0123 \text{ g Na})}{(23 \text{ g/gmol})} = 5.35 \text{ x}10^{-4} \text{ gmol Na}$$

$$\text{Moles of Cl} = \frac{(0.019 \text{ g Cl})}{(35.5 \text{ g/gmol})} = 5.35 \text{ x}10^{-4} \text{ gmol Cl}$$

Based on equal molar composition, the compound is the salt NaCl.

9.6.4 Illustrative Example 4

A 10,000 lb/hr feed of seawater (3.5% salt) is fed to a desalination unit. Calculate the discharge streams, flow rate and concentration, if 0.04% salt is present in the discharge potable water stream and the system provides a 75% production rate.

Solution: A line diagram of the system is provided in Figure 9.2. An overall balance yields:

$$S = W + B;\ S = 0.75W + B$$

$$10{,}000 \text{ lb/hr} = (0.75)(10{,}000 \text{ lb/hr}) + B;\ B = 2{,}500 \text{ lb/hr}$$

A componential salt balance yields:

$$(10{,}000 \text{ lb/hr})(0.035) = (7{,}500 \text{ lb/hr})(0.0004) + (2{,}500 \text{ lb/hr})(x_B)$$

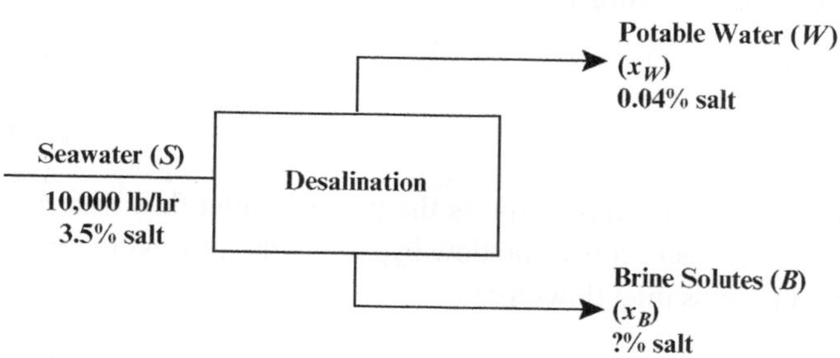

Figure 9.2 Desalination Unit Flow Diagram for Illustrative Example 4.

$$350\,\text{lb/hr} = 3\,\text{lb/hr} + 2{,}500x_B; x_B = (347\,\text{lb/hr})/(2{,}500\,\text{lb/hr}) = 0.139$$

The reject brine solution is 13.9% salt.

9.6.5 Illustrative Example 5

Refer to Illustrative Example 4. Assume the desalination system was designed to produce 5,000 lb/hr of the 0.04% potable water stream. Solve the problem.

Solution: For this system, an overall balance yields:

$$10{,}000\,\text{lb/hr} = (5{,}000\,\text{lb/hr}) + B;\ B = 5{,}000\,\text{lb/hr}$$

A componential salt balance yields:

$$(10{,}000\,\text{lb/hr})(0.035) = (5{,}000\,\text{lb/hr})(0.0004) + (5{,}000\,\text{lb/hr})(x_B)$$

$$350\,\text{lb/hr} = 2\,\text{lb/hr} + 5{,}000x_B; x_B = (348\,\text{lb/hr})/(5{,}000\,\text{lb/hr}) = 0.0.696$$

The reject brine solution is 6.96% salt.

9.6.6 Illustrative Example 6

Determine the percentage of a river's stream flow available to an industry for cooling such that the river temperature does not increase more than 10°F. Fifty percent of the industrial withdrawal is lost by evaporation and the industrial water returned to the river is 60°F warmer than the river.

Solution: Draw a flow diagram representing the process as shown in Figure 9.3. Express the volumetric flow lost by evaporation from the process in terms of that entering the process as:

$$q_{lost} = 0.5q_{in} \tag{9.2}$$

Express the process outlet temperature and the maximum river temperature in terms of the upstream temperature as:

$$T_{out} = T_{up} + 60°F \tag{9.3}$$

$$T_{mix} = T_{up} + 10°F \tag{9.4}$$

Using the conservation law for mass, express the process outlet flow in terms of the process inlet flow. Also express the flow bypassing the process in terms of the upstream and process inlet flows as:

$$q_{out} = 0.5q_{in} \tag{9.5}$$

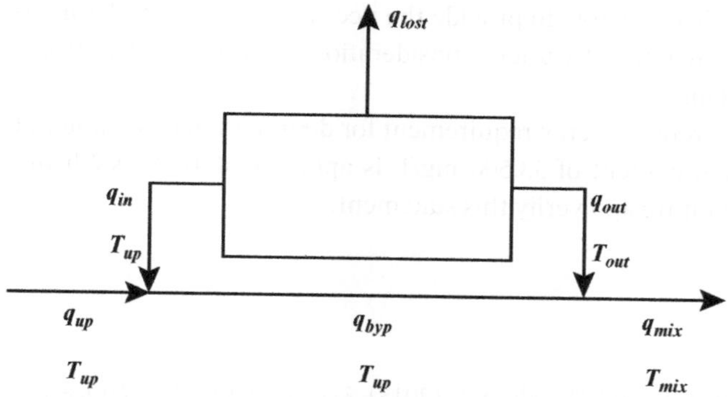

Figure 9.3 Flow Diagram for Illustrative Example 6.

$$q_{byp} = q_{up} - q_{in} \tag{9.6}$$

$$q_{mix} = q_{byp} + q_{out} \tag{9.7}$$

Writing an energy balance, assuming the heat capacity of all streams is equal, and the fluid is noncompressible, i.e. volumetric flow remains proportional to mass flow for all flow streams, the following equation can be written:

$$q_{byp}(C_p)\left(T_{up}\right) + q_{out}(C_p)\left(T_{out}\right) = (q_{mix})(C_p)(T_{mix}) \tag{9.8}$$

Substituting for q_{byp}, q_{out}, T_{out} and T_{mix} using Equations 9.3 through 9.7 yields the following:

$$\left(q_{up} - q_{in}\right)\left(T_{up}\right) + 0.5q_{in}(T_{up} + 60) = (q_{up} - q_{in} + 0.5q_{in})(T_{up} + 10)$$

Simplifying and rearranging yields:

$$10q_{up} = 35q_{in}$$

The percentage of a river's stream flow available to an industry for cooling, q_{in}/q_{up}, is then found to be:

$$q_{in}/q_{up} = 10/35 = 0.286 = 28.6\% \text{ of the upstream flow.}$$

9.7 Problems

1. Describe the early history associated with the desalination process.
2. Discuss the recent advances in desalination technology.
3. Briefly describe the reverse osmosis process.
4. List the variables that can impact the desalination process.

5. Employing nuclear reactors to provide the necessary energy for desalination processes are currently under consideration. Describe nuclear waste disposal problems.
6. The lowest theoretical energy requirement for desalination of seawater at 25°C with a salt content of 33,500 mg/L is approximately 0.7 kWh/m^3. Review the literature and verify this statement.

References

Flynn, A.M., Akashige, T., and Theodore, L. (2018). *Kern's Process Heat Transfer*, 2nd Edition. Salem, MA: Schrivener-Wiley.

Green, D. and Southard, M. (2019). *Perry's Chemical Engineers' Handbook*, 9th Edition. New York, NY: McGraw-Hill.

Skipka, K. and Theodore, L. (2014). *Energy Resources: Availability, Management, and Environmental Impacts*. Boca Raton, FL: CRC Press/Taylor & Francis Group.

Theodore, L. (2011). *Heat Transfer for the Practicing Engineer*. Hoboken, NJ: John Wiley & Sons.

Theodore, L. (2014). *Chemical Engineering: The Essential Reference*. New York, NY: McGraw-Hill.

Theodore, L. and Barden, J. (1995). *Mass Transfer Operations. A Theodore Tutorial*. East Williston, NY: Theodore Tutorials. Originally published by the U.S. EPA/APTI, RTP, NC.

Theodore, L., Dupont, R., and Ganesan, K. (2017). *Unit Operations in Environmental Engineering*. Salem, MA: Scrivener-Wiley.

Theodore, L. and Kunz, (2005). *Nanotechnology: Environmental Implications and Solutions*. Hoboken, NJ: John Wiley & Sons.

Theodore, L. and Ricci, F. (2010). *Mass Transfer Operations for the Practicing Engineer*. Hoboken, NJ: John Wiley & Sons.

Theodore, L., Ricci, F., and Van Vliet, T. (2007). *Thermodynamics for the Practicing Engineer*. Hoboken, NJ: John Wiley & Sons.

Treybal, R. (1951). *Liquid Extraction*. New York, NY: McGraw-Hill.

10

Evaporation

10.1 Introduction

The vaporization of a liquid for the purpose of concentrating a solution has long been in the domain of the chemical engineer and is a common operation in the chemical process industry. The concentration of solutions of nonvolatile solutes through heat-induced vaporization of the solvent (generally water) is called *evaporation*. Evaporation is a relatively expensive operation, and its application in waste treatment is limited primarily to the recovery of valuable byproducts from waste liquids and to the treatment of wastes where no alternate methods are available.

Purifying water via the evaporation route is not nearly as common as concentrating liquid solutions. The ocean/seawater salt concentration is approximately 30–40 g/L. Desalination by evaporation is a viable option, however, for two reasons. First, the operation is not affected by the salt concentration, and secondly, the salt concentration can easily be reduced via evaporation to concentrations below drinking water standards.

The overall desalination process requires the salt water to be heated to the boiling point where water vapor is formed. The "pure" vapor is then condensed to produce potable water. The thermal operating efficiency can be increased by employing a multi-stage evaporator, a type of unit that is addressed later in the chapter (Farag and Reynolds 1996; Flynn et al. 2018; Kern 1950).

The simplest device is an open pan or kettle that receives heat from a coil or jacket or by direct firing underneath the pan. Perhaps the traditional unit is the horizontal-tube evaporator in which a liquid (to be concentrated) in the shell side of a closed, vertical cylindrical vessel is evaporated by passing steam or another hot fluid through a bundle of horizontal tubes contained in the lower part of the vessel. The liquid level in the evaporator is usually less than half the height of the vessel; the empty space permits disengagement of entrained liquid from the vapor passing overhead (Farag and Reynolds 1996; Flynn et al. 2018; Kern 1950).

Introduction to Desalination: Principles, Processes, and Calculations, First Edition. Louis Theodore and R. Ryan Dupont.

Heat exchangers also play a role in evaporation processes in the desalination field. The simplest kind of heat exchanger, which confines both the hot and cold fluids, consists of two concentric tubes, or pipes. When conditions are such that only a few tubes per pass are required, the simplest construction is the double pipe heat exchanger. This consists of special fittings that are attached to standard iron (typically) pipe so that one fluid (usually liquid) flows through the *inside* pipe and the second fluid (also usually a liquid) flows through the *annular* space between the two pipes (Flynn et al. 2018; Kern 1950). Other heat transfer equipment are discussed later in this chapter.

This chapter provides a review of the basics and fundamentals of evaporation through a discussion of the principles of heat transfer and evaporation, classification of heat transfer equipment, classification of evaporators, describing equations, multiple-effect evaporation, thermocompression, and evaporation applications. An application section provides eight illustrative examples of evaporation concepts and implementation for desalination.

10.2 Principles of Heat Transfer and Evaporation

Heat transfer is often only one of several unit operations involved in a process, and the interrelations among the different unit operations involved must be understood since the transfer of heat to and from process fluids is an essential part of many environmental processes including desalination. Heat transfer requires that there be a driving force present for heat to flow. This driving force is the temperature difference between a warmer and colder body.

Practically all of the operations that are carried out in industry involve the production or absorption of energy in the form of heat. The laws governing the transfer of heat and the types of equipment that consider heat flow as one of their main objectives are therefore of great importance. Heat may flow by one or more of the following three basic mechanisms:

1. *Conduction*: When heat flows through a body by the transference of the momentum of individual atoms or molecules without bulk mixing, it is said to flow by conduction. For example, the flow of heat through the wall of a furnace or the metal shell of a boiler takes place by conduction.
2. *Convection*: When heat flows by actual mixing of warmer portions with cooler portions of the same material the heat transfer mechanism is known as *convection*. Convection is restricted to the flow of heat in fluids. Rarely does heat flow through fluids by pure conduction without some convection resulting from the eddies caused by the changes of density of the fluid with changes in temperature. For that reason, the terms conduction and convection are often used together, although in many cases, heat transfer is primarily by convection. For example, the heating of water flowing by a hot surface

is an example of heat transfer due primarily to convection. Heat transfer by convection due to density differences is defined as *natural convection*.

3. *Radiation*: A body emits radiant energy in all directions. If this energy strikes a receiver, a portion of it may be transmitted, a portion may be reflected, and a portion may be absorbed. It is this absorbed portion that represents heat transfer in the form of radiation that is the transfer of energy through space by means of electromagnetic waves. Radiation passing through empty space is not transformed to heat or any other form of energy.

Regarding evaporation, much of the present knowledge of the vaporization and boiling phenomenon is obtained from the work of Jakob and Fritz (1931) and Jakob (1949). When steam flows through a tube which is submerged in a pool of liquid, minute bubbles of vapor form at random points on the surface of the tube. The heat passing through the tube surface where no bubbles form enters the surrounding liquid by convection. Some of the heat in the liquid then flows toward the bubble, causing evaporation from its inner surface into itself. When sufficient buoyancy has been developed between the bubble and the liquid, the bubble breaks loose from the forces holding it to the tube and rises to the surface of the liquid pool. Kelvin postulated that, in order for this behavior to prevail, the liquid must be hotter than the saturation temperature in the incipient bubble. This is possible, since the spherical nature of the bubble establishes liquid surface forces on it so that the saturation pressure inside the bubble is less than that of the surrounding liquid. The saturation temperature of the bubble being lower than that of the liquid surrounding it, heat flows into the bubble. The number of points at which bubbles originate is dependent upon the texture of the tube surface, the number increasing with the roughness. Jakob and Fritz (1931) detected the presence of superheated liquid close to the heating surface and found the difference between the temperature of the superheated liquid and the saturation temperature of the vapor to be less for rough surfaces than for smooth surfaces.

Heat transfer by vaporization without mechanical agitation is obviously a combination of ordinary liquid free convection and the additional convection produced by the rising stream of bubbles. Under very small temperature differences between the tube wall and boiling liquid the formation of bubbles proceeds slowly and the rate of heat transfer is essentially that of free convection (Flynn et al. 2018; Kern 1950). Surface tension as an influence on bubble formation and growth is another variable.

10.3 Classification of Heat Transfer Equipment

Heat exchangers are defined as equipment that effect the transfer of thermal energy in the form of heat from one fluid to another. The simplest exchangers involve the direct mixing of hot and cold fluids. Most industrial exchangers are those in which the fluids are separated by a wall. The latter type, referred to by some as a *recuperator*, can range from a simple plane wall between two flowing fluids to more complex configurations involving multiple passes, fins, or baffles. Conductive and convective heat transfer principles are required to describe and design these units; radiation effects are generally neglected.

As described earlier, heat exchangers are devices used to transfer heat from a hot fluid to a cold fluid and can be classified by their functions. Farag and Reynolds (1996) provide an abbreviated summary of these units and their functions (see Table 10.1). A host of other exchangers can be added to this list, including parallel corrugated plates, plate units, etc. Direct-contact coolers and

Table 10.1 Heat Exchanger Equipment.

Equipment	Function
Chiller	Cools a fluid to a temperature that is obtainable if only water were used as a coolant. It often uses a refrigerant such as ammonia or Freon.
Condenser	Condenses a vapor or mixture of vapors, either alone or in the presence of a non-condensable gas.
Cooler exchanger	Performs a double function: (1) heats a cold fluid and (2) cools a hot fluid. Little or none of the transferred heat is normally lost.
Final condenser	Condenses vapors to a final storage temperature of approximately 100°F. It uses water cooling, which means the transferred heat is often lost to the process.
Forced-circulation reboiler	A pump is used to force liquid through the reboiler (see reboiler).
Heater	Imparts sensible heat to a liquid or a gas by means of condensing steam or some other hot fluid (e.g. Dowtherm).
Partial condenser	Condenses vapors at a temperature high enough to provide a temperature difference sufficient to preheat a cold stream of process fluid. This saves heat and eliminates the need for providing a separate preheater.
Reboiler	Connected to the bottom of a fractioning tower, it provides the reboil heat necessary for distillation. The heating medium may be either steam (usually) or a hot process fluid.
Steam generator	Generates steam for use elsewhere in the plant by using available high-level (temperature) heat, e.g. from tar or a heavy oil.
Superheater	Heats a vapor above its saturation temperature.

(*Continued*)

Table 10.1 (Continued)

Equipment	Function
Thermosiphon reboiler	Natural circulation of the boiling medium is obtained by maintaining sufficient liquid head to provide for circulation (see reboiler).
Vaporizer	A heater which vaporizes all or part of the liquid.
Waste heat boiler	Produces steam; similar to a steam generator, except that the heating medium is a hot gas or liquid produced in a chemical reaction.

specialty condensers can also be included in this list. However, the process industry uses three principal types of heat exchangers:

1. Double-pipe exchanger – the simplest type, used for cooling and heating.
2. Shell and tube exchangers – used for most applications.
3. Finned (extended surfaces) exchanges – used for some applications.

As one might suppose, there is a near infinite variety of heat exchange equipment. These can vary from a simple electric heater in the home to a giant boiler in a utility power plant. A limited number of heat transfer devices likely to be encountered by the practicing environmental engineer have been selected for description in this section. Most of the units transfer heat from one fluid to another fluid, with the heat passing through a solid interface such as a tube wall that separates the two (or more) fluids. The size, shape, and material employed that separate the two fluid is, of course, important. Another potential problem or concern in a heat exchanger is the method of confining one or both of the two fluids.

Heat exchangers used in the environmental, chemical, petrochemical, petroleum, paper, and power industries encompass a wide range of designs that are available from many manufacturers.

Equipment design practice first requires the selection of safe operable equipment. The selection-design process must also seek a cost-effective balance between initial (capital) installation costs, operating costs, and maintenance costs.

10.4 Classification of Evaporators

There are two principal types of tubular vaporizing equipment used in industry: *boilers* and *vaporizing exchangers*. Boilers are directly fired tubular apparatus which primarily convert fuel energy into latent heat of vaporization. Vaporizing exchangers are unfired and convert the latent or sensible heat of one fluid into the latent heat of vaporization of another. If a vaporizing exchanger is used for the evaporation of water or an aqueous solution, it is now fairly conventional to call it an *evaporator*. If it is used to supply the heat requirements at the bottom of a distilling column, whether the vapor formed be steam or not, it is labeled a *reboiler*. When not used for the formation of steam and not a part of a distillation process, a vaporizing exchanger is simply called a *vaporizer*. When an evaporator is used in connection with a power-generating system for the production of pure water or for any of the evaporative processes associated with power generation, it is a *power-plant evaporator*. When an evaporator is used to concentrate a chemical solution by the evaporation of solvent water, it is a *chemical evaporator*. Both classes differ in design. Unlike evaporators, it is the object of reboilers to supply part of the heat required for distillation and not a change in solution concentration, although a change generally cannot be avoided. Very often the term *evaporator* is also applied to a combination of several pieces of equipment, each of which can also be defined as an evaporator (Flynn et al. 2018).

Unfortunately, certain classes of evaporators are still designed as part of an art rather than the rational summation of the individual resistances to heat flow (to be discussed shortly) that is the basis of the science. This is due to the high heat transfer coefficients with which certain classes of evaporators operate, and the difficulty in identifying each of the small individual resistances which make up the overall resistance terms. Alternatively, Davis and Masten (2014) list four types of evaporators: rising film, flash evaporators using waste heat, submerged tube, and atmospheric pressure evaporators. Rising film evaporators are built so that the evaporative heating surface is covered by a liquid film and does not lie in a pool of boiling liquid. Flash evaporators are of similar configuration, but waste heat from a process stream is recirculated through the evaporator to augment the evaporation process. In the submerged tube design, heating coils are submerged in the evaporating fluid. Atmospheric evaporators do not recover the distillate and they do not operate under vacuum.

In an evaporator, a pound of fuel can normally yield about 10 lb of steam, and in a double-effect evaporator (to be discussed shortly), 10 lb of steam will yield a total of approximately 18.5 of pure water. It is little wonder that ships at sea commonly produce their own water requirements from seawater. Seawater contains approximately 3.5% dissolved solids by weight corresponding to 35,000 ppm compared with on average 340 ppm dissolved solids in fresh water. Instead of approximately 90% vaporization, it is customary to vaporize only approximately one third of the feed. The remainder, containing approximately 5% or 5,000 ppm dissolved solids is dumped overboard. Because of the significant blowdown, the use of a vacuum system with a low evaporation temperature is occasionally desirable since low temperatures are also favorable for low rates of scaling. Unlike evaporators in stationary power plants, salt-water evaporators or distillers operate for 600 to 700 hours without scaling (Kern 1950).

10.4.1 Simple Evaporators

Two classes of simple evaporators are shown in Figures 10.1 and 10.2. Figure 10.2 indicates that this form of a simple evaporator includes the potential for the recovery of heat from both condenser and the heat exchanger.

10.4.2 Flash Evaporator

The separation of water from a saline solution can also be achieved by means of a *flash evaporator*. This operation is referred to as a "flash" since the more volatile component of a saturated liquid mixture rapidly vaporizes upon entering a tank or drum which is at a lower pressure and/or higher temperature than the incoming feed. If the feed is a sub-cooled liquid, a pump and heater may be required to elevate the pressure and temperature, respectively, to achieve an effective flash (Figure 10.3). As the feed enters the flash drum, it impinges against mixture. an internal deflector plate the which promotes liquid–vapor separation of the feed mixture. The composition of the feed stream (F) is given by the mole fractions belonging to the set $\{z_i\}$. Similarly, the compositions of the vapor and liquid product streams are given by $\{y_i\}$ and $\{x_i\}$, respectively. Alternatively, some processes require that a vapor stream be cooled, so as to partially condense the least volatile components in the stream. This process is referred to as a *partial condensation*, and the following development may be adapted to such processes. As a result of the flash, the vapor phase will be mostly composed of the more volatile components. Typically, flash

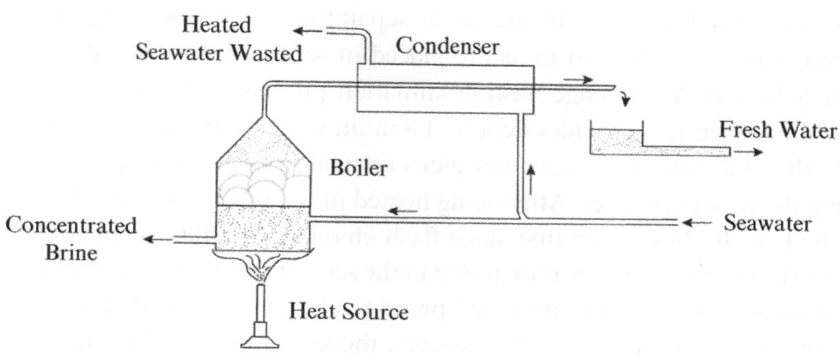

Figure 10.1 Simple Evaporator.

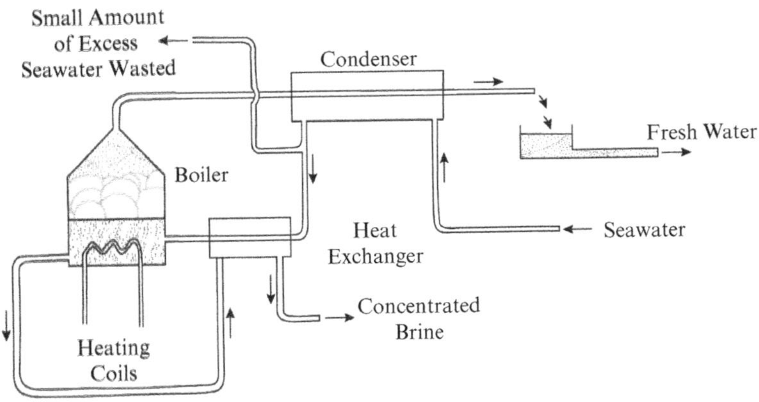

Figure 10.2 Simple Evaporator with Reuse of Heat from Condenser and Heat Exchanger.

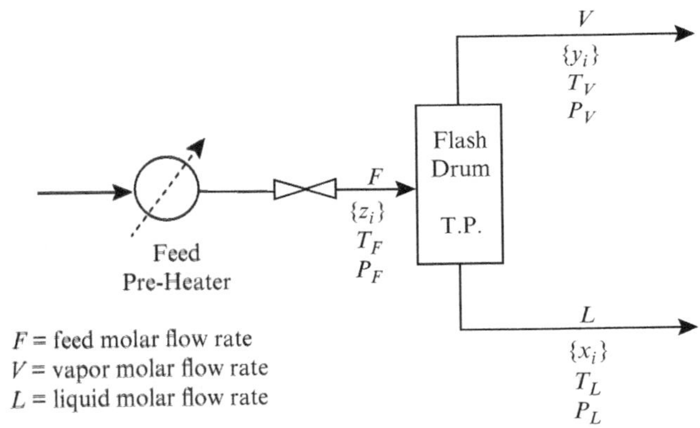

Figure 10.3 Flash Evaporator System.

evaporation is not an efficient means of separation when used only once. However, when several flash units are placed in series, much purer product may be achieved. A two-stage flash desalination process is shown in Figure 10.4, while Figure 10.5 provides details of a multi-stage flash evaporation system. In the first system, the seawater picks up heat in each condenser before entering the heat exchanger. After being heated in the heat exchanger the sea water boils or flashes in the first stage flash chamber and the steam is condensed to fresh water. The vacuum pump in the second stage causes the seawater to flash at a lower temperature and pressure without the addition of heat. In the multi-stage flash evaporation system, the seawater gains heat in each stage in passing through the condenser tubes from Stage 10 to Stage 1. It is then heated to almost boiling by steam in a heat exchanger. As it enters Stage 1 part

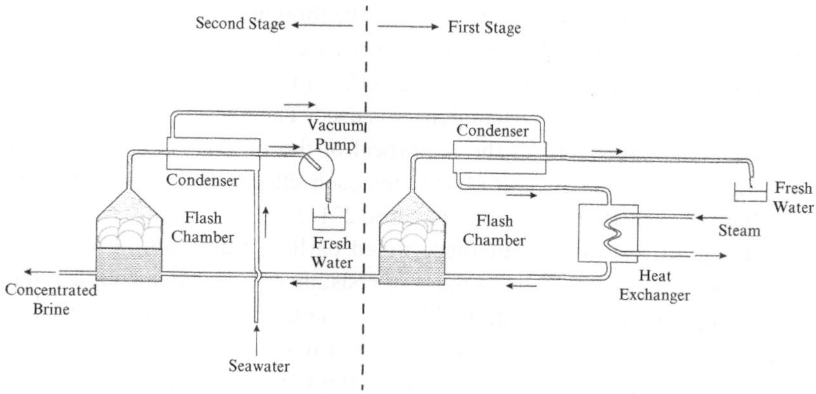

Figure 10.4 Two-stage Flash Evaporation System.

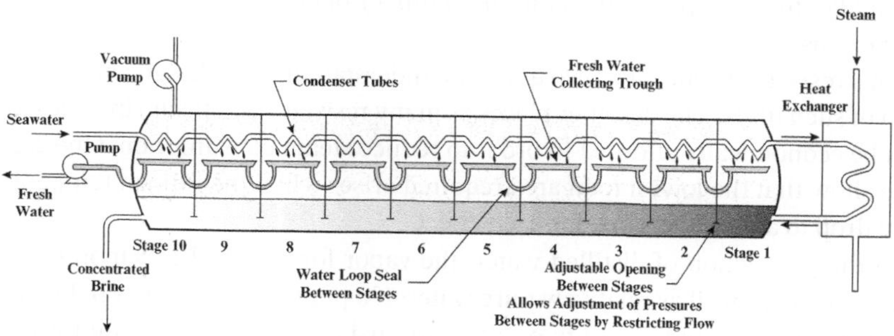

Figure 10.5 Multi-stage Flash Evaporation System.

of the seawater boils or flashes into steam because of the reduction of pressure and is condensed by the condenser tubes. In each successive stage the vacuum increases causing more of the seawater to flash into steam and in each stage the steam is condensed by the condenser tubes and drops into the collecting trough.

10.5 Multiple-effects Evaporators

One of the principal operating expenses of evaporators is the cost of steam for heating. A considerable reduction in those costs can be achieved by operating a battery of evaporators in which the overhead vapor from one evaporator (or "effect") becomes the heating medium in the steam chest of the next evaporator, thus saving both the cost of condensing the vapor from the first unit and supplying heat for the second. Several evaporators may operate in a battery in this fashion.

Basically, a multiple-effect evaporator may be thought of as a series of resistances to the flow of heat. The main resistances are those associated with heat transfer across the heating surface of each effect in the evaporator and across the final condenser if a surface condenser is used. The resistances of the heating surfaces are equivalent to the reciprocal of the product of the area and overall heat transfer coefficient (1/UA) for each effect (Farag and Reynolds 1996; Flynn et al. 2018; Kern 1950; Theodore 2011).

Neglecting all resistances but those noted earlier and assuming they are equal, it can be seen that if the number of resistances (effects) is doubled, the flow of heat (steam consumption) will be cut in half. With half as much to heat, each effect will evaporate about half as much water. But since there are twice as many effects, the total evaporation will be the same. Thus, under these simplifying assumptions, the same output would be obtained regardless of the number of effects (each of equal resistances), and the steam consumption would be inversely proportional to that number of effects or the total heating surface installed.

This resistance concept is also useful in understanding the design and operation of such units. The designer places as many resistances (effects) in series as can be economically afforded in order to reduce steam consumption. One can also show that the lowest total area required arises when the ratio of temperature drop to area is the same for each effect.

In the production of distilled water, the vapor formed in the evaporator is useful steam as well as relatively pure water. If a pound of steam is supplied to an evaporator, it can be used to produce approximately 0.9 lb of steam from a pound of water. The remaining 0.1 lb of water contains the bulk of the impurities, and it is removed from the evaporator as evaporator *blowdown*. The 0.9 lb of vapor from the evaporator can be condensed by partially preheating the evaporator feed, or in a power plant application, it can be mixed directly with cold return condensate before being fed to a boiler.

If, however, the original pound of steam was supplied to a multiple-effect evaporator, the vapor produced in the first evaporator would serve as the heat source in a second evaporator operating at a lower pressure than the first. Thus, an additional utilization could be made of most of the heat. If both evaporators were fed in *parallel* with raw water, about 0.85 lb of pure water would be formed in the first effect and about 0. 75 lb would be formed in the second effect. For each pound of steam supplied, about 1.6 lb of pure water would be produced, more than the 0.9 lb of vapor noted in the previous paragraph. The original pound of steam can also be considered a pound of pure water. When the vapor formed in the first effect is reused as the heating medium in a second effect, it is called a *double-effect* evaporator. When applied to three effects as shown in Figure 10.6, it is a *triple-effect* evaporator and the original pound of steam will produce approximately 2.25 lb of pure water. In this process,

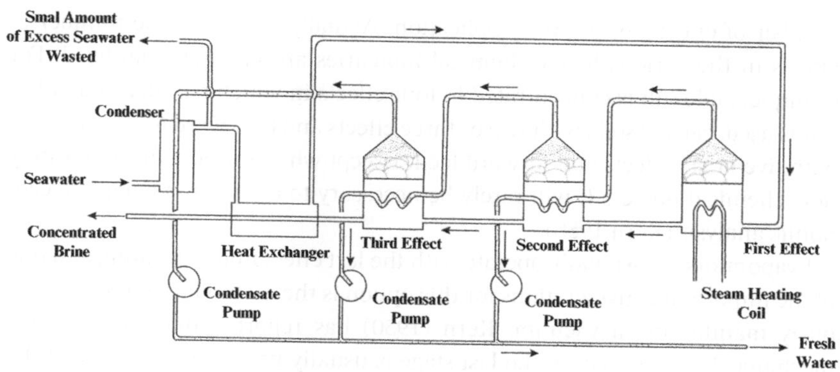

Figure 10.6 Multi-effect Evaporation System.

seawater picks up heat in the condenser and heat exchanger before being heated in the first effect. Steam from the first effect is condensed to fresh water in the second effect. In so doing, heat is given off to the seawater in the second effect which also boils, but at a lower temperature and pressure. This process is repeated in each effect with lower and lower pressure. The last effect uses a separate condenser to condense the fresh water.

In order to maintain temperature differences for heat transfer between the vapor from one effect and the boiling liquid of the next effect, the pressure on each succeeding evaporator must be *lower* than its predecessor. The quantity of blowdown, which is arrived at from experience with water of different impurities, carries with it considerable sensible heat and thereby reduces the amount of vaporization which can be realized in succeeding effects. Due in part to this escape of heat from the system there is a limit to the number of effects which are justifiable. The fixed charges for additional effects ultimately dissipate the savings in energy which result from the use of a large number of effects (Theodore 2014).

In any event, the greater the number of effects the greater the amount of evaporation per pound of steam admitted to the first effect. The operating costs will be less the larger the number of effects used. This is offset, however, by the increased capital cost of the added equipment, and the increased maintenance costs for cleaning and replacement, both of which are considered fixed charges. Supervisory labor will be the same for the operation of any number of effects. The cost of condenser water must also be included, and it, too, will decrease the greater the number of affects employed. The optimum number of effects may be obtained by computing the process requirements with two, three, four, or up to six or eight effects and determining the fixed charges and operating cost resulting from each arrangement. When the total cost is plotted against the number of effects, a minimum will occur corresponding to the optimum number of effects for a given application. Actually, however, the number of effects in the various heavy chemical industries are fairly standardized. For example, table salt is concentrated in four effects in which parallel feed of liquor is used, caustic soda with two or three effects and backward feed, and sugar with five or six effects and forward feed. Except when introducing an entirely new chemical process it will rarely be necessary to carry out a complete economic analysis (Kern 1950).

Evaporators occasionally operate with the last effect under vacuum, and one of the important considerations for this option is the method used for continuously maintaining a vacuum. Kern (1950) has reported that the use of a mechanical compressor on the last stage is usually prohibitive because of the energy required for compression. The thermocompression option is discussed in the next section.

10.6 Thermocompression

The principle of thermocompression is continuously finding wider application in industry. Consider a single-effect evaporator (Flynn et al. 2018; Kern 1950). It is fed with steam and generates vapors which have nearly as much heat content as was originally present in the steam. These vapors are condensed with water as a convenient means of removal, but this is a significant waste of both energy and water. Were it not for the fact that the temperature of the generated vapor is lower than the steam, it would be possible to circulate the vapor back into the steam chest and evaporate continuously without supplying additional steam. But apart from the heat (enthalpy) balance, a temperature difference must exist between the steam and the generated vapors or no heat would have been transferred originally. If the vapors from the evaporator were compressed to the saturation pressure of the steam, however, the temperature of the vapors could be raised to that of the original steam. The cost of supplying the necessary amount of compression is usually small compared with the value of the latent heat in the vapors. The practice of recompressing a vapor to increase its temperature and permit its reuse is *thermocompression*. In this system (Figure 10.7), the vapor compressor takes in water vapor from the boiler-condenser and compresses it to both a higher pressure and temperature. This vapor passes through the coil in the boiler-condenser and can heat sea water while the vapor is condensing into fresh water. The heat exchanger salvages heat from the discharged concentrated brine and fresh water for re-use in the boiler-condenser.

The only energy input required during operation is the mechanical energy for the vapor pump. Where fuel is costly, the compression may be accomplished with a centrifugal compressor (Theodore 2014). Where steam is available at a

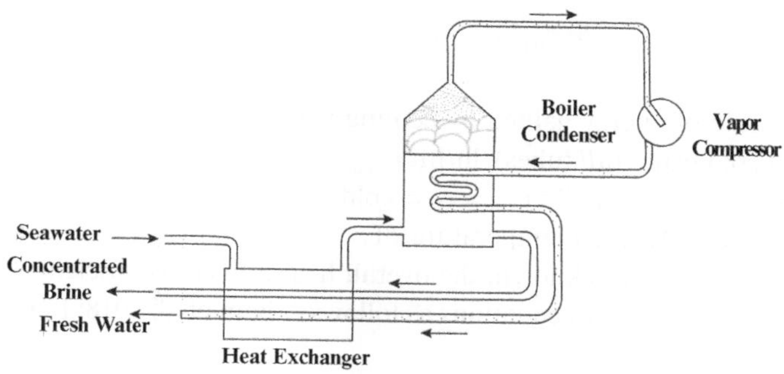

Figure 10.7 Vapor Compression Process.

pressure higher than that required in the evaporator, the recompression can be effected in a steam jet booster. The latter operates on the principle of an ejector and is used to compress vapors instead of a non-condensable gas. Hot concentrated seawater must be discharged at intervals to prevent the buildup of excessive concentrations of salt in the boiler.

The pressure at which steam is discharged from a thermocompressor depends upon pressure and proportion at which the live steam and the vapors are supplied. Kern (1950) and Flynn et al. (2018) have shown that the higher the desired discharge pressure, the greater the percentage of live steam that is required.

10.7 Evaporator Describing Equations

When two process fluids at different temperatures pass one another as in a shell and tube exchanger, heat transfer occurs due to the temperature difference between the two streams. The energy required to accomplish this heat transfer, i.e. the heat exchanger duty, can be determined for both process streams. For a well-insulated exchanger, the two duties should equal one another. The following equation gives the heat exchanger duty of the hot process steam when saturated steam is introduced on the shell side (enthalpy balances between steam and condensate may also be used) (Flynn et al. 2018; Kern 1950):

$$\dot{Q}_h = \dot{m}_s\lambda + \dot{m}_s c_{p,hw}\left(T_{s,i} - T_{s,o}\right) = \dot{m}_s\lambda + \dot{m}_s c_{p,hw}\left(T_1 - T_2\right) \tag{10.1}$$

where $\dot{Q}_h = Q_H$ = rate of energy lost by steam, Btu/hr; $\dot{m}_s$ = total mass flow rate of steam condensate, lb/hr; $\lambda = h_{vap}$ = heat of vaporization, Btu/lb; $c_{p,hw}$ = average heat capacity of condensate (hot water), Btu/lb-°F; $T_{s,i} = T_1$ = steam inlet temperature, °F; and $T_{s,o} = T_2$ = condensate outlet temperature, °F.

The cold process stream duty is calculated similarly:

$$\dot{Q}_c = \dot{m}_c c_{p,cw}\left(t_{c,o} - t_{c,i}\right) = \dot{m}_c c_{p,cw}\left(t_2 - t_1\right) \tag{10.2}$$

where $\dot{Q}_c = Q_C$= rate of energy gained by cooling water, Btu/hr; $\dot{m}_c$ = total mass flow rate of cold water (*all* tubes), lb/hr; $c_{p,cw}$ = average heat capacity of the cold process stream, Btu/lb-°F; $t_{c,o} = t_2$ = cold water outlet temperature, °F; and $t_{c,i} = t_1$ = cold water inlet temperature, °F.

If the heat exchanger duty is known, the overall heat transfer coefficient is calculated from either of the two equations as follows (Flynn et al. 2018; Kern 1950):

$$U_o = \frac{\dot{Q}_h}{A_o \Delta T_{lm}} \tag{10.3}$$

$$U_i = \frac{\dot{Q}_c}{A_i \Delta T_{lm}} \tag{10.4}$$

where U_o= overall heat transfer coefficient based on hot (h) process stream, Btu/ft^2-hr-°F; U_i = overall heat transfer coefficient based on cold (c) process stream, Btu/ft^2-hr-°F; A_o = exchanger transfer area based on outside tube surfaces, ft^2; A_i = exchanger transfer area based on inside tube surfaces, ft^2; and ΔT_{lm} = log mean temperature difference, °F.

Typical values for the *overall* heat transfer coefficient in steam condensers with water in the tubes range from 1,000 to 6,000 W/m^2-K (Theodore 2011). Perry's *Chemical Engineers' Handbook* reports heat transfer coefficients ranging from 400 to 1,000 Btu/hr-ft^2-°R (Green and Southard 2019). These values are approximately the same after conversion of units.

For tubular heat exchangers, as with double pipe heat exchangers, the overall heat transfer coefficient is related to the individual coefficients by the following equation:

$$\frac{1}{UA} = \frac{1}{U_i A_i} = \frac{1}{U_o A_o} = \frac{1}{h_i A_i} + \frac{R_{f,i}}{A_i} + \frac{ln\left(D_o / D_i\right)}{2\pi kL} + \frac{R_{f,o}}{A_o} + \frac{1}{h_o A_o} \tag{10.5}$$

where U = overall heat transfer coefficient, Btu/ft^2-hr-°F; A_i, A_o = inside and outside surface areas ($\pi D_i L$ or $\pi D_o L$), ft^2; A = average of the two areas, ft^2; $U_i, U_o,$ = heat transfer coefficients based on inner and outer tube surfaces, Btu/ft^2-hr-°F;, ft^2; and h_i, h_o = inside and outside heat transfer coefficients, Btu/ft^2-hr-°F; $R_{f,i}$, $R_{f,o}$= fouling factors based on inner and outer tube surfaces; $D_i, D_o,$ = inside and outside pipe diameters, ft; k = thermal conductivity, Btu/hr-ft-°F; and L = tube length, ft.

If the fouling factors. $R_{f,i}$ and $R_{f,o}$ and the tube wall resistance, i.e. the middle terms on the right-hand side of Equation 10.5 are negligible, then the relationship between the overall heat transfer coefficient and the individual coefficients simplifies to:

$$\frac{1}{U_o} = \frac{D_o}{h_i D_i} + \frac{1}{h_o} \tag{10.6}$$

The overall heat transfer coefficient, U, is an average value based on the average duty or $\dot{Q} = (\dot{Q}_h + \dot{Q}_c)/2$, and an average heat transfer area or $A = (A_i + A_o)/2$.

Additional details, including illustrative examples, are available in the literature (Farag and Reynolds 1996; Flynn et al. 2018; Kern 1950; Theodore 2011).

10.8 Illustrative Examples

Eight illustrative examples complete the presentation of materials highlighted in this chapter.

10.8.1 Illustrative Example 1

Which of the following is NOT a heat exchanger?

1. Reboiler
2. Condenser
3. Absorber
4. Superheater

Solution. A reboiler is connected to the bottom of a fractionating tower and provides the reboil heat necessary for distillation. The heating medium, which may be steam or a hot process fluid, transfers heat to the bottoms product of the column. Therefore, a reboiler is classified as a heat exchanger.

A condenser condenses a vapor or mixture of vapors either alone or in the presence of a non-condensable gas. The cooling medium, which may be cooling water or air, absorbs heat from the hot process vapor. Therefore, a condenser is classified as a heat exchanger.

An absorber is a mass transfer device used for separating components from either a liquid or gaseous stream. Therefore, an absorber is NOT a heat exchanger.

A superheater heats a vapor above its saturation temperature. Therefore, a superheater is classified as a heat exchanger.

Therefore, 3 is the correct answer.

10.8.2 Illustrative Example 2

Solar energy plays an important role in several classes of desalination processes. Provide some details on the source of solar energy, i.e. the Sun.

Solution. The Sun is a star that, by the gravitational effects of its mass, dominates the planetary system that includes the Earth. The radiation of its electromagnetic energy furnishes directly or indirectly all of the energy supporting life on Earth, because all foods and fuels are derived ultimately from plants using the energy of sunlight. The total amount of energy emitted by the sun in the form of radiation is remarkably constant, varying by no more than a few tenths of 1 percent over several days. This energy output is generated deep within the sun. Like most stars, the sun is made up primarily of hydrogen (specifically, 71% hydrogen, 27% helium, and 2% other, heavier elements). Near the center of the Sun the temperature is almost 16,000,000 K (about 29,000,000°F) and the density 150 times that of water. Under these conditions the nuclei of individual hydrogen atoms interact, under nuclear fusion. In this process two hydrogen nuclei combine to make one helium nucleus, and energy is released in the form of gamma radiation. This energy is equivalent to that which would be released from the explosion of 100 billion one-megaton hydrogen bombs per second. The nuclear "burning" of hydrogen in the core of the sun extends out to about 25% of the Sun's radius.

10.8.3 Illustrative Example 3

A four-year-old solar evaporation desalinator is no longer delivering the degree of separation required for the process. Rather than replace the unit, you have been asked to recommend what other possible steps can be taken with the existing unit to get it back "on line." Unfortunately, it is not currently possible to completely replace the desalinator. Some other solutions need to be found.

Solution. The issue with the current operation is its inability to perform at its designed level. The first thing that needs to be investigated is what is the current state of the system? How was it originally designed? Were safety margins included in the design that allow for some "tweaking" of the actual performance of the system? What will the materials stand up to? Finally, it is possible that the evaporator needs cleaning and that could increase its performance efficiency.

10.8.4 Illustrative Example 4

Qualitatively describe what an evaporator does.

Solution: This is an open-ended question. Standiford (1963) provides one answer to this example. The requirement for the correct functioning of any evaporator are:

1. It must transfer a great deal of heat, on the order of 1,000 Btu/lb of water evaporated. This, more than anything else, determines the type, size, and cost of each effect of the evaporator.
2. It must efficiently separate the vapor from the residual liquid. What is efficient in one evaporator may be many orders of magnitude different from what is efficient in another (e.g. from a loss of salt value at only a few dollars a ton, to lithium chemicals valued at a dollar a pound, to radioactive waste). Separation may to be important solely because of the value of product lost, pollution problems, or because of fouling or corrosion in the equipment in which the vapor is condensed or in which the condensate is subsequently used.
3. It must make as efficient use of the available heat or mechanical energy as is economically feasible. This means using the vapor evaporated in one part (effect) of the evaporator as the heating steam in another effect that is operating at a lower temperature (as in a multiple-effect evaporator), or compressing the vapor evolved so that it can be used as the heating medium in the same evaporator, or by employing a combination of these. As in most other cases, efficiency is usually gained only as capital cost is increased and, for evaporators, the designer has a wide range to choose from.
4. It must meet conditions dictated by the characteristics of either the liquid being evaporated or the product. In a crystallizing evaporator, crystal size, shape, and purity may be of the utmost importance. If a salting or scaling liquid is to be handled, the evaporator type selected must be capable of dealing with it. Product quality characteristics that may also be important are corrosiveness and degradation at high temperature, long holdup time, or contact with certain metals.

Other considerations are the size of the operation, any foaming characteristics of the liquor, the need for special types of materials of construction such as polished stainless steel (required for many food products), and easy access for cleaning.

10.8.5 Illustrative Example 5

A single-effect evaporator is to concentrate 10,000 lb/hr of a 10% NaOH solution to 75%. The feed enters at 120°F and the evaporator is to operate at an absolute pressure of 14.7 psi. The 75% NaOH solution leaves at the evaporator equilibrium temperature. Assume the solution enthalpies entering and leaving the evaporator are 81 and 395 Btu/lb solution, respectively, and the enthalpy of the saturated steam is 1150 Btu/lb steam. For what heat transfer rate (Btu/hr) should this industrial evaporator be designed?

Solution: This system is shown in Figure 10.8.

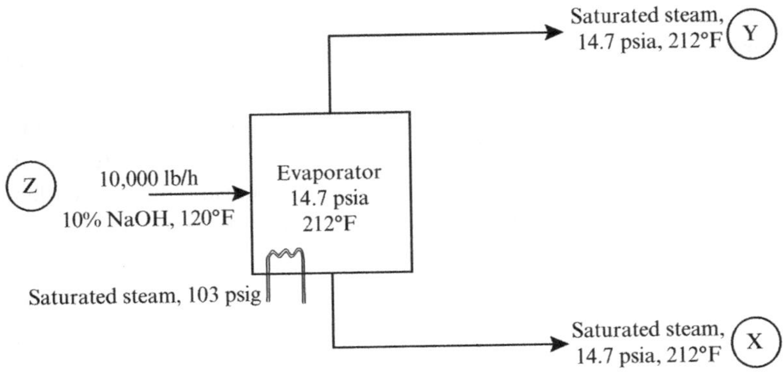

Figure 10.8 Evaporator Flow Diagram for Illustrative Example 5.

On the basis of 1 hour of operation, calculate the flow rate of steam Y, and the 75% NaOH-H_2O solution, X, leaving the evaporator. From a NaOH balance,

$$(10{,}000\,\text{lb/hr})(0.1) = (X\,\text{lb/hr})(0.75);\; X = 1333.3\ \text{lb/hr}$$

From an overall material balance:

$$Z = X + Y; Y = Z - X = (10{,}000 - 1333.3) = 8666.7\ \text{lb/hr}$$

Calculate the evaporator heat required, Q, in Btu/hr.

$$Q = \Delta H \tag{10.7}$$

$$= (8666.7\,\text{lb/hr})(1{,}150\,\text{Btu/lb}) + (1333.3\,\text{lb/hr})(395\,\text{Btu/lb}) - (10{,}000\,\text{lb/hr})(81\,\text{Btu/lb})$$

$$= 9{,}966{,}705\,\text{Btu/hr} + 526{,}654\,\text{Btu/hr} - 810{,}000\,\text{Btu/hr}$$
$$= 9{,}683{,}360\,\text{Btu/hr}$$

10.8.6 Illustrative Example 6

Refer to Illustrative Example 5. Calculate the area requirement of the evaporator if the overall heat transfer coefficient is 500 Btu/hr-ft^2-°F and 103 psig saturated steam (T_s = 340°F) is employed in the steam chest.

Solution: Calculate the area of the evaporator in ft^2 using Equation 10.8 (Kern 1950):

$$Q = UA\,\Delta T;\; A = Q/(U\,\Delta T) \tag{10.8}$$

$$A = (9{,}683{,}360\,\text{Btu/hr})/[(500\,\text{Btu/hr-ft}^2 - {}^\circ\text{F})(340 - 212^\circ\text{F})]$$

$$A == (9{,}683{,}360\,\text{Btu/hr})/(64{,}000\,\text{Btu/hr} - \text{ft}^2) = 151.3\,\text{ft}^2$$

10.8.7 Illustrative Example 7

Refer to Illustrative Example 6. Recalculate the area requirement if atmospheric steam is employed in the steam chest.

Solution: This is an impossible condition since:

$$\Delta T = T_s - T \rightarrow 0 \tag{10.9}$$

and from Equation 10.8, $A \approx \infty$.

10.8.8 Illustrative Example 8

An evaporator is to be fed 5,000 lb of solution containing 2% dissolved solids by weight. The feed, *F*, is at a temperature of 100°F. It is to be concentrated to a solution of 5% solute by weight in an evaporator operating at a pressure of 1 atm in the vapor space. In order to carry out the evaporation, the heating surface is supplied with steam at 5 psig (227°F) and the overall heat transfer coefficient of the evaporator, *U*, is 280 Btu/hr-ft^2-°F. What is the mass of vapor produced, the total mass of steam required, and the surface area required? Neglect enthalpy of solution effects (Badger and Banchero 1955).

Solution. In order to simplify the solution, it will be assumed that the solution is so dilute that its boiling point is the same as the boiling point of water and that its heat capacity and latent enthalpy are the same as that of water. Under these circumstances, the thermal properties of the solution (both feed and product) and of the steam can be taken from the steam tables. This results in the following values for pertinent quantities on a per hour basis (refer to Figure 10.9):

$F = 5{,}000$ lb
$x_F = 0.02$
$T_F = 100$°F
$h_F = 68$ Btu/lb (estimated from steam tables at 100°F)
$x_L = 0.05$
$T_L = 212$°F (at 1 atm)
$h_L = 180$ Btu/lb (estimated from steam tables)
$h_V = 1{,}1\ 56$ Btu/lb (estimated from steam tables at 1.0 atm)

$T_S = 227°F$
$T_V = 212°F$
$h_S = 1{,}156$ Btu/lb (estimated from steam tables at 227°F and 5 psig)
$T_C = 227°F$ (condensate)
$h_C = 195$ Btu/lb (estimated from steam tables)

$$\text{Total solids in feed} = (5000\,\text{lb})(0.02) = 100\,\text{lb}\,(\text{componential balance})$$

$$\text{Total water in feed} = 5{,}000\,\text{lb} - 100\,\text{lb} = 4{,}900\,\text{lb}\,(\text{componential balance})$$

$$\text{Total solids in liquor} = \text{Total solids in feed} = 100\,\text{lb}$$

$$\text{Total lb of liquor} = L = 100\,\text{lb}/(0.05) = 2{,}000\,\text{lb}$$

$$\text{Total lb of vapor} = V = 5{,}000\,\text{lb} - 2{,}000\,\text{lb} = 3{,}000\,\text{lb}\,(\text{overall balance})$$

The following enthalpy (energy) balance (Equation 10.10) results (S representing steam):

$$F\,h_F + S\,h_S = V\,h_V + L\,h_L + S\,h_C \tag{10.10}$$

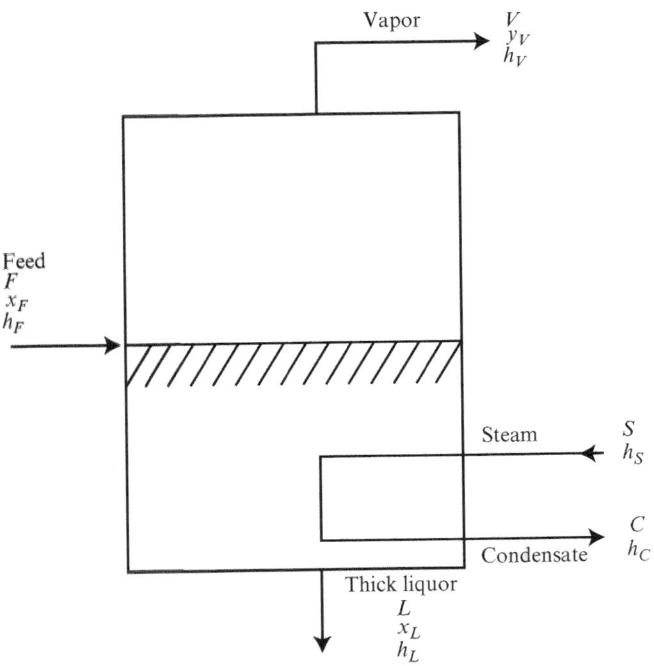

Figure 10.9 Evaporator Flow Diagram for Illustrative Example 8.

Substituting and solving yields the total lb of steam, *S*, required:

$$(5{,}000\,\text{lb})(68\,\text{Btu/lb}) + (S)(1{,}156\,\text{Btu/lb}) = (3{,}000\,\text{lb})(1{,}150\,\text{Btu/lb}) + (2000\,\text{lb})(180\,\text{Btu/lb}) + (S)(195\,\text{Btu/lb});\ S = 3{,}611\,\text{lb}$$

The total heat requirement is:

$$Q = (3{,}611\,\text{lb})(1156 - 195\,\text{Btu/lb}) = 3{,}470{,}000\,\text{Btu}$$

The required area is (assuming all the stated inputs based on 1 hour) is determined from Equation 10.8 as:

$$A = Q/(U\,\Delta T) = 3{,}470{,}000\,\text{Btu/hr}/\left[(280\,\text{Btu/hr} - \text{ft}^2 - {}^\circ\text{F})(227 - 212\,{}^\circ\text{F})\right] = 826\ \text{ft}^2$$

10.9 Problems

1. An evaporator is fed with 5,000 lb/hr of a 20% solution of sodium hydroxide at 100°F. This is to be concentrated to a 40% solution. The evaporator is supplied with saturated steam at 5 psig. Although the unit operates with the vapor space at a pressure of 4 in Hg absolute, the boiling temperature of the solution in the evaporator is 198°F (due to the superheat created by the exposed heating element). The overall heat transfer coefficient is 400 Btu/hr-ft^2-°F. Calculate the steam rate and the required heat transfer area. Assume a basis of 1 hour.
2. Three evaporators are arranged as shown in Figure 10.10. Calculate the flow rate and composition of Stream 7 and the production efficiency for this desalination process.
3. A patent has been submitted that claims a newly developed energy conversion process has unique capabilities. The patent claim indicates that the process can take steam at 212°F and 1 atm and provide heat to a constant temperature reservoir at 300°C while discharging heat to a low temperature reservoir at 60°F. Pertinent data are provided in the flow diagram of Figure 10.11. Is this claim possible?
4. You are tasked with evaluating the design of a proposed solar water heater. The heater works by having water flow through a pipe which is positioned along the focal point of a parabolic mirror (see Figure 10.12). The parabolic mirror has an opening (approximately perpendicular to the sun's rays) of 3 m, and a length of 100 m. On an average day, solar radiation can be assumed to be 1 kW/m^2, and the efficiency of energy transferred (i.e. amount of energy transferred to the water pipe versus incoming solar energy) is approximately

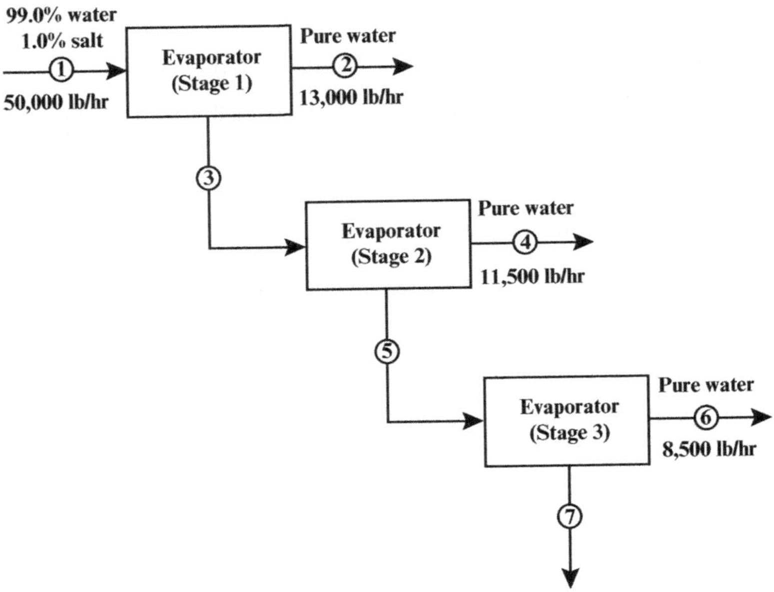

Figure 10.10 Flow Diagram for Three Stage Evaporation Process for Problem 2.

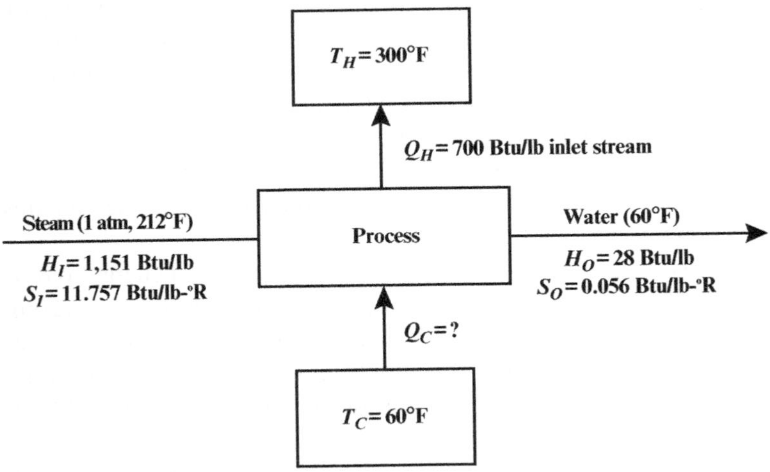

Figure 10.11 Flow Diagram for Energy Conversion Process for Problem 3.

40%. The water to be heated enters the pipe at a temperature of 80°C, an absolute pressure of 2 bar, and a flow rate of 200 kg/h. Assuming that the pressure drop along the pipe is negligible, what is the outlet water temperature? If you find that the exiting stream is not 100% liquid (i.e. if vapor forms

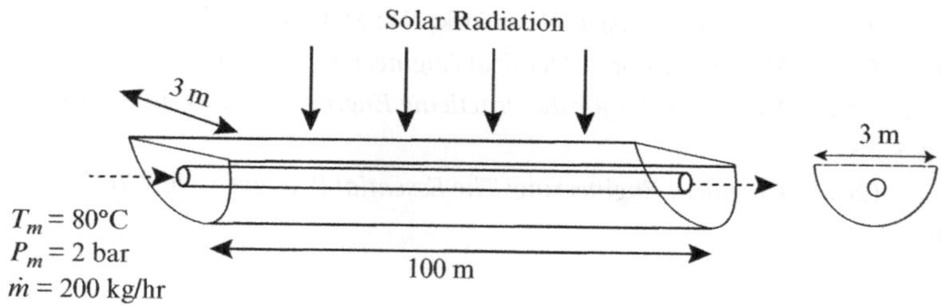

Figure 10.12 Side and Front Views of the Solar Water Heater for Problem 4.

during heating) determine the quality of the stream (the mass fraction of the stream which is vapor).

5. Refer to Problem 4. Assume the feed stream is 1.8% salt in seawater. Estimate the quality of the exit stream if enthalpy of mixing effects can be neglected.
6. Refer to Problem 5. Qualitatively discuss the effect of including enthalpy of mixing effects on the resulting quality of the exit stream.
7. Water discharged from an evaporator is being cooled in a double pipe heat exchanger from $T_1 = 210°F$ to $T_2 = 170°F$. The water flows inside a tube at a rate of 8,000 lb/hr and the tube surface temperature is maintained at 60°F. The overall heat transfer coefficient, $U = 95$ Btu/(hr-ft^2-°F). Calculate the required heat transfer area, A, in ft^2.

References

Badger, W. and Banchero, J. (1955). *Introduction to Chemical Engineering*. New York, NY: McGraw-Hill.

Davis, M. and Masten, S. (2014). *Principles of Environmental Engineering and Science*. New York, NY: McGraw-Hill.

Farag, I. and Reynolds, J. (1996). *Heat Transfer. A Theodore Tutorial*. East Williston, NY: Theodore Tutorials. Originally published by the U.S. EPA/APTI, RTP, NC.

Flynn, A.M., Akashige, T., and Theodore, L. (2018). *Kern, Process Heat Transfer*, 2nd Edition. Salem, MA: Schrivener-Wiley.

Green, D. and Southard, M. (2019). *Perry's Chemical Engineers' Handbook*, 9th Edition. New York, NY: McGraw-Hill.

Jakob, M. (1949). *Heat Transfer*. New York: Wiley.

Jakob, M. and Fritz, W. (1931). Versuche ueber den Verdampfungsvorgang. *Forsch. Ing. - Wes* 2: 435.

Kern, D. (1950). *Process Heat Transfer*. New York, NY: McGraw-Hill.
Standiford, F.C. (1963). Evaporators. *Chemical Engineering* 70 (25): 158–176.
Theodore, L. (2011). *Heat Transfer for the Practicing Engineer*. Hoboken, NJ: John Wiley & Sons.
Theodore, L. (2014). *Chemical Engineering: The Essential Reference*. New York, NY: McGraw-Hill.

11

Reverse Osmosis

11.1 Introduction

Membrane processes are today state-of-the-art separation technologies that have shown promise for future technical growth and wide-scale industrial commercialization. They are used in many industries for process stream and product concentration, purification, separation, and fractionation, as well as for potable water production for marginal water sources. The use of membranes in wastewater treatment systems to replace conventional sedimentation tanks is also expanding. The need for membrane research and development is important because of the increasing use of membrane technology in both traditional and emerging chemical and environmental engineering fields. Membrane processes are increasingly finding their way into the growing engineering application areas of biotechnology, green engineering, specialty chemical manufacture, and biomedical engineering, as well as in the traditional chemical and environmental process industries. Membrane technology is also being considered as either a replacement for or supplement to traditional separations such as sedimentation, distillation, or extraction. Membrane processes can be more efficient and effective since they can simultaneously concentrate and purify, and can also perform separations at ambient conditions in small or existing reactor volumes.

Membranes create a boundary between different bulk gas or liquid mixtures. Different solutes and solvents flow through membranes at different rates; this enables the use of membranes in separation processes. Membrane processes can be operated at moderate temperatures for sensitive components (food, pharmaceuticals, etc.). Membrane processes also tend to have low relative capital and energy costs. Their modular format permits simple and reliable scale-up or retrofitting in existing process tanks.

Key membrane properties include their size rating, selectivity, permeability, mechanical strength, chemical resistance, low fouling characteristics, high capacity, low cost, and consistency. Vendors characterize their equipment with

Introduction to Desalination: Principles, Processes, and Calculations, First Edition. Louis Theodore and R. Ryan Dupont.

ratings indicating the approximate size (or corresponding molecular weight) of components retained by their membranes. Commercial membranes consist of polymers or ceramics. Other membrane types include sintered metal glass and liquid films.

Membrane unit operations are often characterized by the following parameters: (1) driving force utilized, (2) membrane type and structure, and (3) species being separated. Reverse osmosis (RO), Nanofiltration (Theodore 2006), Ultrafiltration (UF), and Microfiltration (MF) all utilize a pressure difference as the driving force to separate a liquid feed into a liquid *permeate* and *retentate*. They are listed in ascending order in their ability to separate a liquid feed on the basis of solute size. RO uses nonporous membranes and can separate down to the ionic level, e.g. seawater in the rejection of dissolved salt. Nanofiltration performs separations at the nanometer range. UF uses porous membranes and separates components of molecular size ranging from the low thousand to several hundred thousand atoms; an example includes components in biomedical processing. MF uses much more porous membranes and is typically employed in the micro- or macromolecular range to remove particulate or larger biological matter from a feed stream (e.g. in the range of 0.05 to 2.5 μm) (Applegate 1984; Schweitzer 1979).

Dialysis membrane processes use a concentration difference as a driving force for separation of liquid feed across a semipermeable membrane, with the major application in the medical field of hemodialysis. *Electrodialysis* separates a liquid employing an electric charge difference as the driving force and is widely used in water purification and industrial processing. *Gas separation* processes can be divided into two categories: gas permeation through nonporous membranes and gas diffusion through porous membranes. Both of these processes utilize a concentration difference across the membranes as the driving force for separation. The gas permeation processes are used extensively in industry to separate air into purified nitrogen and enriched oxygen. Another commercial application is hydrogen recovery in petroleum refineries.

Since membrane separation processes are one of the newer (relatively speaking) technologies being applied in practice, the subject matter is and has been introduced into the engineering curriculum. There are four major membrane processes of interest to the practicing engineer:

1. Reverse osmosis (hyperfiltration) (RO)
2. Ultrafiltration (UF)
3. Microfiltration (MF)
4. Gas permeation (GP)

The four processes have their differences. The main difference between RO and UF is that the overall size and diameter of the particles or molecules in solution to be separated is smaller in RO. In MF, the particles to be separated or concentrated are generally solids or colloids rather than molecules in solution. GP is another membrane process that employs a nonporous semipermeable membrane to "fractionate" a gaseous stream.

The heart of the membrane process is the membrane itself; it is an ultrathin semipermeable barrier separating two fluids that permits the transport of certain species through the membrane barrier from one fluid to the other. As noted, a membrane is typically produced from various *polymers* such as cellulose acetate or polysulfone, but ceramic and metallic membranes are also used in some applications. The membrane is referred to as *selective* since it permits the transport of certain species while rejecting others. The term *semipermeable* is frequently used to describe this selective action.

This chapter will focus on RO because of its widespread use in desalination applications. Although RO receives the bulk of the treatment in the chapter, UF, MF, and GP are addressed in separate sections. The reader should note that the notation and units adopted by this industry are primarily employed in the development to follow. The reverse osmosis process and membrane principles and applications are addressed through a review of the role of osmosis, reverse osmosis equipment, describing equations, other membrane processes, and membrane process applications. Six illustrative examples of concepts and implementation of reverse osmosis and membrane desalination processes are provided at the end of the chapter.

11.2 Introduction to Reverse Osmosis

The most widely commercialized membrane process by far is RO. It belongs to a family of pressure-driven separation operations for liquids that includes not only RO, but also UF and MF.

RO is an advanced separation technique frequently employed in the chemical process industry that may be used when low molecular weight (MW) solutes such as inorganic salts or small organic molecules (e.g. glucose) are to be separated from a solvent (usually but not always water). In normal (as opposed to reverse) osmosis, most water–salt applications, water flows from a less concentrated salt solution to a more concentrated salt solution as a result of the concentration gradient driving force. As a result of the migration of water, an osmotic pressure is created on the side of the membrane to which water flows. In reverse (as opposed to normal) osmosis, the membrane is permeable to the solvent or water and relatively impermeable to the solute or salt. To make water pass through an RO membrane in the desired direction (i.e. away from the concentrated salt solution) in RO applications, a pressure must be applied that is higher than that of the osmotic pressure.

Aside from the classic example of RO for seawater desalination (Dupont et al. 1982), it has found a niche in the food industry for concentration of various fruit juices, in the galvanization industry for concentration of waste streams, and in the dairy industry for concentration of milk prior to cheese manufacturing (Applegate 1984; Brooks 1985; Parkinson 1983).

RO processes are classified into the following two basic categories:

1. *Purification* of the solvent such as in desalination where the permeate or purified water is the product.
2. *Concentration* of the solute such as in concentration of fruit juices where the retentate is the product.

11.3 The Role of Osmosis

Osmosis must be first better explained in order to fully describe RO. Osmosis occurs when a concentrated solution is partitioned from a pure solute or relatively lower concentration solution by a semipermeable membrane. The semipermeable membrane allows only the *solvent* to flow through it freely. Equilibrium is achieved when the solvent from the lower-concentration side ceases to flow through the membrane to the higher-concentration side (thus reducing the concentration) because the mass transfer concentration difference driving force has been reduced. This is shown in Figures 11.1 and 11.2. *Osmotic pressure* is the pressure needed to terminate the flux of solvent through the membrane or the force that pushes on the concentrated side of the membrane (see Figure 11.3). Applying a pressure on the concentrated side halts the solvent flux. *Reverse osmosis* (see Figure 11.4) takes place when an applied

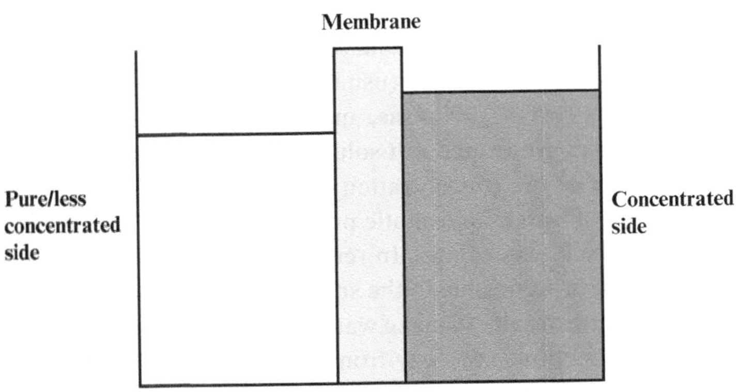

Figure 11.1 Pre-osmosis Equilibrium.

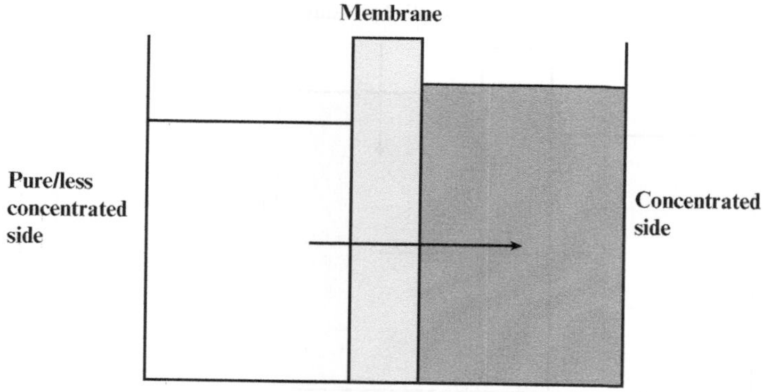

Figure 11.2 Osmosis of a Solvent.

force (pressure) overcomes the osmotic pressure and forces the solvent from the concentrated side through the membrane, leaving the solute on the concentrated side.

One of the authors has provided the following explanation for osmotic pressure (Glaston 1949). This phenomenon can be simply illustrated by tying a semipermeable membrane over the end of an inverted thistle tube which is then partially filled with a concentrated salt solution and dipped into a beaker of water. The water will rise in the tube until the excess hydrostatic pressure will cause the water to flow outward at the same flowrate as the inlet flow caused by the osmosis. This excess pressure must be overcome in reverse osmosis.

As noted earlier, osmotic pressure occurs when two solutions of different concentrations (or a pure solvent and a solution) are separated by a semi-permeable membrane. In simplest terms, it means that the membrane is

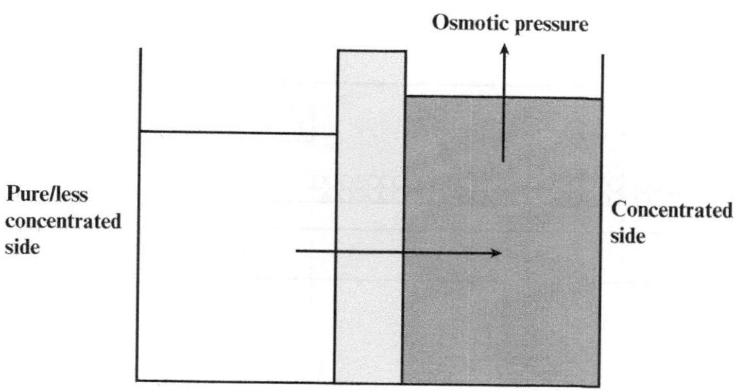

Figure 11.3 Osmotic Pressure.

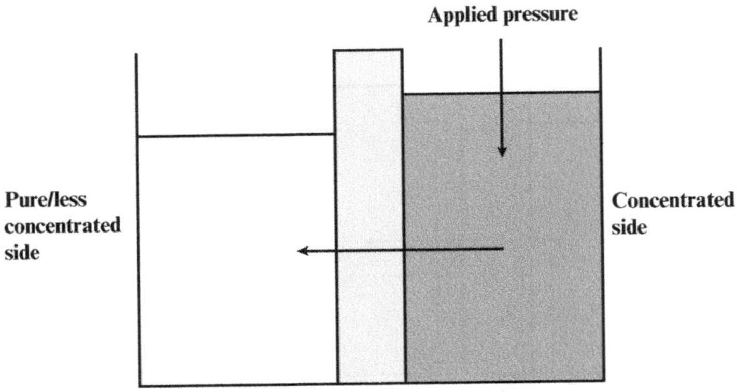

Figure 11.4 Reverse Osmosis.

permeable to the solvent but not to the solute. The solvent molecules in the dilute phase have a higher chemical potential than the molecules in the concentrated phase. This difference in chemical potential causes a flow of solvent molecules from the dilute phase (high chemical potential) to the concentrated phase (low chemical potential). Flow of solvent molecules into the concentrated solution continues until osmotic equilibrium is reached, i.e. until the chemical potential of the solvent molecules are equal.

Summarizing, Figure 11.5 provides a more detailed pictorial representation of what happens on the surface of a membrane in RO. A concentrated solution enters as the feed and is separated with the assistance of the membrane and the filtered solvent exits as the permeate. The retentate is the solvent that is not cleaned and is more concentrated than the feed solution. A step-by-step explanation (see Figure 11.5) of this phenomenon follows:

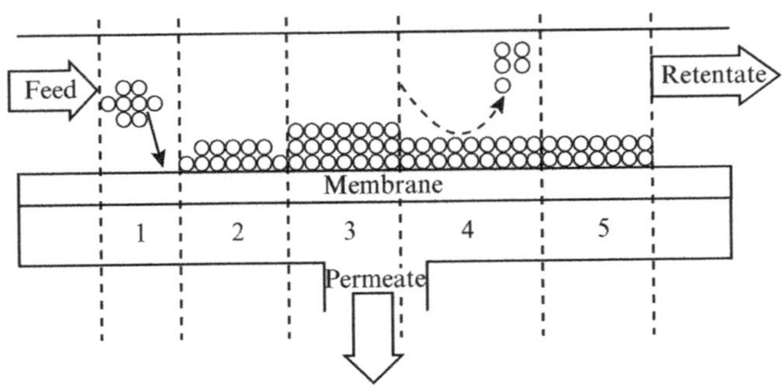

Figure 11.5 Diagram of Steps Taking Place in Membrane Separation Processes.

1. The feed enters and the solution is forced to the surface of the membrane.
2. Some of the solvent passes through the membrane (not shown because it is smaller than solute) and some passes on to the retentate side.
3. The solute builds up a layer on the surface of the membrane, causing the flux to decrease (which reduces the quantity of feed being purified).
4. The feed solution that is still flowing comes into contact with the solute buildup on the membrane surface and removes some or even all of the solute on the membrane and re-entrains it in the flow to become the retentate. Downstream, the re-entrained solute can then be re-trapped on the surface of the membrane.
5. The whole process can be repeated with a small film on the membrane where solute has been trapped already or on a totally clean part downstream.

In the process, the retentate becomes more concentrated with the solute because the solute that is removed and trapped on the membrane becomes re-entrained by the tangential force of the feed solution that is still passing through the unit. The solvent is effectively forced through the membrane to make permeate (pure solvent) by the design of the filter itself.

RO is the most selective of the three membrane processes described earlier that are used in industry for (primarily) liquid purification: MF, UF, and RO. The three processes are shown schematically in Figure 11.6. In MF, the

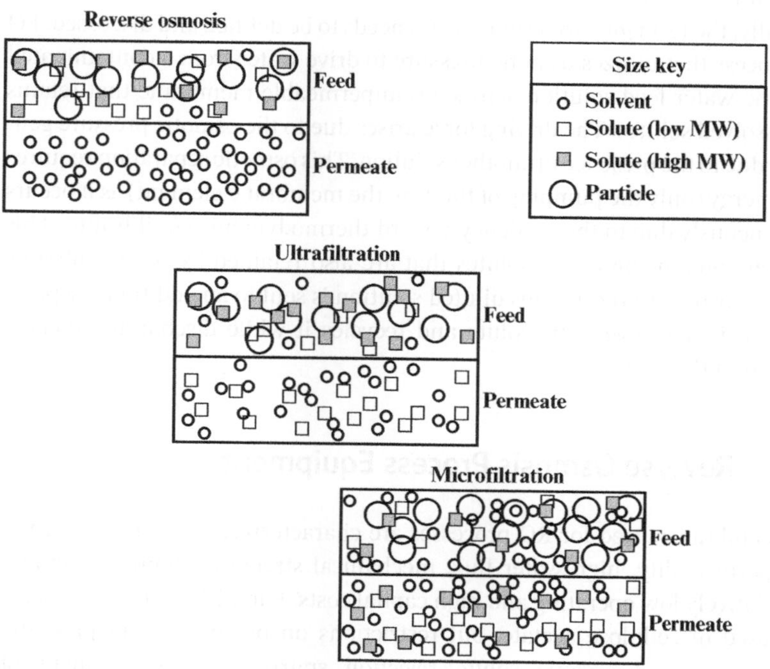

Figure 11.6 Different Membrane Separation Processes.

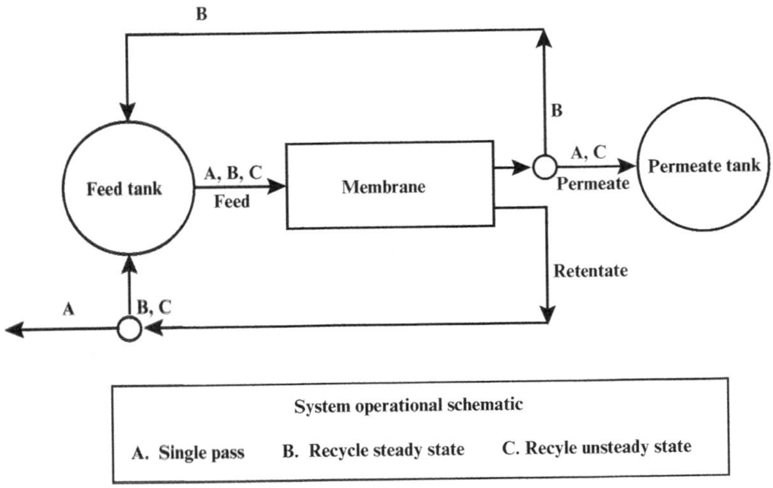

Figure 11.7 Process Flow Diagrams for the RO Process. (Adapted from Slater and Paccione 1987).

particles to be concentrated are generally solids or colloids rather than molecules in solution. Various process flow schematics of RO systems are provided in Figure 11.7.

Finally, the term *forward osmosis* (FO) needs to be defined and discussed. FO is a process that utilizes osmotic pressure to drive water from a contaminated or saline water feed solution across a semipermeable membrane that retains the dissolved solutes. The driving force arises due to the osmotic pressure generated due to the presence of another solution. This osmotic separation requires little energy (only the pumping of fluids to the membrane element) as it occurs spontaneously due to the tendency toward thermodynamic equilibrium. This solution contains dissolved solutes that are also retained by the membrane. After the osmosis process, this diluted solution is sent to an additional separation unit that recovers the solute and recycles it while discharging potable water from the unit.

11.4 Reverse Osmosis Process Equipment

The membranes used for RO processes are characterized by a high degree of semi-permeability, high water flux, mechanical strength, chemical stability, and relatively low operating and high capital costs. Early RO membranes were composed of cellulose acetate, but restrictions on process stream pressure, temperature, and organic solute rejection spurred the development of noncellulosic and composite materials. RO membranes are now designed for high salt retention, high permeability, mechanical robustness (to allow module fabrication and withstand operating conditions), chemical robustness (for fabrication materials, process fluids, cleaners, and sanitizers), low extractables, low fouling characteristics, high capacity, low cost, and consistency. The predominant RO membranes currently used in water applications include cellulose polymers, thin composites consisting of aromatic polyamides, and cross-linked polyether urethane urea.

RO membranes may be configured or designed into certain geometries for system operation: plate and frame, tubular, spiral wound (composite), and hollow fiber (Brooks 1985). In the *plate-and-frame* configuration, flat sheets of membrane are placed between spacers with heights of approximately 0.5 to 1.0 mm. These are, in turn, stacked in parallel groups. *Tubular* units are also commonly used for RO. This is a simpler design in which the feed flows inside of a tube whose walls contain the membrane. These types of membranes are usually produced with inside diameters ranging from 12.5 to 25 mm and generally produced in lengths of 150 to 610 cm. There is also the *hollow fine-fiber* (HFF) arrangement used in approximately 70% of worldwide desalination applications. Millions of hollow fibers are oriented in parallel and fixed in epoxy at both ends. A feed stream is sent through a central distributor where it is forced out radially through the fiber bundle. As the pressurized feed contacts the fibers, the permeate is forced into the center of each hollow fiber. The permeate then travels through the hollow bore until it exits the permeator. A *spiral-wound* cartridge is occasionally employed in this configuration. Here, the solvent is forced inward toward the product tube while the concentrate remains in the space between the membranes. A flat film membrane is made into a "leaf." Each leaf consists of two sheets of membrane with a sheet of polyester tricot in between which serves as a collection channel for the water product. Plastic netting is placed between each leaf to serve as a feed channel. Each leaf is then wrapped around the product tube in a spiral fashion.

11.5 Describing Equations

The membrane operation for the purification of seawater that incorporates a selective barrier can be simply described using the line diagram provided in Figure 11.8. This membrane operation typifies the case in which a feed stream (seawater) is separated by a semipermeable membrane that rejects salt but selectively transports water. A purified stream (the permeate) is therefore produced while at the same time, a concentrated salt stream (the retentate) is

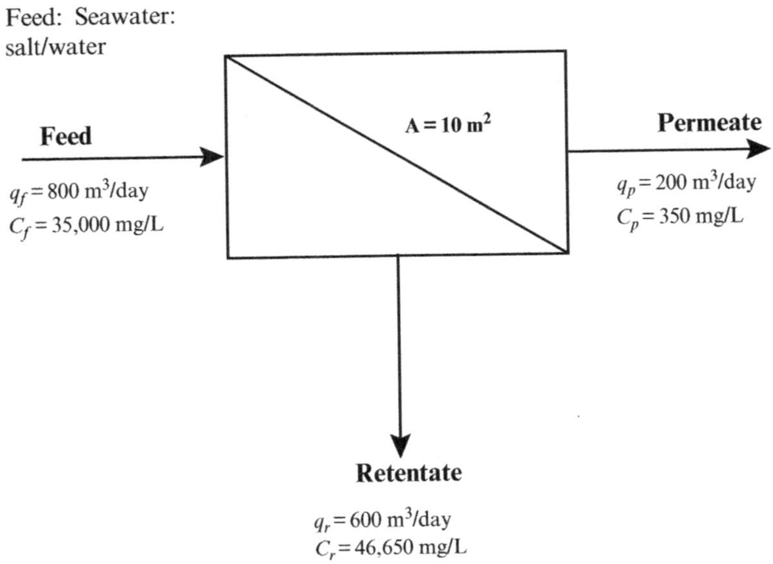

Figure 11.8 Seawater Desalination by RO.

discharged. With reference to Figure 11.8, a simple material balance can be written on the overall process flows and for that of the solute as:

$$q_f = q_r + q_p \tag{11.1}$$

$$C_f q_f = C_r q_r + C_p q_p \tag{11.2}$$

where q = volumetric flow rates, C = solute concentration, and subscripts f, r, and p refer to the feed, retentate, and permeate, respectively.

As noted earlier, one important characteristics of the RO process is the osmotic pressure. Osmotic pressure is related to both the solute concentration and the temperature of the solution as described in the Van't Hoff equation:

$$\pi = iC_sRT \tag{11.3}$$

where π = osmotic pressure, psi; i = Van't Hoff factor, dimensionless; C_s = solute concentration, mol/L; R = universal gas constant, (L-atm)/(mol-K); and T = absolute temperature, K.

The Van't Hoff factor, i, in Equation 11.3 factors in the number of ions in solution. For example, NaCl separates into two ions, Na^+ and Cl^-, therefore making the Van't Hoff factor equal to 2. Closer inspection of Equation 11.3 reveals that the Van't Hoff equation is analogous to the ideal gas law.

The change in osmotic pressure, $\Delta\pi$, across the membrane in this operation must be overcome in order to achieve RO. This is described in the following equation:

$$\Delta\pi = \pi_f - \pi_p \tag{11.4}$$

where π_f = osmotic pressure in the feed, psi; and π_p = osmotic pressure in the permeate, psi.

This change in osmotic pressure, $\Delta\pi$, can also be calculated using the concentrations of both the feed and the permeate as well as an empirical coefficient denoted as Ψ. This formula is:

$$\Delta\pi = \psi(C_f - C_p) \tag{11.5}$$

where Ψ = a constant, L-psi/g; C_f = feed concentration, g/L; and C_p = permeate concentration, g/L.

The permeate flux, J_p, is an important characteristic of an RO system. It is related to the permeate flow as well as to the area of the membrane. This is represented in the following equation:

$$J_p = \frac{q_p}{A_m} \tag{11.6}$$

where J_p, = permeate flux, gal/(day-ft^2); q_p = permeate flow, gal/day; and A_m = membrane surface area, ft^2.

The flux can be determined experimentally by measuring each incremental volume of permeate, ΔV, collected in time period Δt, and dividing by the surface area of the membrane. In water-based processes such as desalination, the permeate naturally consists mostly of water. Therefore, the permeate flux can be considered to be equal to *water flux*. Equation 11.7 defines this water flux:

$$J_p = J_w = A_w\left(\Delta P - \Delta\pi\right) = \frac{K_s}{t_m}\left(\Delta P - \Delta\pi\right) = \frac{\Delta P}{\mu R_m} \tag{11.7}$$

where J_w, = water flux, gal/(day-ft^2); A_w = water permeability coefficient, gal/(day-ft^2-psi); ΔP = pressure drop, psi; K_s = permeability constant, gal/(day-ft-psi); t_m = membrane thickness, ft; μ = solution viscosity, lb-sec/in^2; and R_m = membrane resistance coefficient, 1/ft. The reader should note that $\left(\Delta P - \Delta\pi\right)$ in Equation 11.7 represents the driving force across the membrane, here in terms of pressure. This driving force term can also be represented by concentration (as indicated later) or an electric field.

The water permeability coefficient can be experimentally determined by obtaining data on the unit with pure water; this eliminates the $\Delta\pi$ term since both sides of the membrane contain pure water. Water permeability coefficients supplied by a manufacturer are generally provided at 25°C. As temperature varies water diffusivity and viscosity vary, affecting the water flux rate through these membranes. Required membrane area correction factors, A_{WT}/A_{W25}, the area required at temperature T versus that at 25°C, have been reported by Kaup (1973) to vary as follows: T = 10°C, 1.58; T = 15°C, 1.34; T = 20°C, 1.15; T = 25°C, 1.0; T = 30°C, 0.84.

Another important factor is the *solute flux*, J_s, which can be determined by:

$$J_s = \frac{q_p C_p}{A_m} = \frac{\dot{m}_p}{A_m} = J_p C_p \tag{11.8}$$

where J_s, = solute flux, g/(day-ft^2); and $\dot{m}_p$ = permeate mass flow rate, g/d.

The solute flux can also be related to the solute concentration by utilizing the solute permeability factor. This relationship is given by:

$$J_s = B_s\left(\Delta C_s\right) \tag{11.9}$$

where B_s = solute permeability coefficient, gal/(day-ft^2); and ΔC_s = change in solute concentration, g/L, with appropriate unit conversions to yield proper units for J_s.

The selectivity of a membrane for a given solute can be expressed as the *percent rejection* (%R). Percent rejection represents a membrane's effectiveness and is a measure of the membrane's ability to selectively allow certain species to permeate and others to be rejected. This is an important characteristic or consideration when selecting a membrane for a separation process. The percent rejection represents the percentage of solute that is *not* allowed to pass into the permeate stream, and is given by:

$$\%R = \left(\frac{C_f - C_p}{C_f}\right) \text{x}100 \tag{11.10}$$

Rejections are sometimes written in terms of the average concentration of the feed and the concentrate since the feed becomes the concentrate as water moves through the membrane unit. The percent rejection of a given solute will vary from the entry point of the system to the final membrane unit. Percent rejection values are high, i.e. > 95%, for most solutes except for gases and small organic molecules.

Finally, the solvent recovery, Y, is a measure of the quantity of solvent that is allowed to pass through the membrane. This is defined as the quotient of the permeate flow divided by the feed flow, as shown by Equation 11.11.

$$Y = \frac{q_p}{q_f} \tag{11.11}$$

The concentration of solute in the permeate is a function of the water flux, which is dependent on the osmotic pressure, i.e. the concentration of the solute on the feed-concentrate side of the membrane. Solute concentration is also a function of the solute flux which is dependent on the solute concentration difference across the membrane.

The concentrate concentration is important with respect to fouling and scaling which must be considered in both the design and operation of an NF or RO system. If the solubility of salts is exceeded there is a potential for precipitation on the membrane surface. In some cases, as with calcium carbonate, pH adjustment with an acid dosage to the feed can control the precipitation. With other compounds such as calcium sulfate and silica, polymeric anti-scalants are needed to control scaling. Fouling can be caused by the accumulation of natural organic matter on the membrane surface, and periodic off-line cleaning of the membranes is needed.

The actual concentrate concentration, C_c, can be predicted by combining Equations 11.1, 11.2, 11.10 and 11.11, and expressing C_c in terms of Y and $\%R$ as shown in Equation 11.12.

$$C_c = \frac{\left[1 - Y\left(1 - \%R\right)\right] C_f}{1 - Y} \tag{11.12}$$

11.6 Other Membrane Separation Processes

Ultrafiltration, microfiltration, and gas permeation are briefly reviewed in this section. The reader is referred to the literature (Applegate 1984; Schweitzer 1979; Slater et al. 1992; Theodore and Ricci 2010) for more details on these processes. It should be noted that these three processes rarely find application in desalination.

11.6.1 Ultrafiltration

UF is a membrane separation process that can be used to concentrate single solutes or mixtures of solutes. Transmembrane pressure (the membrane pressure drop) is the main driving force in UF operations, and separation is achieved via a sieving mechanism. The UF process can be used for the treatment or concentration of oily wastewater, for pretreatment of water *prior* to RO, and for the removal of bacterial contamination (pyrogens). In the food industry, UF is used to separate lactose and salt from cheese whey proteins, to clarify apple juice, and to concentrate milk for ice cream and cheese production (Kosikowski 1986; Maubois 1980). The most energy-intensive step in ice cream production is concerned with the concentration of skimmed milk, where membrane processes are more economical for this step than vacuum evaporation (Garcia et al. 1989). UF processes are also used for concentrating or dewatering fermentation products as well as purifying blood fractions and vaccines.

UF membranes are rated in terms of their molecular weight cutoff, thereby separating proteins and other biochemicals according to their molecular weight differences. Thin, mechanically strong, flexible, nonadsorbing and flat-textured, UF membranes are available in a wide variety of cutoff sizes. For example, the YM-30 membrane will retain any material with a MW > 30,000. Most UF membranes have an asymmetric structure with a thin selective membrane supported by a thicker, porous structure, which is the case for the YM-30 membrane.

UF may be regarded as a membrane separation technique where a solution is introduced on one side of a membrane barrier while water, salts, and/or other low-molecular-weight materials pass through the unit under an applied pressure. As noted, these membranes separation processes can be used to concentrate single solutes or mixtures of solutes. The variety in the different membrane materials allows for a wide temperature-pH processing range.

The main economic advantage of UF is a reduction in both design complexity and energy usage since UF processes can simultaneously concentrate and purify process streams. The fact that no phase change is required leads to highly desirable energy savings. A major disadvantage is the high capital cost that might be required if low flux rates for purification demand a large system. However, UF processes are usually economically sound in comparison to other traditional separation techniques.

11.6.2 Microfiltration

MF is employed in modern biochemical and biological separation processes. For example, in cell harvesting, microfiltration can be used instead of centrifugation or pre-coat rotary vacuum filtration to remove yeast, bacteria, or mycelial organisms from fermentation broth. Both MF and UF are used for cell harvesting. MF is used to retain cells and colloids, while allowing passage of macromolecules into the permeate stream. UF is used to concentrate macromolecules, cells, and colloidal material, while allowing small organic molecules and inorganic salts to pass into the permeate stream. Pore sizes in MF are around 0.02 to 1.0 μm in diameter as compared with 0.001 to 0.02 μm (300 to 300,000 Daltons) for UF (ranges vary slightly depending on the source) (Mulder 1996; Porter 1990; Wankat 1990).

Similar types of equipment are used for MF and UF, but membranes with larger pore sizes are installed in MF applications. MF and UF belong to a group of separation processes that depend on pressure as the driving force for separation. MF processes operate at lower pressures than UF, but at higher pressures than conventional particulate filtration. Ideal membranes possess high porosity, a narrow pore size distribution, and a low binding capacity. When separating microorganisms and cell debris from fermentation broth, a biological cake is formed. Principles of cake filtration apply to MF systems except that the small size of the yeast particles results in a cake with a relatively high resistance to flow and a relatively low filtration rate.

In *dead-end filtration*, feed flow is perpendicular to the membrane surface, and the thickness of the cake layer on the membrane surface increases with filtration time. Consequently, the permeation rate decreases with time as well. *Cross-flow filtration*, on the other hand, features feed flowing parallel to the membrane surface which is designed to decrease formation of a cake by *sweeping* previously deposited solids from the membrane surface and returning them to the bulk feed stream. Cross-flow filtration is far superior to dead-end filtration for cell harvesting because the biological cake is highly compressible, which causes the accumulated layer of biomass to rapidly blind the filter surface in dead-end operation. Therefore, MF experiments are often conducted using cross-flow filtration because of the advantages that this filtration mode offers (Slater and Hollein 1993; Slater et al. 1989).

11.6.3 Gas Permeation

Gas separation processes are generally considered as relatively new and emerging technologies because they are not included in the curriculum of many traditional chemical and environmental engineering programs (Clements et al. 1986; Davis and Sandall 1990; Permea Inc 1986; Slater et al. 1989, 1990, 1992; Wankat 1990).

Gas Permeation (GP) is the term used to describe a membrane separation process in which a gaseous feed stream is fractionated into permeate and nonpermeate streams. The non-permeating stream is typically referred to as the *nonpermeate* in gas separation terminology, although it is defined as the retentate in liquid separation. Transport separation occurs by a solution diffusion mechanism. Membrane selectivity is based on the relative permeation rates of the components through the membrane. Each gaseous component transported through the membrane has a characteristic individual permeation rate that depends on its ability to dissolve and diffuse through the membrane material. The mechanism for transport is based on solubilization and diffusion; the two describing relationships on which the transport is based are Fick's law (diffusion) and Henry's law (solubility) (Slater et al. 1990). Gas membranes generally work because individual gases differ in their solubility and diffusivity through the nonporous polymers they are made of. As noted earlier, gas separation membranes separate gases from other gases. Some gas filters, which remove liquids or solids from gases, are MF membranes.

The use of membranes in gas separation processes was commercialized by Monsanto in the early 1970s with the development of the hollow fiber Prism system for which the Monsanto Company won the 1981 Kirkpatrick Chemical Engineering Achievement Award. The hollow fiber membrane allowed, for the first time, the practical use of membranes in large-scale gas separation and purification processes. Several other firms, including UOP, Air Products and Chemicals, Dow, DuPont, and Grace produce these gas permeation membrane units (Slater et al. 1992).

Gas permeation systems have gained popularity in both traditional and emerging engineering areas (Ismail et al. 2015). These systems were originally developed primarily for hydrogen recovery. There are presently numerous applications of gas permeation in industry, and other potential uses of this technology are in various stages of development. Applications today include gas recovery from waste gas streams, landfill gases, and ammonia and petrochemical products. Gas permeation membrane systems are also employed in gas generation and purification, including the production of nitrogen and enriched oxygen gases (Slater et al. 1990), and enrichment of biogas (Beil and Beyrich 2013).

11.7 Applications

Reverse osmosis is widely utilized today by a host of industries for a surprising number of operations. Aside from the classic example of RO for seawater desalination, it has found a niche in the food industry for concentration of fruit juices, in the galvanic industry for concentration of waste streams, and in the dairy industry for concentration of milk prior to cheese manufacturing. (Applegate 1984; Brooks 1985; Parkinson 1983). A more novel use of reverse osmosis is the production of low-alcohol beer by breweries in Denmark, France, and Germany.

The wine and juice industries have applications for hyperfiltration as well. Flavor tests have shown positive results indicating the potential of membrane processing for improving taste. Using membranes with pore sizes controlled within a specific range has resulted in bitterness and "off-flavors" being removed from finished wine products as well as from grapefruit and orange juice. After using an RO process, a taste-testing panel found a low quality

Chenin Blanc to have a significantly better taste (Schweitzer 1979). However, this is a difficult application as desirable and undesirable taste elements have a similar molecular size along with steric and polar characteristics that are also similar.

Another important application of hyperfiltration is the dialysis. This technique is used in patients who suffer from kidney failure and can no longer filter waste products (urea) from the blood. In general, RO equipment used for dialysis can reduce ionic contaminants by up to 90%. In this process, the patient's blood flow in a tubular membrane while a dialysate flows counter currently on the outside of the feed tube. The concentrations of undesirables such as potassium, calcium and urea are high in the blood and low or absent in the dialysate. Although this treatment successfully mimics the filtration capabilities of the kidney, it cannot replace its endocrine functionality (Slater 1991).

Water covers around 70% of Earth's surface, but 97.5% of it is unfit for human consumption. With the world facing a growing freshwater shortage from which the United States may not be spared, one method of producing freshwater is desalination. The major application of RO is water desalination. Some areas of the world that do not have a ready supply of freshwater have chosen to desalinate seawater or brackish water using RO to generate potable drinking water. Because no heating or phase change is required, the RO process is both a relatively low energy and economical water purification process. A typical saltwater RO system consists of an intake, a pretreatment component, a high-pressure pump, membrane apparatus, remineralization, and pH adjustment components, as well as a disinfection step. A pressure difference driving force of about 1.0 to 6.5 mPA is generally required to overcome the osmotic pressure of saltwater (Schweitzer 1979).

11.8 Illustrative Examples

Six illustrative examples complete the presentation of materials on reverse osmosis and membrane separation processes highlighted in this chapter.

11.8.1 Illustrative Example 1

Verify that the quantities provided in Figure 11.8 satisfy both a componential and overall material balance.

Solution. An overall balance yields (in m^3/d):

$$q_f = q_r + q_p;\ 800\,\text{m}^3/\text{d} = 600\,\text{m}^3/\text{d} + 200\,\text{m}^3/\text{d} = 800\,\text{m}^3/\text{d}\,\surd$$

A componential balance on the salt gives (in mg/d):

$$(35{,}000\,\text{mg/L})\left(800\,\text{m}^3/\text{d}\right)\left(1{,}000\,\text{L/m}^3\right) =$$
$$(46{,}650\,\text{mg/L})\left(600\,\text{m}^3/\text{d}\right)\left(1{,}000\,\text{L/m}^3\right) + (350\,\text{mg/L})\left(200\,\text{m}^3/\text{d}\right)\left(1{,}000\,\text{L/m}^3\right)$$
$$2.8\times10^{10}\,\text{mg/d} = 2.799\times10^{10}\,\text{mg/d} + 0.006\times10^{10}\,\text{mg/d} = 2.8005\times10^{10}\,\text{mg/d}\,\checkmark$$

Both balances are satisfied.

11.8.2 Illustrative Example 2

With reference to Figure 11.8, calculate the permeate flux and the membrane selectivity.

Solution. For the flux, using Equation 11.6, substitution yields:

$$J_p = \frac{q_p}{A_m} = \frac{200\ \text{m}^3/\text{d}}{10\ \text{m}^2} = 20\ \text{m}^3/\text{m}^2/\text{d}$$

For membrane selectivity, expressed as percent rejection, %R, Equation 11.10 is used as follows:

$$\%R = \left(\frac{C_f - C_p}{C_f}\right)\text{x}100 = \frac{(35{,}000 - 350\ \text{mg/L})}{35{,}000\ \text{mg/L}}(100) = 99\%$$

11.8.3 Illustrative Example 3

Consider the seawater desalination example in Illustrative Examples 1 and 3. If the applied pressure gradient across a membrane is 500 psi and the membrane thickness is 10 μm, determine the permeability of the membrane in m^2/sec-Pa.

Solution. Employing Equation 11.7, the membrane permeability constant, K_s, can be determined as follows after rearrangement:

$$K_s = \frac{J_p t_m}{(\Delta P - \Delta\pi)} \tag{11.13}$$

Substitution yields:

$$K_s = \frac{\left(20\ \text{m}^3/\text{m}^2/\text{d}\right)\left(10\ \text{x}10^{-6}\ \text{m}\right)}{(500\ \text{psi})(6894.76\ \text{Pa/psi})} = \left(5.8\ \text{x}\ 10^{-11}\text{m}^2/\text{d-Pa}\right)\left(1\ \text{d}/86{,}400\ \text{sec}\right)$$
$$= 6.71\ \text{x}\ 10^{-16}\text{m}^2/\text{sec-Pa}$$

11.8.4 Illustrative Example 4

A new membrane material (#CSS-1) is to be evaluated for its solute and solvent permeability. A small test cell is utilized with a 5.0 cm diameter circular membrane. The test solution contains 6,000 mg/L of NaCl in water at 25°C. Assume that the following relationship holds for osmotic pressure of NaCl in water.

At an operating pressure gradient of 750 psi, the permeate flow rate is 0.0152 cm^3/sec = 0.0152 g/sec, and the permeate solute concentration is 150 mg/L. Assuming there is no concentration polarization and that the operating conditions remain constant, determine the water flux in g/cm^2-sec and the solute flux in g/cm^2-sec.

Solution: The membrane surface area is:

$$A_m = \frac{\pi(5\ \text{cm})^2}{4} = 19.63\text{cm}^2$$

Employing Equation 11.6 for water as the permeate yields:

$$J_w = \frac{q_p}{A_m} = \frac{(0.0152\ \text{g}/\text{sec})}{(19.63\ \text{cm}^2)} = 7.74 \text{ x } 10^{-4}\text{g/cm}^2\text{-sec}$$

Employing Equation 11.8, the solute flux is determined as:

$$J_s = \frac{q_p C_p}{A_m} = \frac{(0.0152\text{cm}^3/\text{sec})(150\text{mg}/\text{L})(1\text{L}/1000\text{cm}^3)}{(19.63\ \text{cm}^2)}$$

$$= \frac{(0.00228\text{mg}/\text{sec})}{(19.63\text{cm}^2)} = 1.16\text{x}10^{-4}\text{mg/cm}^2\text{-sec}$$

$$J_s = 1.16 \times 10^{-7}\text{g/cm}^2\text{-sec}$$

11.8.5 Illustrative Example 5

Refer to Illustrative Example 4. Calculate the percent solute rejection, i.e. the membrane selectivity.

Solution: The percent rejection is given by Equation 11.10, and substitution yields:

$$\%R = \left(\frac{C_f - C_p}{C_f}\right)\text{x}100 = \frac{(6{,}000\ \text{mg}/\text{L} - 150\ \text{mg}/\text{L})}{(6{,}000\ \text{mg}/\text{L})}\text{x}100$$

$$= \frac{5{,}850\ \text{mg}/\text{L}}{6{,}000\ \text{mg}/\text{L}}\text{x}100 = 97.5\%$$

11.8.6 Illustrative Example 6

A 10-year old reverse osmosis system is no longer delivering the degree of separation required for the process. Rather than replace the unit, you have been asked to recommend what other possible steps can be taken for the existing unit to get back "on line." Unfortunately, it is not currently possible to completely replace the unit. Some other solutions need to be found.

Solution: The issue with the current operation is its inability to perform the separation at its designed level. The first thing that needs to be investigated is what is the current state of the process? How was it originally designed? Were safety margins included in the design that allow for some "tweaking" of the operation of the process? How much change in pressure (drop) is allowed? What will the materials of construction stand up to?

A possibility for the process is changing the flow rate of the feed water. It also may be possible to change some units in the process. It is possible after 10-years of service that some equipment is not working at its optimum settings. An examination should be made of their contribution of the failure to desalinate the seawater. The composition and conditions of the feed stream also need to be examined. Are they the same as what the process was originally designed to separate? Has the process been altered upstream in such a way that the separation is no longer feasible? Perhaps the feed stream has changed in such a manner that the original process is no longer capable of separating it. Finally, is it possible that the process needs cleaning and additional maintenance that could increase the separation efficiency?

11.9 Problems

1. A UF system is being used to concentrate a seawater slurry to 10% by weight solids prior to entering a desalination unit. If 1 ft^2 of membrane area is required to treat 5 lb/hr of solids to the desired final slurry concentration, what membrane area, in ft^2, is required for a slurry flowrate of 6,000 lb/min?
2. You have been requested to calculate the required membrane area for a proposed RO desalination plant. The following design information has been provided:
   Permeate (pure water) design flow rate, $q_p = 50\ m^3/hr$
   Membrane design pressure drop, $\Delta P = 175$ kPa
   Membrane resistance coefficient, Rm = $3.9 \times 10^{12}\ m^{-1}$
   Operating temperature = 6°C
   Seawater viscosity @ 6°C = 0.00155 Pa-sec.
3. A reverse osmosis unit is to demineralize 100,000 gal/d of treated tertiary effluent. The following data are available for this membrane system: the

water permeability coefficient, A_w = 0.035 gal/(day-ft-psi) @ 25°C; the pressure drop across the membrane, P = 425 psi; the change in osmotic pressure, $\Delta\pi$, across the membrane = 45 psi; and the lowest operating temperature will be 10°C. Determine the water flux at 25°C and the corresponding membrane area required to meet production requirements at the lowest operating temperature expected for the source water.

4. A reverse osmosis unit is to demineralize 200,000 gal/d of brackish water. The following data are available for this membrane system: the water permeability coefficient, A_w = 0.030 gal/(day-ft-psi) @ 25°C; the pressure drop across the membrane, P = 380 psi; the change in osmotic pressure, $\Delta\pi$, across the membrane = 45 and psi; the lowest operating temperature will be 15°C. The membrane area/unit volume of equipment = 2,500 ft^2/ft^3. Determine the water flux at 25°C and the corresponding membrane area required to meet production requirements at the lowest operating temperature expected for the source water. Also determine the space (ft^3) required for the equipment.
5. A low-pressure RO membrane system treating a brackish groundwater is operating at 80% recovery. The rejection for both calcium and sulfate is 95%. The feed water contains 100 mg/L of calcium and 200 mg/L of sulfate, and the K_{sp} for $CaSO_4$ is 2.5×10^{-5}. Determine the concentration of calcium and sulfate in the concentrate and determine if the solubility of $CaSO_4$ has been exceeded, posing serious scaling potential.
6. Using the RO membrane system described in Problem 3 treating a brackish groundwater, determine the expected permeate concentration of chloride if the feed concentration is 4,500 mg/L. The solute permeability coefficient, B_s = 0.74 gal/(day-ft^2). Also determine the percent rejection of the membrane for Cl^-.

References

Applegate, L. (1984). Membrane separation processes. *Chemical Engineering* 91 (6): 64–89.

Beil, M. and Beyrich, W. (2013). Chapter 15 – Biogas upgrading to biomethane. In: *The Biogas Handbook, Science, Production and Applications* (ed. A. Wellinger, J. Murphy, and D. Baxter). Cambridge, UK: Woodhead Publishing, Limited. https://doi.org/10.1533/9780857097415.3.342.

Brooks, K. (1985). Membranes push into separations. *Chemical Week* 138 (1): 21–24.

Clements, L., Otten, M., and Bhat, P. (1986). *Laboratory membrane gas separator-a new teaching tool.* Paper 53b presented at the AIChE Annual Meeting, Miami Beach, FL.

Davis, R. and Sandall, O. (1990). A membrane gas separation experiments for the undergraduate laboratory. *Chemical Engineering Education* 87: 10–21.

Dupont, R.R., Eisenberg, T.N., and Middlebrooks, E.J. (1982). *Reverse Osmosis in the Treatment of Drinking Water*. Water Quality Series UWRL/Q-82/05. Logan, UT: Utah Water Research Laboratory. https://digitalcommons.usu.edu/water_rep/505.

Glaston, S. (1949). *Textbook of Physical Chemistry*, 2nd Edition. New York, NY: D. Van Nostrand.

Ismail, A.F., Khulbe, K.C., and Matsuura, T. (2015). *Gas Separation Membranes: Polymeric and Inorganic*. New York, NY: Springer.

Kaup, E.C. (1973). Design factors in reverse osmosis. *Chemical Engineering* 4: 46–55.

Kosikowski, F. (1986). New Cheese-Making Procedures Utilizing Ultrafiltration. *Food Technology* 40 (6):71–77.

Maubois JL. (1980) Recent Developments of Membrane Ultrafiltration in the Dairy Industry. In: *Ultrafiltration Membranes and Applications. Polymer Science and Technology*, (ed Cooper A.R.) 13: 305–318.

Mulder, M. (1996). *Basic Principles of Membrane Technology*, 2nd Edition. Boston, MA: Kluwer Academic.

Parkinson, G. (1983). Reverse osmosis: Trying for wider applications. *Chemical Engineering* 90 (5): 26–31.

Permea Inc. (1986). *Prism Separators*, Bulletin PERM-6-008. St. Louis, MO: Permea Inc.

Porter, M.C. (1990). *Handbook of Industrial Membrane Technology*. Park Ridge, NJ: Noyes Publications.

Schweitzer, P. (1979). *Handbook of Separation Techniques for Chemical Engineers*. New York, NY: McGraw-Hill.

Slater, C.S. and Hollein, H. (1993). Educational initiatives in teaching membrane technology, desalination. *Chemical Engineering Education* 90: 625–634.

Slater, C.S., Hollein, H., Antonechia, P.P., Mazzella, L.S., and Paccione, J. (1989). Laboratory experiences in membrane separation processes. *International Journal of Engineering Education* 5: 369–378.

Slater, C.S. and Paccione, J.D. (1987). A reverse osmosis system for an advanced separation process laboratory. *Chemical Engineering Education* 22: 138–143.

Slater, C.S., Vega, C., and Boegel, M. (1990). Membrane gas separation experiments for a chemical engineering laboratory. *American Society for Engineering Education (ASEE) Annual Conference Proceedings*, 648–650. Washington, D.C.: ASEE.

Slater, C.S., Vega, C., and Boegel, M. (1992). Experiments in gas permeation membrane processes. *International Journal of Engineering Education* 9: 1–7.

Theodore, L. (2006). *Nanotechnology: Basic Calculations for Engineers and Scientists*. Hoboken, NJ: John Wiley & Sons.

Theodore, L. and Ricci, F. (2010). *Mass Transfer Operations for the Practicing Engineer*. Hoboken, NJ: John Wiley & Sons.

Wankat, P.C. (1990). *Rate-Controlled Separation*. Boston, MA: Chapman & Hall.

12

Crystallization

12.1 Introduction

Throughout the history of the chemical industry, crystals have been produced via crystallization methods that range between something as simple as allowing vats of hot concentrated solution to cool to those as complex as continuous, carefully controlled, multi-step processes which yield a crystal product of a particular size or size distribution, shape, moisture content, and purity. Today, consumer demands require crystal products to meet rigid specifications that can also include specific color, odor, particle size *and* particle size distribution, and caking characteristics. Most of the development of crystallization processes has been for use in water desalination, and on a laboratory scale, and the process has been used in the treatment of ammonia nitrate wastewater, paper mill bleach solutions, plating liquors, and arsenal red water. Freeze crystallization also has the potential for the removal of 1 to 10% total dissolved solids (TDS) in other aqueous waste streams.

Crystallization is an important mass transfer operation that is often employed in the preparation of a pure product. In the process, a crystal usually separates out as a substance of definite composition from a solution of varying composition. Any impurities in the liquid (often referred to as the *mother liquor*) are carried in the crystalline product only to the extent that they adhere to the surface or are *occluded* (retained) within the crystals that may have grown together during or after the crystallization operation.

The separation of a solid from a solution onto a crystal occurs only if there is a state of imbalance involving a mass driving force, i.e. if a differential in chemical potential (or concentration) between the bulk of the liquid solution and the crystal interface exists. This effectively means that the solution must be supersaturated.

Although crystallization is ordinarily thought of as the deposition of a solid, crystalline phase from a liquid phase by cooling, evaporation, or both, the same principles apply to crystal formation by precipitation caused by the addition of a third substance. This other component may either react to form a precipitate or simply decrease the solubility of the precipitated material. One may also view crystallization as a phase equilibrium application. These solid crystals are usually formed from a homogenous liquid phase. If one starts with an unsaturated solution formed by dissolving some solid in liquid, more solid can be dissolved

Introduction to Desalination: Principles, Processes, and Calculations, First Edition. Louis Theodore and R. Ryan Dupont.

until the solution becomes *saturated*. More solid can still be dissolved in the solution causing it to become *supersaturated*. At this point, solids begin to deposit out of solution in a process often referred to as *nucleation*. Crystal growth continues until the solution reaches the previously attained saturation (equilibrium) point.

There are several different ways that crystallization can occur. The four most often encountered in practice are:

1. Cooling
2. Evaporation
3. Cooling and evaporation (also denoted adiabatic evaporation)
4. Salting out

Process (1) is most commonly employed, provided the solubility of the component being crystallized decreases with decreasing temperature.

The problems concerning, crystallization which are most frequently encountered by the practicing environmental engineer are:

1. Yield of a given product
2. Purity of the product
3. Energy requirements for cooling, evaporation, etc
4. Shape of the individual crystals
5. Size of the crystals
6. Uniformity or size distribution of the crystals
7. Rate of production of the desired crystals
8. Caking

It should also be noted that when crystallization begins, heat is evolved (liberated). The amount of heat evolved is defined as the heat (or enthalpy) of crystallization.

This chapter introduces the reader to the concepts and applications of the crystallization process to water desalination and wastewater treatment through discussion of the crystallization process, crystal physical characteristics, crystallization equipment, describing equations, and design considerations. An application section provides seven illustrative examples related to the general subject of crystallization.

12.2 The Crystallization Process

The crystallization process essentially involves three steps:

1. Formation of crystals
2. Crystal growth
3. Separation of the crystals from residual liquid

A simple schematic of this overall process is provided in Figure 12.1.

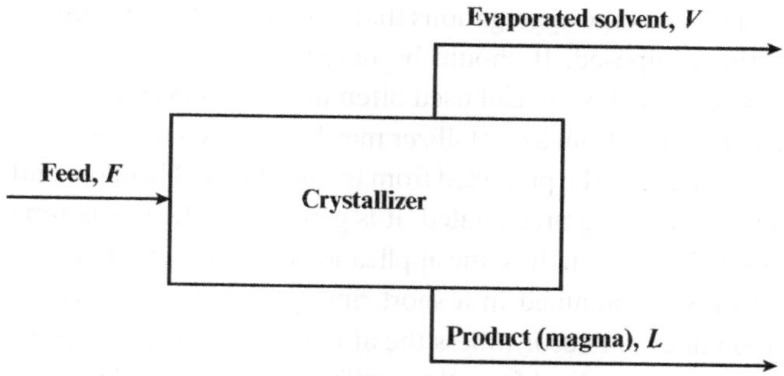

Figure 12.1 Generalized Diagram of the Crystallization Process.

There are various processes that can accomplish this phase separation while meeting the desired design requirements associated with a crystallization operation and/or the rate of crystallization. As with many mass transfer (as well as heat transfer) calculations, the crystallization rate is expressed in terms of an overall coefficient which takes into account all individual resistances. The coefficient is almost always determined from experimental data and/or experience. Details on product yield, product purity, crystal size, and/or size distribution must also be specified. Each of these topics is discussed below with the latter topic receiving additional treatment in the next section.

The rate of crystallization involves two distinct steps:

1. The rate of formation of new crystals, or nucleation
2. The rate of precipitation on crystals already present, usually referred to as *crystal growth*

The mechanism of nucleation is essentially unknown. However, the process of crystal growth has been studied extensively, with some researchers suggesting that crystal growth occurs at erratic rates and that the solution concentration at the crystal surface is not uniform.

It is common knowledge that the solubility of small crystals is greater than that of larger crystals. Therefore, material may be *simultaneously* depositing on larger crystals while smaller crystals are dissolving when both are exposed to the same solution. As a result, it is difficult in practice to maintain uniform conditions while growing crystals. Thus, it is usually impossible to avoid locally varying conditions in the solution.

Prior to crystallization, it is common practice to inject small crystals, known as *seeds*, into the solution so that solids will be deposited more easily. Since the solubility of crystals of this size is less than that of submicroscopic crystals, the crystallization unit may be operated in such a manner as to retain the supersaturated solution with respect to the seeds, but not supersaturated with respect to crystals of the size of the aggregations that will inevitably form under the operating conditions imposed. It should be noted that seeding is rarely used in continuous processes; it is instead used often in batch processes.

Predetermination of the yield from a crystallizer may be obtained from a material balance. The yield can usually be predicted from the solubility (Theodore and Rici 2010) of the solid phase being precipitated. It is general practice to assume equilibrium or a saturated condition. In some applications, crystallization occurs slowly and equilibrium is not attained in a short time period. In many cases, controlled precipitation does not occur unless the aforementioned seed crystals are present. If solubilities are obtained from the equilibrium diagrams discussed earlier, the prediction of the equilibrium yield can be made by applying the usual *lever arm ratio* principle (Theodore and Rici 2010) between the phases which are known to be present at the specified operation temperature.

The solubilities usually reported in the literature are expressed in terms of mass of "salt" per 100 units of mass of pure solvent. The prediction of a yield of a non-hydrated salt from a solution is simple since the amount of solvent present during crystallization is generally constant and the quantity remaining at the operating temperature is known (Brown 1950).

Impurities in crystallization operations are introduced primarily from the mother liquor that is not completely removed from the solid product. The extent of occlusion and the completeness of washing are important factors in the determination of the purity of a crystalline product. The agglomeration of crystals into larger crystals, which can occur during the growth of the crystals, makes washing more difficult and accordingly results in a lower purity of product. However, agitation decreases the tendency of the crystals to agglomerate.

Purification techniques have become increasingly important in recent years. Mechanisms by which the impurities discussed earlier can be incorporated into crystalline products include adsorption of impurities onto crystal surfaces, solvent entrapment in cracks, crevices and agglomerates, and inclusion of pockets of liquid. The recommendation for producing high purity crystals is to maintain supersaturation at a low level so that large crystals are formed. The characteristics of an impurity can also determine whether it is positioned on the surface or in the interior of the host crystal.

Although crystallization is often thought of as solely a separation process, it is a filtration step at the end of the process that accomplishes the actual crystal separation. The residual liquid (*mixed magma underflow*) from a crystallizer typically contains approximately 30% by mass solids and can be thickened further before filtration by being passed through a gravity settler. Such pre-thickening allows dewatering of the "mother liquor," which allows either more time for the final wash step or a shorter filter cycle.

The filter cycle almost always includes a washing step at the end of the full cycle. This permits a thorough drain before the crystals are discharged. It should also be noted that the system's throughput and performance is often limited by the liquid–solid separation step. In effect, this then becomes the critical step in the overall operation (Brown 1950).

Perhaps the most troubling problem is associated with the final product. Crystalline materials have a tendency to bind together or cake on storage. Most crystalline products are required in a free-flowing form, as with sugar and table salt, or to be capable of being distributed uniformly over surfaces (e.g. fertilizers). It is also important for the crystalline product to remain in a particulate state. Caking may also require a crushing operation before the crystals can be used.

Numerous problems can arise during the operation, design, and analysis of a crystallizer. Ten of the most important problems are listed here:

1. Crystal-size distribution (CSD)
2. Purity
3. Fouling
4. Vapor release and location
5. Scale-up
6. Stability of operation
7. Liquid-solid separation
8. Space requirements
9. Capital costs
10. Operating and maintenance costs

With reference to the crystallization process, the average size to which crystals are grown is usually dictated by the manner in which they are to be used. As noted earlier, fine crystals have less opportunity for agglomeration and accordingly occlude less mother liquor; however, fine crystals have a greater surface area and offer greater difficulty of complete removal of mother liquor entrained on the crystals. In addition, producing crystalline material of a given size range (a specific upper and lower limit) is difficult to accomplish.

It should be noted that the removal rate of crystals of a given size can greatly alter a crystallizer's ability to continue the production of crystals having that same size and/or size distribution. This effect is particularly pronounced in a separation unit where the crystals of a desired size distribution range are preferentially removed and the under-sized particles recycled to the crystallizer. This can shift the distribution to a smaller and narrower size range.

Finally, two of the goals of the design and practicing engineers who work on the CSD problem are to be able to:

1. Understand CSD so as to analyze and modify what a particular operating crystallizer can and cannot do in the way of producing acceptable crystal size and crystal size ranges.
2. Include the effects of CSD in the design of new units, i.e. to design an original process that is not yet operating (Brown 1950).

12.3 Crystal Physical Characteristics

The three principle physical characteristics of crystals are size, shape, and density. By knowing these three properties it is possible to determine not only how crystals will behave in a crystallization process, but also how they will behave in their later use. Although density is often not treated as an important factor, consider, for example a ping-pong ball and a golf ball. Under a microscope, they will appear almost equal in size; however, if both were tossed into a moving fluid stream they would behave quite differently. Even though the size and shape are similar, the density is quite different and the behavior of the two objects is far from being similar. This is one of the reasons that many object to physical sizing alone for crystal characterization.

A common method of specifying large crystal sizes is to designate the screen mesh that has an aperture corresponding to the crystal diameter. Since various screen scales are in use, confusion may result unless the screen scale involved is specified. The *screen mesh* generally refers to the number of screen openings per unit length or area. The aperture for a given mesh will depend on the wire size employed. The clear space between the individual wires of a screen is termed the *screen aperture*. As indicated earlier, the term *mesh* is applied to the number of apertures per linear inch (e.g. a 10-inch mesh screen will have 10 openings per inch and the aperture will be 0.1 inch minus the diameter of the wire).

The Tyler and the U.S. Standard Screen Scales (Table 12.1) are the most widely used in the United States. The screens are generally constructed of wire mesh cloth, with the diameters of the wire and the spacing of the wires being specified. These screens form the bottoms of metal pans about 8 inches in diameter and 2 inches high, the sides of which are fashioned so that the bottom of one sieve nests snugly on top of the next (Reynolds et al. 2004; Theodore 2008).

Size distributions are often characterized by a mean diameter. Although numerous "means" have been defined in the literature, the most common are the arithmetic mean and the geometric mean. The arithmetic mean diameter of a number of crystals is simply the sum of each of the diameters divided by the number of diameters measured. The geometric mean diameter is the nth root of the product of n number of diameters in the sample. In addition to the arithmetic and geometric means, a size distribution may also be characterized by the median diameter. The median diameter is that diameter for which 50% of the crystals are larger in size and 50% are smaller in size. Another important characteristic is the measure of central tendency. It is sometimes referred to as the *dispersion* or *variability*. The most common term employed is the *standard deviation*. These terms are discussed shortly (Shaefer and Theodore 2007).

One basic way of summarizing data is by the computation of a central value. The most commonly used central value statistic is the arithmetic average, or the *mean*. This statistic is particularly useful when applied to a set of crystal

Table 12.1 Tyler and U.S. Standard Screen Scales.

Tyler mesh	Aperture, µm	U.S. mesh	Aperture, µm
400	37	400	37
325	43	325	44
270	53	270	53
250	61	230	62
200	74	200	74
170	88	170	88
150	104	140	105
100	147	100	149
65	208	70	210
48	295	50	297
35	417	40	420
28	589	30	590
20	833	20	840
14	1,168	16	1,190
10	1,651	12	1,680
8	2,362		
6	3,327		
4	4,699		
3	6,680		

size data having a fairly symmetrical distribution. The mean is an important statistic in that it summarizes all the data in the set and because each crystal is taken into account in its computation. The formula for computing the arithmetic mean, $\overline{X}$, is:

$$\overline{X} = \frac{X_1 + X_2 + X_3 + \ldots X_n}{n} = \frac{\sum_{i=1}^{n} X_i}{n} \tag{12.1}$$

where X_i = any individual measurement, i; and n = total number of measurements.

The most commonly used measure of dispersion, or variability, of sets of data is the *standard deviation*, *s*. Its defining formula is given by the expression:

$$s = \sqrt{\frac{\sum_{=1}^{n} (X_i - \overline{X})^2}{n-1}} \tag{12.2}$$

(Note: The terms represent the *sample* standard deviation. In statistical circles, the symbol σ represents another, but nearly similar, standard deviation. Unfortunately, σ has often been employed to represent the sample standard deviation (Shaefer and Theodore 2007).

A typical size analysis method of representation employed in the past is provided in Table 12.2. These numbers mean that 40% of the crystals by mass are greater than 5 mm in size, 27% are less than 5 mm but greater than 2.5 mm, 20% are less than 2.5 mm but greater than 1.5 mm, and the remainder (13%) is less than 1.5 mm Another form of representing data is provided in Table 12.3.

Frequency distribution curves are usually plotted on regular coordinate (linear) paper. The curve describes the amount of material (crystals) contained within each size range. A plot of percent mass versus crystal size (d) on a linear scale gives a curve with a peak at the preferential size. Such a curve is shown in Figure 12.2. This figure shows a normal probability distribution that is symmetric about the preferential size. This curve is only occasionally encountered for crystal size distribution; however, this curve may occasionally be approached in some applications. Size data can also be plotted as a cumulative plot. The size for each size range is plotted on the ordinate. The cumulative percent by weight (frequency) is plotted on the abscissa. The cumulative percent by weight can be given as a cumulative percent less than

Table 12.2 Early Crystal Size Distribution (CSD) Representation.

-	> 5.0	mm	40%
< 5.0	> 2.5	mm	27%
< 2.5	> 1.5	mm	20%
< 1.5	-	mm	13%
			100%

Table 12.3 Crystal Size Distribution (CSD) in Arithmetic Increments.

Size range, mm	Percent in size range
0-2	10
2-4	15
4-6	30
6-8	30
8-10	10
> 10	5

stated size (%LTSS) or cumulative percent greater than stated size (%GTSS). The cumulative percent by weight can be plotted on either a linear percentage or a probability percentage scale. The size range (ordinate) is usually a logarithmic scale.

Frequently, the cumulative distribution is plotted on special coordinate paper called log-probability paper. The size of each size range is plotted on a logarithmic ordinate. The percent by weight larger than crystal size d is plotted on the probability scale as the abscissa. If the distribution is lognormal, the distribution curve plots out as a straight line. It should be noted that one can just as easily plot percent mass less than sized (%LTSS) on the abscissa. The geometric mean value of a lognormal distribution can be read directly from a plot similar to that represented in Figure 12.3. The geometric mean size is the 50% size on the plot.

As discussed previously, the geometric standard deviation is a measure of the dispersion or spread of a distribution. The geometric standard deviation is the root mean square deviation about the mean value and can be read directly from a plot such as shown in Figure 12.3. For a lognormal distribution (which plots d maximum versus percent mass larger than d), the geometric standard deviation is given by:

$$\sigma = \sigma_{gm} = \frac{50\%\ size}{84.13\%\ size} \tag{12.3}$$

or

$$\sigma = \sigma_{gm} = \frac{15.87\%\ size}{50\%\ size} \tag{12.4}$$

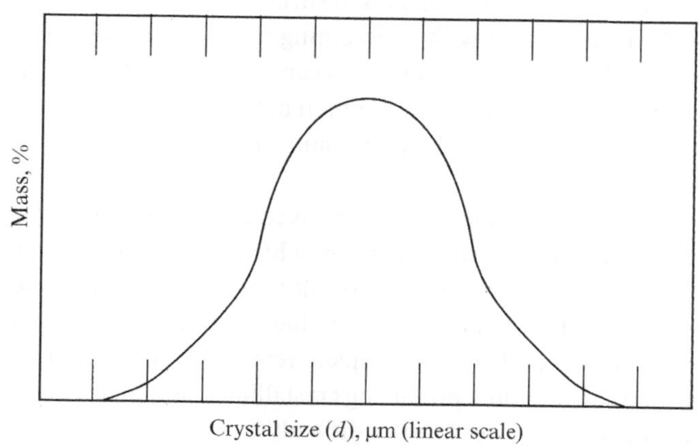

Figure 12.2 Normally Distributed Crystal Size Distribution (CSD).

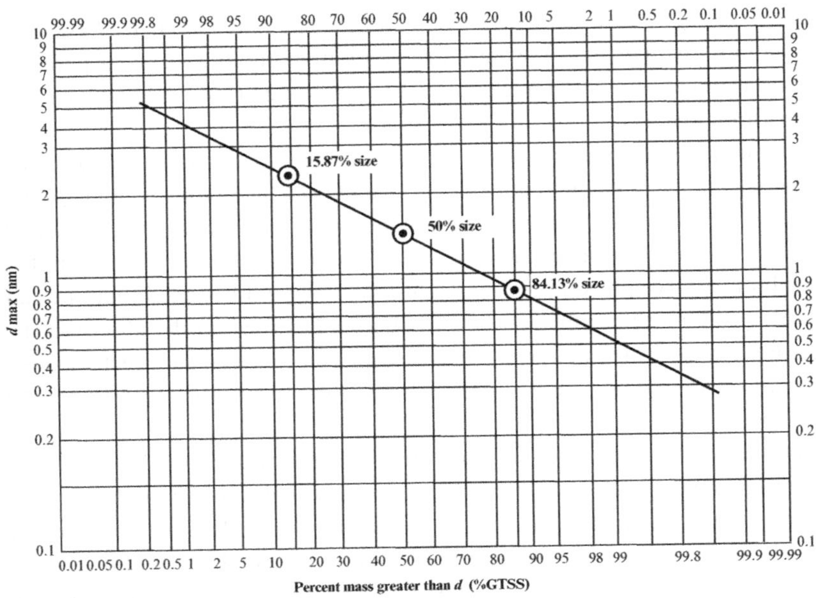

Figure 12.3 Log Normal Crystal Size Distribution (CSD).

Thus, all one must do is determine the 50% size and the 84.13% size from the plot and divide them to determine the geometric standard deviation of the crystal distribution.

It is safe to say that crystals produced via the crystallization process have a distribution of sizes that can vary over large ranges, and the crystal size distribution (CSD) is expressed as a number or mass distribution in terms of the size. Other crystal factors include appearance, purity, the solid–liquid separation process, and other properties involving shape and surface area.

However, the problem associated with maintaining crystal size and crystal size distribution during the crystallization process cannot be overestimated. It is fair to say that both of these problems are common in this approach to water desalination. Additional design considerations and approaches to address them are provided later in Section 12.6.

With reference to the crystallization process, the average size to which crystals are grown is usually dictated by the manner in which they are to be used. As noted earlier, fine crystals have less opportunity for agglomeration and accordingly occlude less mother liquor; however, fine crystals have a greater surface area and offer greater difficulty of complete removal of mother liquor entrained on the crystals. In addition, producing crystalline material of a given size range (a specific upper and lower limit) is difficult to accomplish.

It should be noted that the removal rate of crystals of a given size can greatly alter a crystallizer's ability to continue the production of crystals having that same size and/or size distribution. This effect is particularly pronounced in a separation unit where the crystals of a desired size-distribution range are preferentially removed and the under-sized particles recycled back to the crystallizer. This can shift the distribution to a smaller and narrower size range.

Finally, two of the goals of the design and practicing engineers who work on the CSD problem are to be able to:

1. Understand CSD so as to analyze and modify what a particular operating crystallizer can and cannot do in the way of producing acceptable crystal size/and crystal size ranges, and
2. Include the effects of CSD on the design of new units, i.e. to design an original process that is not yet operating (Randolph 1970)

12.4 Crystallization Equipment

Equipment and equipment selection procedures are usually based on maximum throughput capacities for a new crystallization process. The reason for this is to enable the equipment to perform satisfactorily under the most extreme operating conditions. Material and energy balances based on these conditions are also required before the individual equipment is selected. Most importantly, this procedure is also influenced by the manner in which crystallization occurs. As noted in the Introduction, the methods to accomplish this are crystallization via cooling, via evaporation, cooling and evaporation (adiabatic evaporation), and a salting out process. Details on the equipment that can accomplish these tasks are provided later.

Perhaps the simplest type of equipment for crystallization is a tank in which natural cooling is allowed to lower the temperature of a solution. A hot concentrated solution with the solute is simply poured into the tank where it cools by natural convection. However, there are several drawbacks when employing this practice. This type of crystallizer is inefficient relative to the quantity produced per unit volume or per unit time because of the slow cooling rate. It offers no control over size or size distribution of crystals and it also favors the formation of large crystals that are subject to occlusion of the mother liquor since the crystals tend to grow together. This agglomeration phenomenon can be reduced with agitation. This class of equipment is employed when precipitating small quantities of a material and where the size range of the product is not critical.

Crystallization can also be achieved by cooling a hot, concentrated solution on a surface. The cooling, in turn, results in solute solubility that favors solid-phase precipitation. This type of equipment is available in several designs. The most popular of these in earlier years was the *Swenson-Walker crystallizer*. This unit contains a water-jacketed open trough with a semi-cylindrical bottom. Hot, concentrated solution enters at one end of the trough and the crystals that accumulate on the cooling surface are lifted and pushed along by a spiral stirrer to the other end of the crystallizer. Crystals are removed at the end of the crystallizer by an inclined, spiral flight conveyor that lifts them onto a draining board or a conveyor which carries the crystals to centrifuges or other drying and/or separation operations that may be required. This, as well as other units in this category, are usually employed for systems with a *steep solubility* slope for which a relatively large yield of crystals can be achieved with a modest drop in temperature.

The *vacuum crystallizer* accomplishes cooling via a flashing process. It is a device that cools the solution by the evaporation of a portion of the solvent. The energy for vaporization is obtained from the sensible enthalpy (heat) of the feed. The unit is usually evacuated to a low pressure by steam jet ejectors. The flashing of the feed due to the lower pressure results in both the cooling and concentration of the solution. One of the problems that can arise with this type of crystallization unit is that the operating pressure required to attain the desired crystallization is so low that the evolved vapor cannot be condensed by any cooling water available. It has the advantage, however, of vaporizing some of the solvent to bring about additional yield.

In an *evaporative crystallizer*, crystallization is caused by evaporating the solvent from the feed, which can either be a weak, unsaturated solution or a hot, concentrated mixture. These classes of crystallizers are employed when the solute has a small, low slope temperature-solubility curve. The operation of the evaporator seldom allows much flexibility as to the size, size distribution, or shape of crystals produced. Some classifying action does occur in that fine crystals are usually carried in the circulating mother liquor until they attain a size that will result in their precipitation out of the circulating stream into the solids-removal equipment (Brown 1950). These units are normally steam heated although the heating can also be accomplished by passing hot gases through the solution. If a traditional evaporator is employed, the classic equation for heat transfer may be applied (Reynolds et al. 2004):

$$Q = UA\Delta T \tag{12.5}$$

where Q is the rate of heat transfer, U is the overall heat transfer coefficient (the resistance to heat transfer), A is the available area for heat exchange, and ΔT is the temperature difference driving force across the heat transfer area.

The *salting-out crystallizer* uses a third component that does not chemically react to induce supersaturation, but instead displaces the solubility of the solute in the mixture. The term salting-out is employed because sodium chloride

(NaCl) is usually added as the third component although any substance that depresses the solubility of the original solute can be added (as the third component). In some applications, the process is referred to as an *antisolvent crystallization*. Examples include alcohols and ketones.

Reactor crystallizers are another category of units but are rarely employed. In such units, a mass transfer step (e.g. gas absorption), occurs in the mother liquor. This results in supersaturation and provides additional mass transfer from the liquid to the solid phase in the form of crystallization (Randolph 1970).

Auxiliary equipment for a crystallization process can include slurry pumps, vacuum pumps, mixers, condensers, steam traps, and entrainment separators. These latter units are not different from than those described earlier for other mass transfer operations (Theodore 2008).

12.5 Describing Equations

Overall and componential material balance calculations in the chemical process industry include transient effects that can account for process upsets, start-ups, shutdowns, and so on. The describing equations for these time-varying (unsteady-state) systems are differential. The equations usually take the form of a first-order derivative with respect to time, where time is the independent variable. However, calculations for most crystallization processes assume steady-state conditions with the ultimate or final design based on the worst-case or maximum flow conditions. This greatly simplifies the calculations since the describing equations are no longer differential, but are algebraic.

The development of mass (and energy to follow) balances follow traditional procedures. Consider, as with other mass transfer operations, material and energy balances and any heat transfer equations that can be written for the various crystallization processes. The material balances can provide the process yield, i.e. the mass of crystals formed from a given mass of solution. The effect of evaporation or of cooling must be included if applicable.

As an example, the evaporation of solvent from a crystallizer is pictured in Figure 12.4. A solute componential and total material balance (on a mole basis) may be written:

$$F = V + L + P;\ \text{total} \tag{12.6}$$

$$x_F F = V(0) + x_L L + x_C P;\ \text{componential} \tag{12.7}$$

where F is the feed, V is the quantity of evaporation, L is liquid (filtrate) withdrawal, P is the crystal production, and x represents the solute content of a

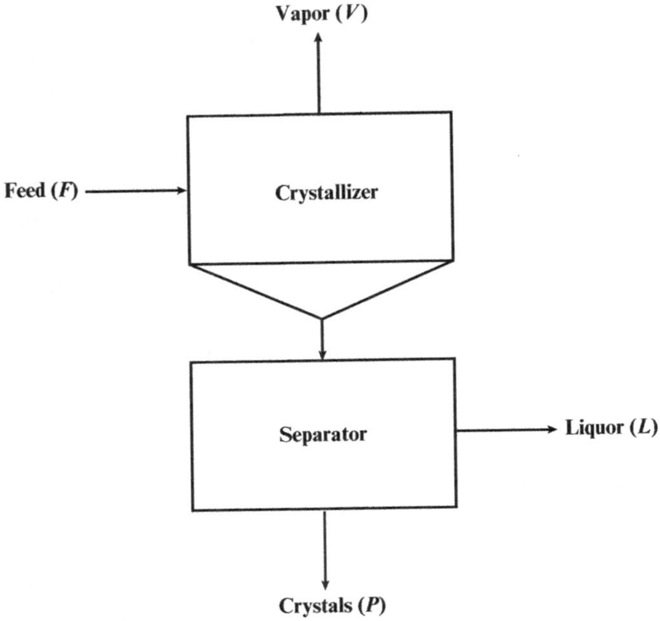

Figure 12.4 Line Diagram of an Evaporative Crystallizer.

stream in units consistent with the flow rate. There is also an equilibrium expression relating x_L and x_C.

Once the material balance is completed, one may then proceed directly to the energy calculations, some of which can play a significant role in crystallizer design. As with material balance calculations, energy calculations for crystallizers almost always are based on steady-state conditions. One of the principal jobs of a practicing engineer involved with crystallizers is to account for the energy that flows into and out of the unit and to determine overall energy requirements. This is accomplished by performing energy balances on the unit.

As noted earlier, crystallization operations may be carried out either by cooling a solution or by evaporative concentration, or by both. The cooling of a solution may be accomplished by the removal (transfer) of heat of both the sensible heat and the heat evolved during crystallization of the product by (to) cooling water or air. If the solution is crystallized by evaporation, the required heat may be supplied by the sensible heat of solution as in vacuum crystallizers, or it may be supplied from an external source. In the former case, the heat of vaporization of the solvent from the solution may be safely assumed equal to the heat of vaporization of the pure solvent. In the latter case, the heat of solution (data for which is rarely available) may be assumed equal to the heat of solution at infinite dilution (Theodore and Rici 2010). Obviously, for crystallization problems requiring energy balances, e.g. where evaporation occurs or when the temperature of an adiabatic crystallizer is unknown, enthalpy data must be provided or made available.

Referring once again to Figure 12.4, one may also write an energy balance:

$$F\widehat{H}_F + \dot{Q} = V\widehat{H}_V + L\widehat{H}_L + P\widehat{H}_C \tag{12.8}$$

where F, V, L, P refer to the feed, vapor, liquid and product phase mass flow rates, respectively; $\dot{Q}$ is the rate of heat flow; and $\widehat{H}_F$, $\widehat{H}_V$, $\widehat{H}_L$, $\widehat{H}_C$ are the changes in enthalpy for the feed, vapor, liquid and crystal phases, respectively. The enthalpies in this equation can be determined, provided the temperatures of the product streams are known. In addition, the operation is usually either adiabatic or may be safely assumed to be adiabatic, so that $\dot{Q} = 0$.

Several excellent examples of calculating crystallizer energy balances are available in the literature (Chopey 1994).

12.6 Design Considerations

As with any mass transfer device, there are usually five conceptual steps to be considered with the design of equipment. These are:

1. Identification of the parameters that must be specified
2. Application of fundamentals underlying theoretical equations or concepts
3. Enumeration, explanation, and application of simplifying assumptions
4. Possible use of correction factors for "non-ideal" behavior
5. Identification of other factors that must be considered for adequate equipment specification

Since design calculations are generally based on the maximum throughput capacity for the proposed crystallizer or process, these calculations are never completely accurate. It is usually necessary to apply reasonable safety factors when setting the final design for the crystallizer. Safety factors vary widely, particularly with crystallizers, and are a strong function of the accuracy of the data involved, calculational procedures, and past experience. Nonetheless, simple qualitative design procedures are provided shortly.

Assuming a crystallizer is the mass transfer operation of choice, the class of crystallizer is normally selected first. For high feed rate (e.g. greater than 50,000 lb/hr), a continuous, as opposed to a batch process is normally selected. Overall process decisions are based on capital, operating and maintenance costs, space, location, production rate, crystal characteristics, size and size distribution of crystals, physical and chemical characteristics of the feed liquor and slurry, corrosion issues, environmental concerns, etc.

Regarding a new design, scale-up problems can be expected when the design is based on pilot-plant or laboratory units. Even after the equipment is fully constructed, significant startup problems are the norm. Finally, it should be noted that the scaling-up of crystallization equipment is probably more difficult than that for any of the other mass transfer operations.

If a pilot or small-scale crystallizer produces a satisfactory product, then the design of the larger crystallizer must simulate a number of different conditions obtained in the smaller unit. The four most important conditions are:

1. Identical flow characteristics of liquids and solid particles
2. Identical degrees of supersaturation in all equivalent regions of the crystallizer
3. Identical initial seed sizes (if applicable) and magma densities
4. Identical contact times between growing crystals and supersaturated liquor

A step-by-step procedure for the design of a single solute forced circulation crystallization unit can be used as follows::

1. Production rate of crystals, P
2. Feed (liquid) mass fraction
3. Desired CSD and appropriate residence time to achieve it
4. Feed temperature
5. Design temperature and corresponding mother liquid saturation mass fractions
6. Centrifuge wash water ratio
7. Crystal bulk density, ρ_B
8. Magma density, ρ_M
9. Liquor density, ρ_L

Specifications (5) through (9) are generally drawn from experience. Marnell (2009) also provides a detailed design example for a crystallizer treating a urea-water feed stream.

As discussed earlier, the design of crystallization equipment can be difficult. Knowledge of phase equilibrium and physical properties does not allow one to accurately predict the behavior of the actual process. Bench-scale and pilot-plant experiments on the actual stream are usually recommended and some equipment vendors require a pilot-plant test before designing the equipment. Even with such testing, operational adjustments must usually be made on the actual commercial installation before the equipment will operate to yield an acceptable product.

12.7 Illustrative Examples

Seven illustrative examples complete the presentation of crystallization technology highlighted in this chapter.

12.7.1 Illustrative Example 1

Define the terms *occlusion* and *magma*.

Solution. Occlusion is a word that describes a process that "prevents the passage of." It is employed with reference to the mother liquor retained within crystals. The word *magma* is simply "a suspension of precipitated matter in a water substance."

12.7.2 Illustrative Example 2

Discuss the crystallization separation process relative to the presence (or absence) of a eutectic (a mixture of two or more components which usually do not interact to form a new chemical compound but, which at certain ratios, inhibit the crystallization process of one another resulting in a system having a lower melting point than either of the components).

Solution. In cases where a complete solid solution occurs with no eutectic, it is possible to separate a binary mixture into its components. However, such systems are the rare exception; eutectic formation is the general rule. In most cases, the degree of separation attainable is therefore limited by the eutectic composition. To accomplish a complete separation, an additional step is required.

12.7.3 Illustrative Example 3

The size of a crystal is specified as Tyler 8–14 screen mesh. Determine its size.

Solution. To calculate the size of the crystal in the problem statement, refer to the Tyler screen information in Table 12.1. An 8-mesh opening = 2,362 μm, while a 14-mesh opening = 1,168 μm. Since an 8 by 14 mesh indicates that an object will pass through an 8-mesh, but not pass (be captured) through a 14-mesh screen, one size cannot be specified for the crystal. The crystal is in the size range 1,168–2,362 μm and the average or "mean" arithmetic size is 1,765 μm.

12.7.4 Illustrative Example 4

A tiny spherical crystal has a diameter of 100 nanometers (nm). Calculate the volume (cm^3) and surface area (cm^2) of the crystal.

Solution. The volume (V) of the crystal is:

$$V = \frac{\pi D^3}{6} = 0.524(100)^3 = 0.524\,x10^6 \text{nm}^3 \tag{12.9}$$

Since there are 10^7 nm/cm the volume becomes

$$V = 0.524\,x10^6\left(10^{-7}\right)^3 = 0.524\,x10^{-13}\ \text{cm}^3$$

The surface area of the crystal, A, is given by:

$$A = \pi D^2 = 3.14(100)^2 = 3.14\,x10^4 \text{nm}^2 \tag{12.10}$$

Converting to units of cm^2 yields:

$$A = 3.14\,x10^4\left(10^{-7}\right)^2 = 3.14\,x10^{-10}\ \text{cm}^2$$

12.7.5 Illustrative Example 5

Compare a 10 μm spherical crystal to one with a diameter of 100 mm Include the ratio of volumes and surface area in the calculation.

Solution. The ratio of diameters (RD) is:

$$RD = \frac{100\ mm}{10\mu m} = \frac{(100)(1000)}{10} = 10{,}000$$

Since the volume and surface area are a function of the $(\text{diameter})^3$ and $(\text{diameter})^2$, respectively, the volume ratio (VR) and surface area ratio (SAR) are:

$$VR = (10{,}000)^3 = 10^{12}$$

$$SAR = (10{,}000)^2 = 10^8$$

Table 12.4 Crystal Size-separation Efficiency Data for Illustrative Example 6.

d, mm	Efficiency, %	Number of Crystals
1.0	0	10,000
100	100	1

12.7.6 Illustrative Example 6

The crystal size separation efficiency information for a crystallizer desalination unit is provided in Table 12.4. Calculate the overall separation efficiency for the filter on a: (a) number basis, (b) mass basis, (c) volume basis.

Solution. (a) on a number basis, the efficiency, E_N is:

$$E_N = \frac{(10{,}000)(0)+(1)(100)}{(10{,}000+1)} = \frac{100}{10{,}001} = 0.01\%$$

(b) on a mass basis, the efficiency, E_M, is calculated knowing that the mass of a crystal is proportional to the volume of the crystal, or to its diameter cubed, and assuming the crystal density is the same regardless of particle size. This results in the following:

$$E_M = \frac{(10{,}000)(1.0)^3(0)+(1)(100)^3(100)}{(10{,}000)(1.0)^3+(1)(100)^3} = \frac{10^8}{10^6+10^4} = 99\%$$

(c) on a volume basis, the efficiency, E_V, is the same as the E_M value since the mass of a crystal is directly proportional to its volume. Therefore, $E_V = 99\%$.

12.7.7 Illustrative Example 7

As noted earlier in the chapter, crystallization is employed heavily as a separation process in the inorganic chemistry industry, particularly where salts are recovered from aqueous media. In the production of organic chemicals, crystallization is also used to recover product, to refine intermediate chemicals, to remove undesired salts, and to recover "pure" water. The feed to a crystallization system consists of a solution from which solute is crystallized through one or more of a variety of processes. The solids are normally separated from the crystallizer liquid, washed, and discharged to downstream equipment for further processing.

Potassium nitrate is obtained from an aqueous solution of 20% KNO_3, During the process, the aqueous solution of potassium nitrate is evaporated, leaving an outlet stream with a concentration of 50% KNO_3, which then enters a crystallization unit where the outlet product is 96% KNO_3 (anhydrous crystals) and 4% water. A residual aqueous solution that contains 0.55 g of KNO_3 per gram of water also leaves the crystallization unit and is mixed with the fresh solution of KNO_3 at the evaporator inlet. Such recycling provides a means of preventing pollution by minimizing the loss of valuable raw materials. The flow diagram in Figure 12.5 depicts the process.

What is the mass flowrate, *M*, in kilograms/hr of the recycled material, R, when the feedstock flowrate is 5,000 kg/hr?

Solution. Using Figure 12.5, an overall material balance on KNO_3 leads to:

$$0.96C = 0.2(5{,}000)$$

This equation may be solved for the rate of product leaving the crystallizer:

$$\text{C} = 1{,}000/0.96 = 1{,}042\,\text{kg/hr}$$

An overall mass balance around the crystallizer may also be written:

$$\text{M} = \text{R} + \text{C} = \text{R} + 1{,}042$$

Since the mass concentration of the KNO_3 in R is (0.55/1.55), the equation for the KNO_3 mass balance around the crystallizer is:

$$0.5M = 0.96(1{,}042) + (0.55/1.55)R = 1{,}000 + 0.355R$$

These two equations may be solved simultaneously for the amount of KNO_3 leaving the evaporator in terms of the amount of recycle. Solving the second equation for M yields:

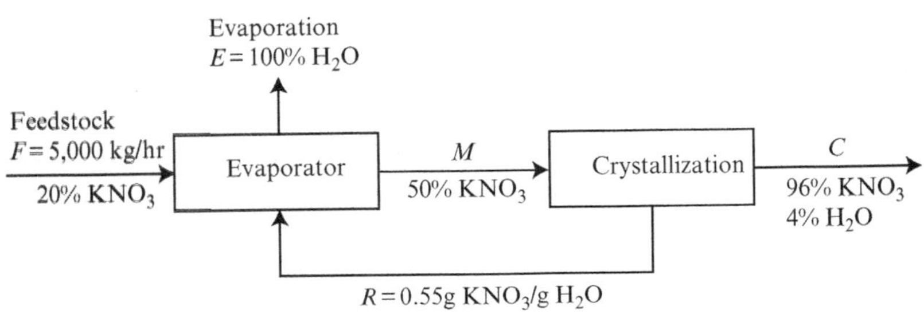

Figure 12.5 Example Crystallizer Process Flow Diagram for Illustrative Example 7.

$$M = 2{,}000 + 0.71R$$

Therefore,

$$2{,}000 + 0.71R = R + 1{,}042$$

and the new mass flow rate of the recycle stream is

$$R = 3{,}301\,\text{kg/hr}$$

The reader is left with the exercise of calculating the rate of "pure" water, *E*, production.

12.8 Problems

1. The following crystal sizes (in mm) in a sample were recorded from a crystallizer: 22, 10, 8, 15, 13, and 18. Find the median, the arithmetic mean, the geometric mean, and the standard deviation of these crystal sizes.
2. Calculate the overall efficiency of a crystallization separation unit using the data provided in Table 12.5.
3. Using the process diagram in Figure 12.6 that will produce 2,000 lb/hr of $NaNO_3$ crystals from a 90% (by mass) $NaNO_3$-NaCl mixture, write pertinent material balance equations around the cooler and crystallizer.
4. Write pertinent material balance equations around the heater and evaporator for the process provided with system data in Figure 12.6.
5. Combining the results of Problems 3 and 4, obtain the flow rates for each of the eight streams noted in Figure 12.6.
6. Complete a material balance for the process pictured in Figure 12.7. The following information is provided:
   Production rate: 3,125 kg/hr
   Feed composition: 75% urea, 25% water
   Magma density: 1,170 kg/m^3
   Magma crystal concentration: 450 kg crystals/m^3 magma
   Liquor density: 1,340 kg/m^3
   Magma recirculation rate: 567,900 kg/hr
   Centrifuge wash water ratio: 0.12 kg H_2O/kg urea

Also, size the crystallizer, Design information for this unit is provided here:
   Average residence time, τ (based on *M*): 3.38 h
   Maximum (with safety factor) vapor velocity,

$$v = C_f\left(\frac{\rho_l - \rho_v}{\rho_v}\right)^{0.5};C_f = 0.0122\,\text{m/s} \tag{12.11}$$

Table 12.5 Crystal Size Distribution and Separation Information.

Crystal size range, mm	% Crystals in size range	% Less than specific size	% Greater than specific size	Mean diameter, mm	Removal efficiency, E_i, %
0–5	2	2	98	2.5	0.6
5–10	2	4	96	7.5	7.7
10–15	1	5	95	12.5	19.2
15–20	4	9	91	17.5	38.3
20–30	6	15	85	25.0	56.2
30–50	11	26	74	40.0	79.1
50–100	13	39	61	75.0	98.9
100 +	61	-	-	100 +	100

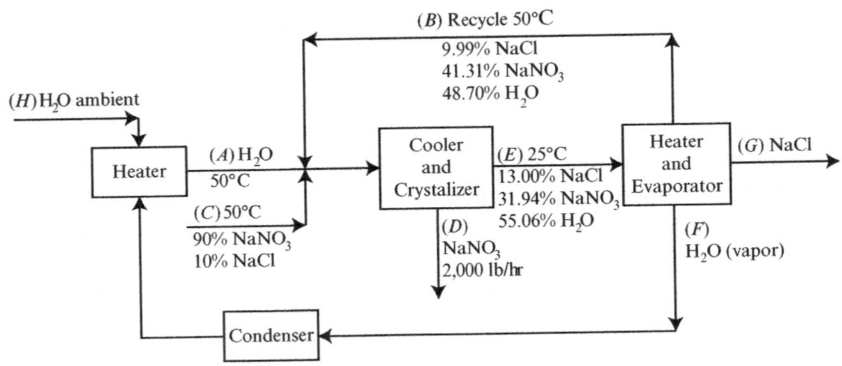

Figure 12.6 Crystallizer Process Flow Diagram with System Data.

Crystallizer diameter, D, is based on the maximum vapor velocity, v
Crystallizer magma cylinder height, $H = 1.15$ day
Freeboard height (for droplet disengagement): 2 m

7. Quick-size a magna heat exchanger if $Q = 3{,}400{,}000$ kJ/hr and $\Delta T_{lm} = 45$°C. Assume U = 200 Btu/hr-ft^2-°F. Use the standard heat transfer equation, Equation 12.5.

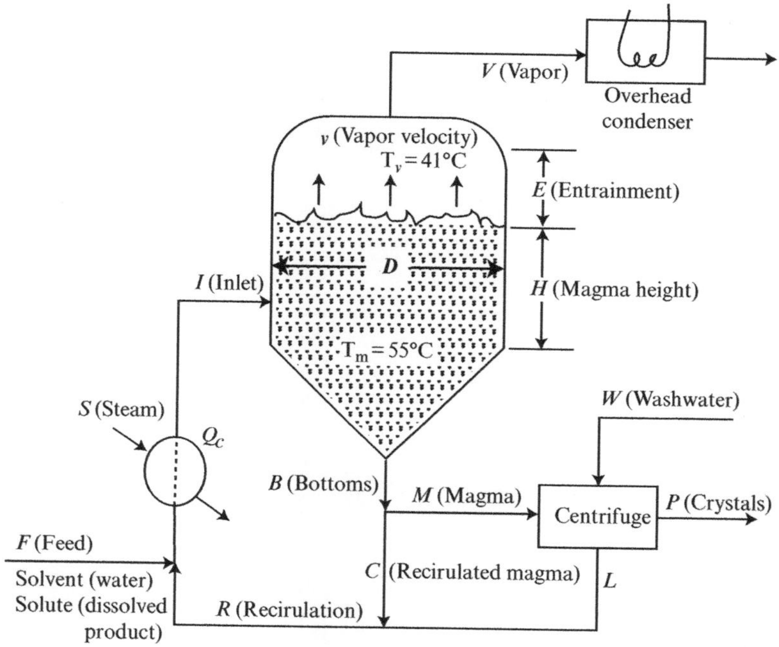

Figure 12.7 Forced Circulation Crystallizer Flow Diagram with System Data for Problem 6.

References

Brown, G. and Associates. (1950). *Unit Operations*. Hoboken, NJ: John Wiley & Sons.

Chopey, N. (1994). *Handbook of Chemical Engineering Calculations*, 2nd Edition. New York, NY: McGraw-Hill.

Randolph, A. (1970). Crystallization. *Chemical Engineering* May 4.

Reynolds, J., Jerris, J., and Theodore, L. (2004). *Handbook of Chemical and Environmental Engineering Calculations*. Hoboken, NJ: John Wiley & Sons.

Shaefer, S. and Theodore, L. (2007). *Probability and Statistics, Applications for Environmental Science*. Boca Raton, FL: CRC Press/Taylor & Francis Group.

Theodore, L. (2008). *Air Pollution Control Equipment Calculations*. Hoboken, NJ: John Wiley & Sons.

Theodore, L. and Rici, F. (2010). *Mass Transfer Operations for the Practicing Engineer*. Hoboken, NJ: John Wiley & Sons.

13

Traditional Desalination Processes

13.1 Introduction

With the world facing a freshwater shortage from which the United States will not be spared, one method for producing potable water that has been around for centuries is *desalination*. The last three chapters provided details on the role evaporation, membranes, and crystallizers can play in the desalination process. This chapter introduces the reader to a number of desalination processes using these basic principles for the separation of potable water from seawater or saline brine mixtures. Details of each of four processes are included in the material to follow, along with a discussion of hybrid systems and other non-commercial desalination technologies. An application section provides seven illustrative examples related to traditional water desalination options.

13.2 The GADUTH Solar Evaporation Systems

Two solar evaporator systems are described in this section. The first provides for energy recovery from a conventional solar system, while the second is based on a greenhouse solar system that includes a unique energy recovery component.

13.2.1 The GADUTH Solar Evaporation Process

The GADUTH Solar Evaporation Process is shown in Figure 13.1. In this process which provides desalination through evaporation, preheated seawater (1) is fed to an evaporator (2) which is heated by solar energy (3). The resulting steam produced (4) is fed to a condenser (5) where the overhead vapors (4) are condensed using heated seawater (6) from the heat exchanger to produce pure water (7). The heated seawater (8) from the evaporator is then directed to a heat exchange where the concentrated brine (9) releases some of its thermal energy to the cool seawater feed stream (10) and is discharged from the process. The preheated seawater feed from the heat exchanger gains additional heat when it is directed to the condenser

Introduction to Desalination: Principles, Processes, and Calculations, First Edition. Louis Theodore and R. Ryan Dupont.

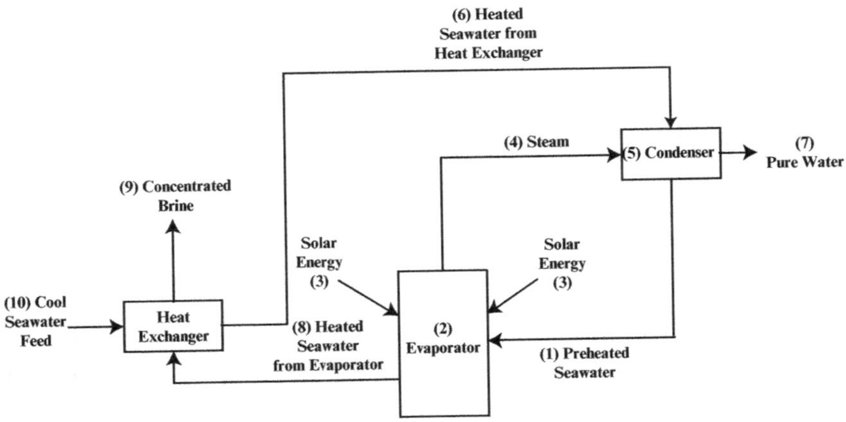

Figure 13.1 The GADUTH Solar Evaporation Process.

(5) where the overhead vapors from the evaporator are condensed. This warm preheated seawater (1) is then fed to the evaporator (2), completing the cycle.

An enthalpy balance (Flynn et al. 2018; Theodore 2016; Theodore et al. 2017) around the process indicates that the energy provided by the sun is transferred to the concentrated brine solution and pure water discharge streams. In effect,

$$H_{solar} + H_{feed} = H_{conc.\ brine} + H_{pure\ water} \tag{13.1}$$

or

$$H_3 + H_{10} = H_9 + H_7 \tag{13.2}$$

Additional energy conservation measures could be included in the process subject to economic considerations.

13.2.2 The GADUTH Greenhouse Solar Evaporator (Ganiaris et al. 2018)

In this process (see Figure 13.2a), solar radiation passes through the glass in a greenhouse (2) and heats seawater in a large pan (3) to a temperature that exceeds that of the surrounding air. The steam vapor from the heated seawater (4) condenses on the cooler inside of the glass (5) and flows by gravity into fresh water collectors (6). Concentrated brine is continuously removed from the pan (7). The glass roof of the house is constructed in a manner that allows the fresh water to condense on the inside and seawater feed (8) to flow downward outside (or inside) the glass at spaced intervals. In this manner, a certain percentage of the glass receives unobstructed sunrays while the feed water is heated in the remaining glass during its approach to the basin pan in the greenhouse. A side view of the greenhouse design is provided in Figure 13.2b.

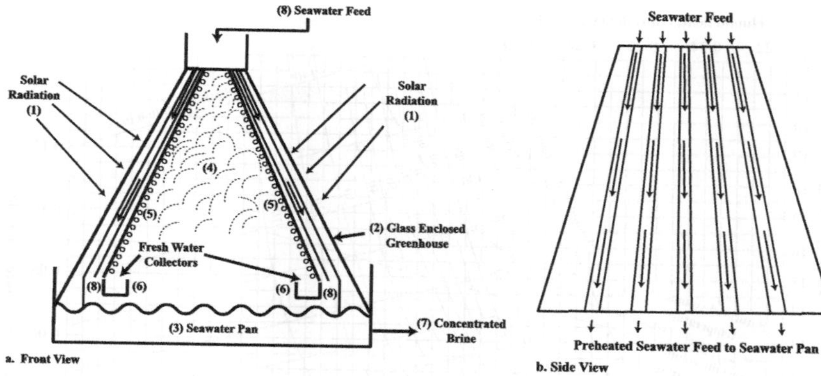

Figure 13.2 The GADUTH Greenhouse Solar Evaporator.

13.3 The Theodore Humidification/Dehumidification Processes

As noted in Chapter 5, in many unit operations it is necessary to perform calculations involving the properties of mixtures of air and water vapor. Such calculations often require knowledge of:

1. The amount of water vapor carried by air under various conditions
2. The thermal properties of such mixtures
3. The changes in enthalpy content and moisture content as air containing some moisture is brought into contact with water or wet solids, and other similar processes

This section reviews the properties of mixtures of air and water vapor, the mechanisms of humidification (and drying) processes, and the role humidification/dehumidification plays in two desalination processes.

Figures 13.3 and 13.4 contain the psychrometric charts from thermodynamic properties drawn from the literature. A General Electric chart is also provided in Figure 13.5. The following are some helpful points on the use of these psychrometric charts:

1. Heating or cooling at temperatures above the dew point (temperature at which the vapor begins to condense) corresponds to a horizontal movement on the chart. As long as no condensation occurs, the absolute humidity remains constant.
2. If the air is cooled, the system follows the appropriate horizontal line to the left until it reaches the saturation curve and follows it thereafter.

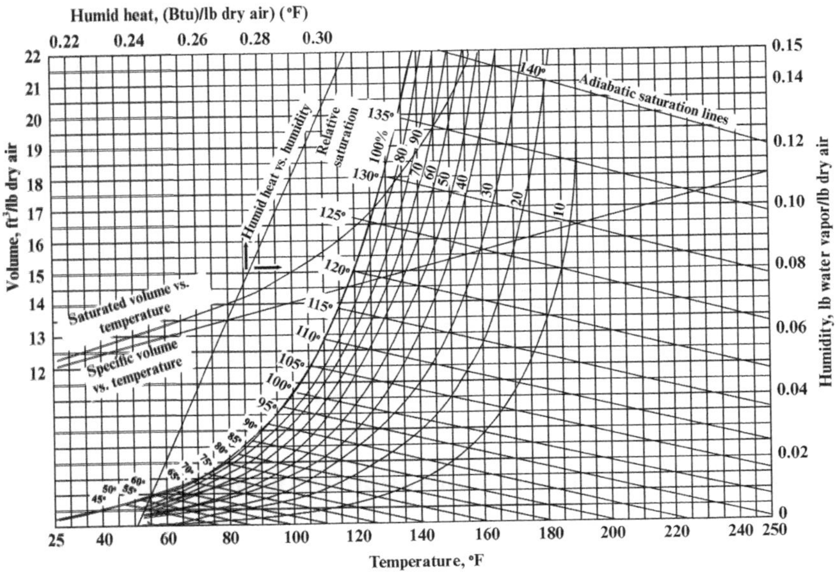

Figure 13.3 Psychrometric Chart, Low Temperatures. Barometric Pressure of 1 atm.

3. In problems involving use of the humidity chart, it is convenient to choose the mass of air on a dry basis since the chart uses this basis.

Guidelines for the use of psychrometric charts provided by in the General Electric psychrometric chart in Figure 13.5 include:

1. Sensible heating and cooling of air are represented on the psychrometric chart by a straight horizontal line between the dry-bulb temperature limits of the process. These processes are distinguished by a change in dry-bulb temperature, relative humidity, wet-bulb temperature, total heat, specific volume, and by no change in moisture conlent, dew-point temperature, and vapor pressure of the air (Figure 13.5a).
2. Cooling and dehumidifying of air are represented on the psychrometric chart by a straight line drawn between the initial condition of the air and the point on the 100 percent line corresponding to the temperature of the cooling surface. This applies only when the surface temperature is below the initial dew point. The final condition of the air will depend on the total heat extracted from the air. This process is distinguished by a change in all properties of the air (Figure 13.5b).
3. Humidifying of air, with no temperature changes, is represented by a straight vertical line along the dry-bulb temperature line of the air between the moisture content limits of the process. This process is distinguished by an increase in relative humidity, wet-bulb temperature, total heat, specific

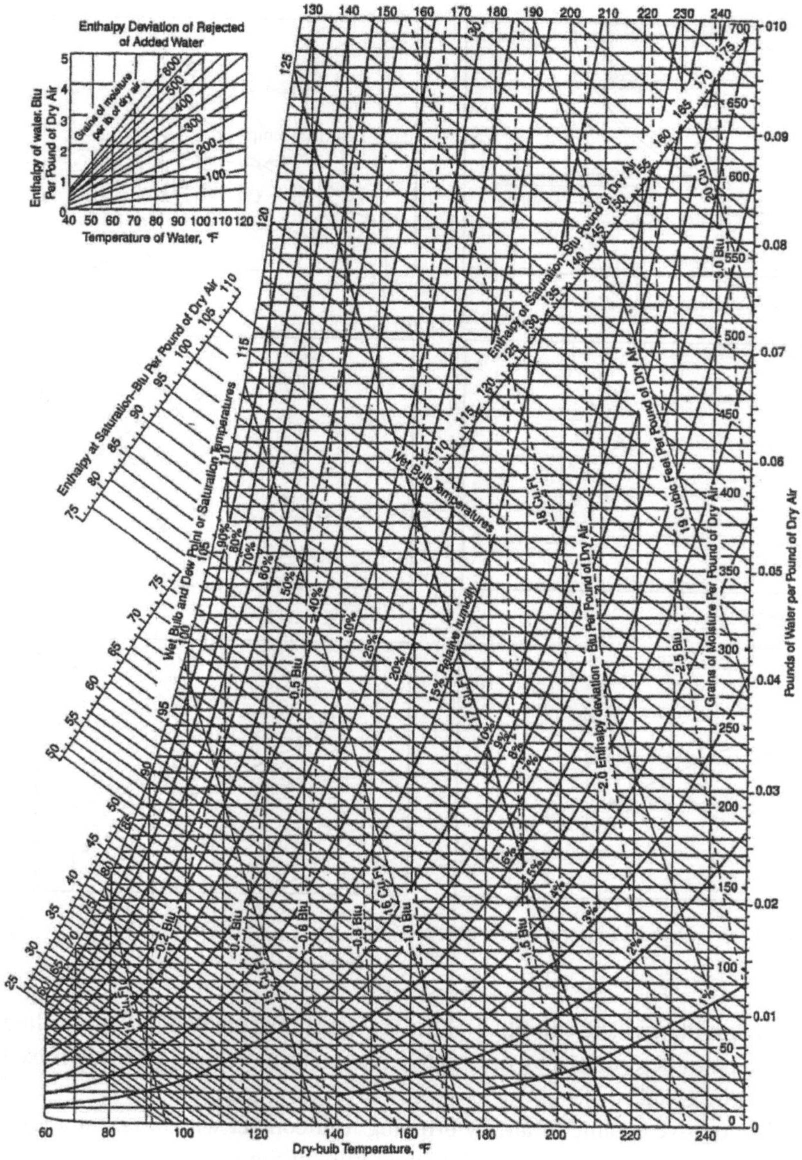

Figure 13.4 Psychrometric Chart, High Temperatures. Barometric Pressure of 1 atm.

volume, moisture content, dew-point temperature and vapor pressure of the air (Figure 13.5c).

4. Mixing of air at one condition with air at some other condition is represented by a straight line drawn between the points representing the two

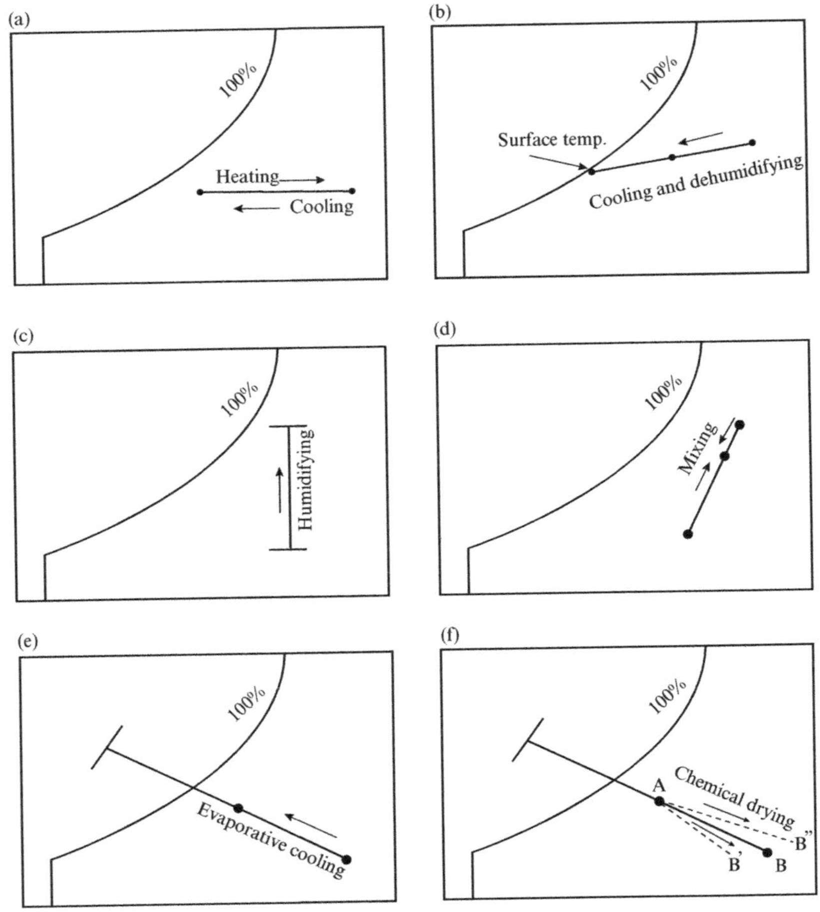

Figure 13.5 General Electric Psychrometric Chart.

air conditions. The condition of the resultant mixture will fall on this line at a point determined by the relative mases (lever rule) of air being mixed (Figure 13.5d).

5. Evaporative cooling of air, by bringing it in contact with water at a temperature equal to the wet-bulb temperature of the air, is represented by a straight line drawn along the wet-bulb temperature line of the air between the limits of the process. In this process, the total heat of the air remains unchanged because the sensible heat extracted from the air is returned as latent heat by an increase of moisture content. This process is distinguished by a change in dry-bulb temperature, relative humidity, specific volume,

moisture content, dew-point temperature, vapor pressure, and by no change in wet-bulb temperature (Figure 13.5e).

6. Chemical drying of air is represented by a straight line along the wet-bulb temperature between the limits of the process (AB) only in case the drying is purely by adsorption (the drying agent does not dissolve in the water extracted from the air) and only in case the drying agent does not retain an appreciable amount of the heat of vaporization liberated when the water is condensed on the surface of the adsorbent. In case an appreciable amount of this heat is retained by the adsorbent, the process takes place on a line below the wet-bulb temperature (AB'). If the drying agent is soluble in water (such as calcium chloride) the drying process is above (AB") or below (AB') the wet-bulb temperature, depending on whether heat is liberated or absorbed when the agent is dissolved in water (Figure 13.5f).

The evaporation of water into air for the purpose of increasing the air humidity is known as *humidification*. Closely allied to this is the evaporation of water into air for the purpose of cooling the water. *Dehumidification* consists of condensing water from air to decrease the air humidity. All of these processes are of considerable industrial importance and involve the contacting of air and water accompanied by heat transfer.

The cooling of water in a cooling tower is accomplished by bringing the water into contact with unsaturated air under such conditions that the air is humidified and the water is brought approximately to the wet-bulb temperature (Theodore and Rici 2010). When air at the wet-bulb temperature enters the tower it receives sensible heat from the hot water and its temperature is raised so that it no longer is saturated. Water then evaporates continuously into the air as it travels upward in the tower. Generally, the temperature of the water sent to a cooling tower rarely exceeds 120°F and is usually less. The lowest possible temperature to which water can be cooled in a cooling tower is the wet-bulb temperature of the air. The difference between the outlet water temperature from the tower and the wet-bulb temperature is defined as the *approach temperature*. When water is present in air, it is possible to extract the water by cooling the air–water mixture below the mixture's dew point temperature. If the temperature is reduced further, water is forced out of the air–water mixture as condensation (dew) is formed, i.e. dehumidification. The emphasis in recent years has been to employ *mechanical-draft* equipment that include spray columns and spray chambers, as well as open surface spray ponds and cooling ponds. Other methods of dehumidification also include adsorption and absorption processes (Theodore and Rici 2010).

Two solar evaporator systems relying on humidification/dehumidification principles are described here.

13.3.1 The Theodore HUMidification Process (THUMP)

THUMP (Theodore and Flynn 2018) involves three steps: heating the brackish/sea water by any of a host of means, contacting the heated water with air by any of a host of means that results in a hotter humidified air, and condensing the saturated air by any of a host of means that separates the resulting condensed water from a less saturated air. The user also has the option of recycling the hot discharge air from the cooler. Figure 13.6 shows a schematic of THUMP.

13.3.2 The GADUTH Dewdrop Process

The GADUTH Dewdrop Process (Figure 13.7) uses a humidification followed by dehumidification approach where relatively dry air (1) is employed as a carrier gas to evaporate water from a feed of seawater (2). This evaporation process decreases the temperature of both the air and feed while saturating the air (3) with water from the feed. This is the humidification part of the process. The saturated air is then cooled in a condenser (4), a dehumidification process, where water in the carrier air condenses as dew to produce pure water (5) via recycled cooled concentrated brine solution (6), that also yields dry air (7) and warm brine (8) discharge streams. Various forms of energy recovery can be included in the GADUTH Dewdrop process, subject to economic considerations.

13.4 Hybrid Systems

This section is concerned with hybrid systems that involve: (1) two or more different devices or processes connected in series, or (2) a process that utilizes two or more separation mechanisms to enhance pure water capture. The

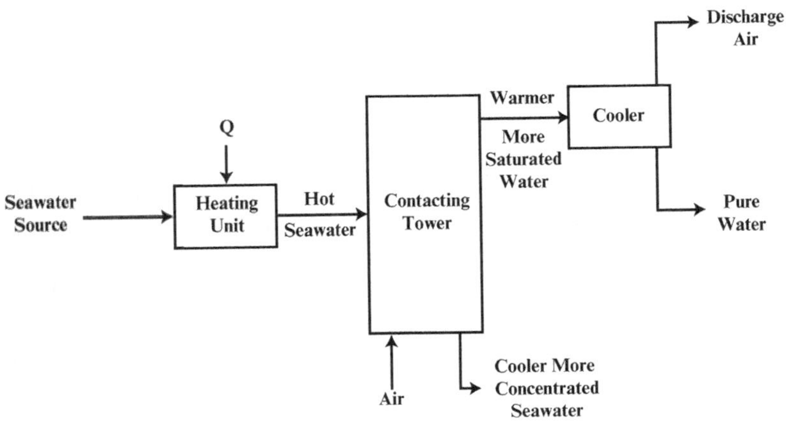

Figure 13.6 The Theodore Humidification Process (THUMP).

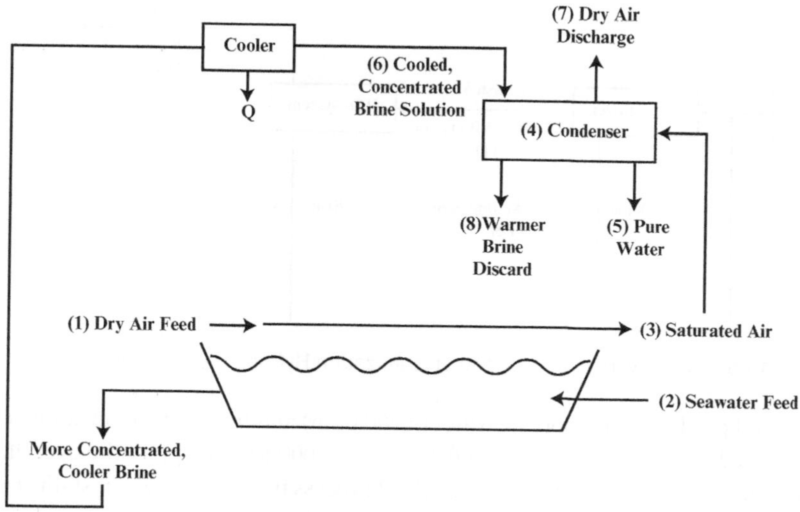

Figure 13.7 The GADUTH Dewdrop Process.

hybrid systems involving two or more processes in series are usually treated from an engineering and calculational standpoint, sequentially, with the outlet of one process becoming the inlet to the next process in series. The second type of hybrid system utilizing two or more different separation mechanisms is highlighted in the example desalination systems discussed shortly.

Reverse Osmosis (RO) has been commercialized successfully for the desalination of both sea water (35,000 ppm) and brackish water. However, the process has certain limitations:

1. The cost of equipment and operation is substantially higher for sea water than brackish water.
2. RO is not adaptable to very large plants since the filtration area (membrane) is proportional to capacity and the pumps required to deliver high pressure (1,000 psi) are limited in size; thus, there are no economies of scale.
3. An RO system becomes extremely costly for high salinity (i.e. Red Sea≈ 40,000 to 45,000 ppm) feed or other wastewaters from mining operations, whereas the cost of freeze concentration (crystallization) does not increase proportionately with the higher salinity feedwater.

The freeze desalination process also has certain limitations:

1. The ice crystals must be separated from the brine and washed in order to produce potable water (<500 ppm).
2. These unit operations ("wash column" or centrifuge) are costly and difficult to scale up.

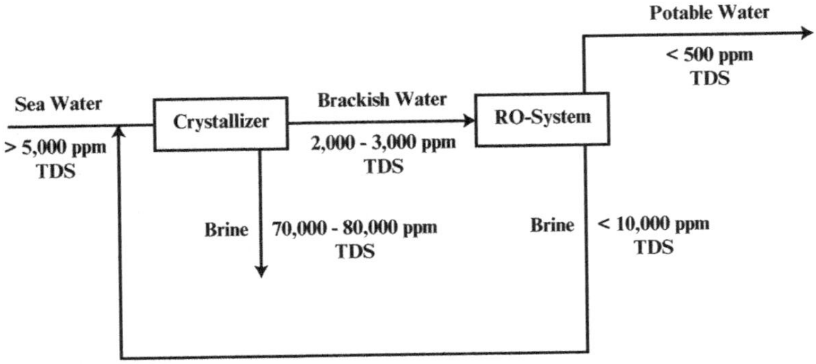

Figure 13.8 The Reverse Osmosis and Crystallization Hybrid I (ROACH I) Process.

If the end product is not potable water (<500 ppm) and freeze crystallization is used to reduce the salinity from 35,000 ppm to 2,000 to 3,000 ppm, the overall cost would be substantially lower. A hybrid process utilizing freeze crystallization for pre-treatment of the feed (sea water) followed by a reverse osmosis system (ROACH I, Figure 13.8) could overcome the limitations of both processes. If the feed is brackish water (<5,000 ppm) the process can be reversed (ROACH II, Figure 13.9) with freeze crystallization being used for post-treatment of RO system effluent. Both ROACH flow diagrams contain some data and information from initial studies conducted by GADUTH.

A combination of thermal desalination and membrane desalination processes is also another possible configuration for hybrid desalination systems. Two such thermal/membrane processes in current operation include the following:

1. Multistage Flash with Reverse Osmosis (MSF-RO). The process draws waste steam from a power plant to preheat sea water prior to the MSF process. The RO plant uses electricity generated by the power plant, operating

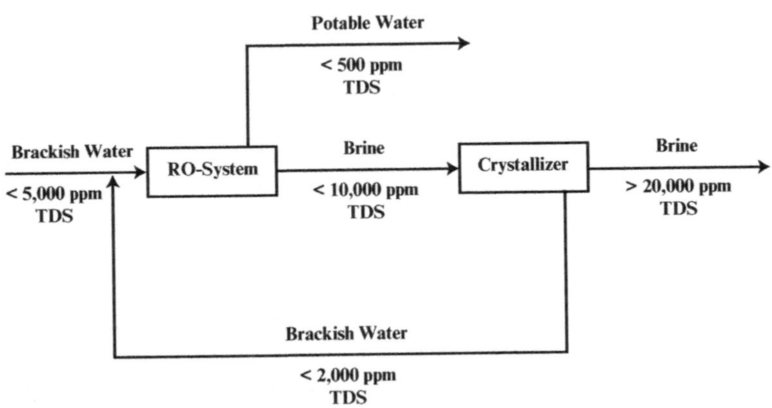

Figure 13.9 The Reverse Osmosis and Crystallization Hybrid II (ROACH II) Process.

during periods of reduced demand to optimize the overall efficiency of the hybrid process. One of the largest MSF-RO hybrid sea water desalination plants is located in Ras Al-Khair, Saudi Arabia. The plant uses a gas turbine for power production (2,400 MW capacity), has a total capacity of 228 MGD, and consists of 8 MSF units coupled with 17 RO units. A total of 70% of the total production is provided by the MSF plant, while 30% is provided by RO. The plant is operated by the Saline Water Conversion Corporation (SWCC).

2. Multi-Effect Evaporation with Reverse Osmosis (MED-RO). The largest hybrid seawater desalination plant utilizing MED-RO is located in the United Arab Emirates, and is called Fujairah 2. This plant has actively operated since May 2011, having a capacity of 100 MGD produced by MED and 30 MGD produced by RO. Gas turbines powering the system produce 2,000 MW of electric power.

13.5 Other Non-commercial Desalination Processes

Five other desalination processes have been investigated by a number of researchers but have been found to be of no current commercial value. These include the following processes:

1. Hydrates. Low molecular weight hydrocarbons, such as propane, combine with water to form insoluble inclusion compounds, known as gas hydrates. When the feed water is saltwater, gas hydrates are salt free. This process is very similar to freezing with a secondary refrigerant.
2. Solvent Extraction. Highly substituted 5- and 6-carbon secondary and tertiary amines or low molecular weight glycol and glycol ethers can dissolve salt-free water without losing their identity as a separate phase.
3. Ion Exchange. Resin beds pick up salt from saline water feed. Ion exchange beds are then regenerated by strong counter ion solutions. This process is used routinely for water softening and demineralization of low salinity water sources.
4. Thermal Diffusion. This process uses a temperature gradient to separate pure water from a brine solution at lower temperatures than required for water evaporation.
5. Polarization. This process uses electrical potential to concentrate ions near the surface of electrodes. The voltage used in this process is lower than it is required for electrolysis.

It should be noted that one or more of these processes could become commercially viable at some time in the future with changes in energy costs, technology improvements, changes in the value of the finished product (potable water), etc.

13.6 Illustrative Examples

Seven illustrative examples complete the presentation of traditional desalination technologies highlighted in this chapter.

13.6.1 Illustrative Example 1

Quantitatively explain the effect of the wet-bulb temperature on how a sling psychrometer works.

Solution. The wet-bulb temperature (T_{wb}) is the temperature at which a thermometer with a wet wick wrapped around the bulb stabilizes. As water evaporates from the wick to the ambient air, the bulb is cooled; the rate of cooling depends on how humid the air is. No evaporation occurs if the air is saturated with water; hence, the wet-bulb temperature and the dry-bulb temperature are the same when the air is at 100% relative humidity.

13.6.2 Illustrative Example 2

Determine the key properties of humid air at a dry-bulb temperature of 160°F and a wet-bulb temperature of 100°F.

Solution. Refer to the psychrometric chart in Figure 13.4. If the air were to be cooled until moisture just begins to condense, the dew point would be reached. This is represented by a horizontal line at constant absolute humidity intersecting the saturation curve at a dew point of approximately 87.5°F.

The relative humidity is approximately 14% (interpolating between 10% and 15% relative humidity lines). The absolute humidity is the horizontal line extended to the right, it intersects the ordinate at an absolute humidity of 0.0285 lb H_2O/lb BDA.

The humid volume is approximately 16.3 ft^3 moist air/lb BDA (interpolating between 16 and 17 ft^3 moist air volume).

The enthalpy for saturated air at a T_{wb} of 100°F is 71.8 Btu/lb BDA. For the unsaturated air, the enthalpy deviation is -1.0 Btu/lb BODA; therefore, the actual enthalpy for the moist air at a T_{wb} of 100°F and a T_{db} of 160°F is 70.8 Btu/lb dry air.

13.6.3 Illustrative Example 3

A flue gas (MW = 30 lb/lbmol, dry basis) is being discharged from a scrubber at 180°F (dry bulb) and 125°F (wet bulb). The gas flow rate on a dry basis is

10,000 lb/hr. The absolute humidity at the dry-bulb temperature of 180°F and wet-bulb temperature of 125°F is 0.0805 lb H_2O/lb dry air.

1. What is the mass flow rate of the wet gas?
2. What is the actual volumetric flow rate of the wet gas?

Solution. Curves showing the relative humidity (ratio of the mass of the water vapor in the air to the maximum mass of water vapor the air can hold at that temperature, i.e. if the air were saturated) of humid air appear on the psychrometric chart. (See Figure 13.10). The curve for 100% relative humidity is also referred to as the saturation curve. The abscissa of the humidity chart is air temperature, also known as the dry-bulb temperature (T_{db}). The wet-bulb temperature (T_{wb}) is another measure of humidity; it is the temperature at which a thermometer with a wet wick wrapped around the bulb stabilizes. As water evaporates from the wick to the ambient air, the bulb is cooled; the rate of cooling depends on how humid the ambient air is. No evaporation occurs if the air is saturated with water; hence T_{wb} and T_{db} are the same. The lower the humidity, the greater the difference between these two temperatures. On a psychrometric chart, constant wet-bulb temperature lines are straight with negative slopes. The value of T_{wb} corresponds to the value of the abscissa at the point of intersection of this line with the saturation curve.

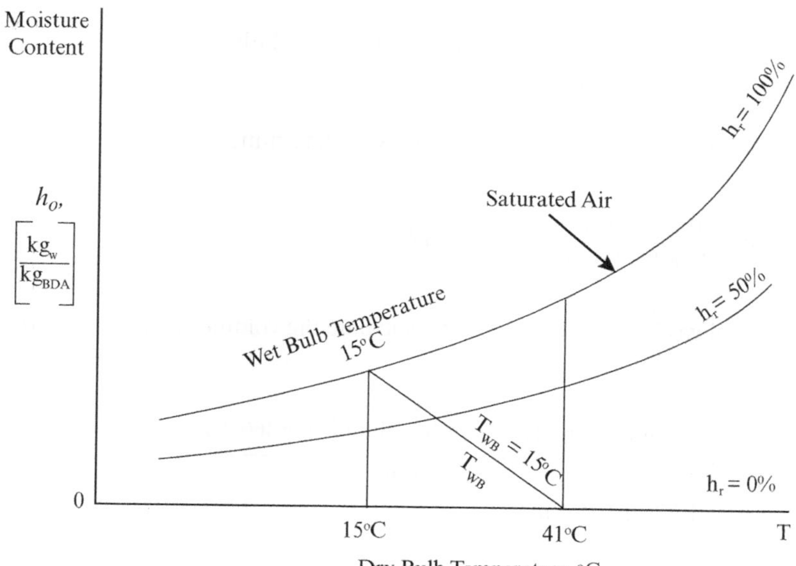

Figure 13.10 Diagram of Psychrometric Chart for Illustrative Example 3.

Calculate the flow rate of water in the air. Note that both the given flowrate and humidity are on a dry basis.

$$\text{Water flow rate} = (0.0805)(10{,}000) = 805\,\text{lb/hr}$$

Calculate the total flow rate by adding the dry gas and water flow rates:

$$\text{Total flowrate} = 10{,}000 + 805 = 10{,}805\ \text{lb/hr}$$

The moles of water and dry gas are thus:

$$\text{Moles gas} = (10{,}000\,\text{lb/hr})/(30\,\text{lb/lbmol}) = 333.3\,\text{lbmol/hr}$$

$$\text{Moles water} = (805\,\text{lb/hr})/(18\,\text{lb/lbmol}) = 44.7\,\text{lbmol/hr}$$

Calculate the mole fraction of water vapor using the above-mentioned units:

$$y_{water} = \frac{44.7\ \text{lbmol / hr}}{(44.7 + 333.3\ \text{lbmol / hr})} = 0.12$$

The average molecular weight of the mixture then becomes:

$$\begin{aligned} \text{MW} &= (1.0 - 0.12)(30\,\text{lb/lbmol}) + (0.12)(18\,\text{lb/lbmol}) \\ &= 28.6\,\text{lb/lbmol} \end{aligned}$$

The molar flow rate of the wet gas can now be determined as:

$$\dot{n} = \frac{10{,}805\ \text{lb / hr}}{28.6\ \text{lb / lbmol}} = 378\ \text{lbmol / hr}$$

The ideal gas law may be applied to calculate the volumetric flowrate of the wet gas as:

$$\begin{aligned} q = \frac{\dot{n}RT}{P} &= \frac{(378\text{lbmol / hr})(0.73\text{atm-ft}^3\text{ / lbmol-°R})(460 + 180\,°\text{R})}{1.0\text{atm}} \qquad (13.3) \\ &= 1.77 \times 10^5\ \text{ft}^3/\text{hr} \end{aligned}$$

13.6.4 Illustrative Example 4

Potassium nitrate is obtained from an aqueous solution of 25% KNO_3. During a hybrid desalination process the aqueous solution of potassium nitrate is evaporated, leaving an outlet stream with a concentration of 50% KNO_3, which then enters a crystallization unit where the outlet product is 96% KNO_3 (anhydrous crystals) and 4% water. A residual aqueous solution that contains 0.65 g of KN per gram of water also leaves the crystallization unit and is mixed with the fresh solution of KNO_3 at the evaporator inlet. Such recycling provides a means of preventing pollution by minimizing the loss of valuable raw material. The flow diagram in Figure 13.11 depicts the process.

What is the mass flow rate, *M*, in kg/hr of the recycled material, *R*, when the feedstock flow rate is 3,000 kg/hr?

Solution. Using Figure 13.11, an overall material balance on KNO_3 leads to:

$$0.96C = 0.25(3{,}000 \text{ kg/hr})$$

This can be solved for the rate of product leaving the crystallizer as:

$$C = (750 \text{kg/hr}) / 0.96 = 781.25 \text{kg/hr} \tag{13.4}$$

An overall mass balance around the crystallizer can also be written as:

$$M = R + C = R + 781.25 \text{kg/hr}$$

Since the mass concentration of the KNO_3 in *R* is, the equation for the KNO_3 mass balance around the crystallizer is:

$$\begin{aligned} 0.5M &= 0.96(781.25 \text{kg/hr}) + (0.65/1.65)R = 750 + 0.394R \\ M &= 1{,}500 + 0.788\,R \end{aligned} \tag{13.5}$$

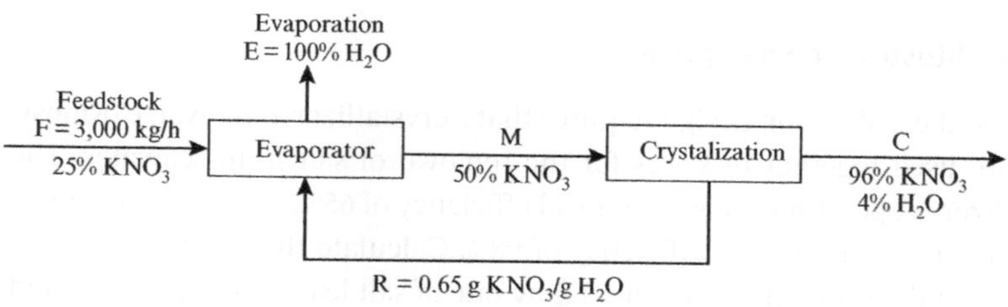

Figure 13.11 Process Flow Diagram for Illustrative Example 4.

Equations 13.4 and 13.5 may be solved simultaneously for the amount of KNO_3 leaving the evaporator in terms of the amount of recycle. Solving yields:

$$M = 1{,}500 + 0.788R = R + 781.25\,\text{kg/hr}$$

$$0.212R = 718.75\,\text{kg/hr}; R = 3{,}390\,\text{kg/hr}$$

13.6.5 Illustrative Example 5

The temperature and dew point of the air entering a dryer are 130 and 60°F (328 and 289 K), respectively. Using a humidity chart (Figure 13.4), find the following properties at adiabatic saturation of the air: its humidity, its percentage humidity, its adiabatic saturation temperature, and the humidity at adiabatic saturation.

Solution. In Figure 13.4, humidity is the ordinate (along the right side of the graph) of the point on the saturation line (the 100% humidity line) that corresponds to the dew point and the intersection with the gas temperature, the latter being read from the abscissa along the bottom of the figure. In the present case, the humidity is found to be 0.015 lb water/lb of dry air (0.015 kg water/kg of dry air).

The percent humidity is determined by first locating the dry-bulb temperature, that is 130°F, along the abscissa. From there a perpendicular line is drawn to intersect the 0.015 lb humidity line, and the percentage humidity line (interpolating a line if necessary) that passes through that intersection. In this case, the intersection is between the 10% and 15% humidity lines, and the percentage humidity can be interpolated to be 12%.

To determine the adiabatic saturation temperature, the adiabatic cooling line should be located. These lines have negative slope and pass through a point where this line intersects the 100% humidity line. This abscissa at the 100% humidity line is the adiabatic saturation temperature. In this case, the abscissa and the adiabatic saturation temperature are 82°F (301 K).

The humidity at adiabatic saturation is the ordinate, along the right side of the graph. Its value is 0.024 lb water/lb dry air (0.024 kg water/kg of dry air).

13.6.6 Illustrative Example 6

A proposed desalination design requires that a crystallizer and reverse osmosis units are used together in series for the removal of salt from seawater. The crystallizer is operating at a salt removal efficiency of 65%, while the RO unit is operating at a salt removal efficiency of 98%. Calculate the mass flow rate of salt leaving the crystallizer, the mass flow rate of salt leaving the RO unit, and the overall fractional efficiency of salt removal through the desalination system if 76.0 lb of salt enters the process every hour.

Solution. The hybrid desalination system is pictured in Figure 13.12. As indicated in Figure 13.12, M_i is defined as the mass flow rate of salt into each unit, with $M0$ being the mass flow rate into the crystallizer, M_1 being the mass flow rate into the RO unit, and M_2 being the mass flow rate out of the combined hybrid process. Based on the problem statement, $M_0 = 76$ lb/hr.

By definition,

$$M_1 = (76 \text{ lb/hr})(1 - 0.65) = 26.6 \text{ lb/hr}$$

In addition,

$$M_2 = (26.6 \text{ lb/hr})(1 - 0.98) = 0.532 \text{ lb/hr}$$

The overall efficiency, E, of the hybrid desalination system can then be calculated as follows:

$$E = \frac{M_o - M_2}{M_0} \text{x}100 = \frac{(76 \text{ lb/hr} - 0.532 \text{ lb/hr})}{76 \text{ lb/hr}} \text{x}100 = 99.3\% \text{ removal}$$

13.6.7 Illustrative Example 7

A dehumidification unit in a large factory draws in moist air al 500 lb/hr. Moist air enters the unit at a temperature of 82.4°F and a wet-bulb temperature of 75°F. The exit stream of the unit is 60.8°F and saturated with water. Determine the amount of water (in lb/hr) condensed in the dehumidifier pictured in Figure 13.13.

Solution. This problem is meant to serve as an example of a basic material balance problem. As with all material balance problems, it is first recommended that a system diagram be drawn as shown in Figure 13.13. In order to solve this problem, one must make use of the relative humidity and the psychrometric charts. As noted earlier, the relative humidity is the ratio of the partial

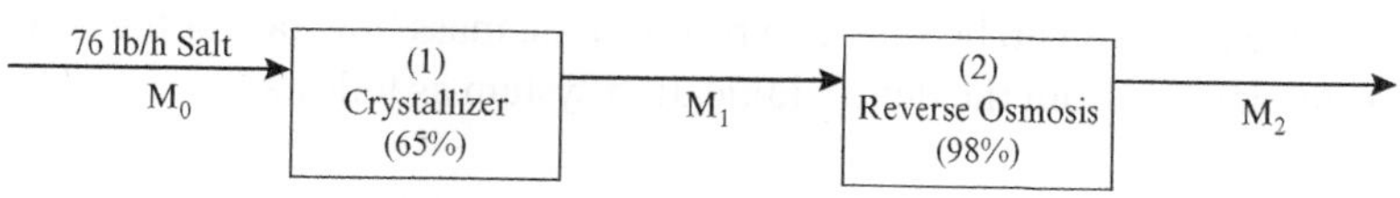

Figure 13.12 Process Flow Diagram of Hybrid Desalination System for Illustrative Example 6.

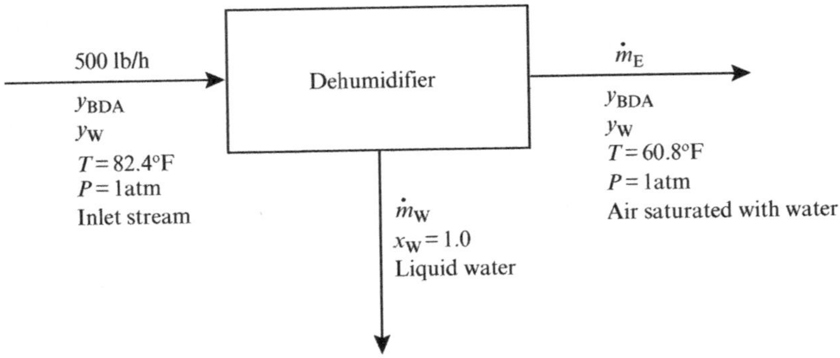

Figure 13.13 Process Flow Diagram of Dehumidifier System for Illustrative Example 7.

pressure of water in the gas phase to the vapor pressure of water at a temperature T. Since the outlet stream is saturated with water, its relative humidity is 100%.

The relative humidity of the inlet stream, however, must be determined via Figure 13.4 by locating the appropriate intersection of T_{db} and T_{wb}, for the inlet conditions. Using the temperature values given in the problem statement, the relative humidity of the influent air from Figure 13.4 is determined to be 74%.

At the inlet temperature, the vapor pressure of water, p_A', is 0.0372 atm. Using the definition of relative humidity, Y_R, the partial pressure of water in the inlet stream, p_A, can be determined as:

$$Y_R = \frac{p_A}{p_A'} = 0.74 = \frac{p_A}{0.0372};\ p_A = (0.74)(0.0372\ \text{atm})$$
$$= 0.0276\ \text{atm}$$

At the exit conditions, the vapor pressure of the water, p_A', is 0.0179 atm, and since the outlet is at 100% saturation, the partial pressure of the water is also 0.0179 atm. Knowing that the partial pressure of the gas and the gas mole fraction are equivalent, these partial pressure values can be used in the componential balance for water through the system as follows:

$$500\ \text{lb / hr} = \dot{m}_w + \dot{m}_E, \text{total material balance}$$

$$(0.0276)(500\ \text{lb / hr}) = \dot{m}_w + (0.0179)(\dot{m}_E), \text{water balance}$$

These two equations can be solved to result in the mass flow rate of the condense water coming from the dehumidification system as follows:

$$\dot{m}_E = 500\ \text{lb / hr} - \dot{m}_w$$

$$(13.8 \text{ lb/hr}) = \dot{m}_w + (0.0179)(\dot{m}_E)$$

$$(13.8 \text{ lb/hr}) = \dot{m}_w + (0.0179)(500 \text{ lb/hr} - \dot{m}_w)$$

$$(13.8 \text{ lb/hr}) = \dot{m}_w + (8.95 \text{ lb/hr} - 0.0179\ \dot{m}_w)$$

$$4.85 \text{ lb/hr} = 0.9821\dot{m}_w;\ \dot{m}_w = 4.94 \text{ lb/hr}$$

13.7 Problems

1. Discuss some normal, everyday domestic applications involving dehumidification.
2. For air with a dry- and wet-bulb temperature of 80°F and 65°F, respectively, calculate the following properties of the air: relative humidity in %, moisture content in lb/lb BDA, and the dew point in °F.
3. A wet gas (molecular weight = 29 lb/lbmol, dry basis) is being discharged from a cooling unit at 170°F dry bulb and 120°F wet-bulb temperature. The gas flow rate is 20,000 lb/hr on a dry basis. What is the mass flow rate of the wet gas?
4. Refer to Problem 3. What is the actual volumetric flow rate of the wet gas?
5. Salt is obtained from an aqueous solution of 5% NaCl. In the process, the aqueous salt solution is first evaporated, leaving an outlet stream with a concentration of 20% NaCl, which then enters a crystallization unit where the outlet product is 50% salt and 50% water. A residual aqueous solution that contains 0.15 lb salt/lb water also leaves the crystallization unit and is mixed with the feed solution at the evaporator inlet. What is the mass flow rate, *M*, in lb/hr of the recycled material, *R*, when the aqueous salt feed is 10,000 lb/hr?
6. An air stream at 90°F, 14.7 psia, with a relative humidity of 60% is flowing into a water spray chamber where enough water is added to reach saturation. The outlet temperature and pressure of the saturated stream are 80°F and 12 psia, respectively. After the spray chamber, the stream is heated and compressed (H/C) to 30 psia and has a final relative humidity of 30%. What is the mole ratio of the amount of water added (i.e. absorbed by the air stream) to the amount of incoming moist air in the spray chamber? Assume a basis of 100 lbmol of feed to solve the problem.

7. Refer to Problem 6. What is the dew point temperature of the moist air leaving the compressor/heater unit and what is the final vapor pressure of the moist air?

References

Flynn, A.M., Akoshige, T., and Theodore, L. (2018). *Kern's Process Heat Transfer*, 2nd Edition. Salem, MA: Scrivener-Wiley.

Ganiaris, N., Dupont, R.R., and Theodore, L. (2018). © *The GADUTH Greenhouse Solar Evaporator*. East Williston, NY.

Theodore, L. (2016). *Chemical Engineering: The Essential Reference*. New York, NY: McGraw-Hill.

Theodore, L., Dupont, R., and Ganesan, K. (2017). *Unit Operations in Environmental Engineering*. Salem, MA: Scrivener-Wiley.

Theodore, L. and Flynn, A.M. (2018). © *The Theodore Humidification Process*. East Williston, NY.

Theodore, L. and Rici, F. (2010). *Mass Transfer Operations for the Practicing Engineer*. Hoboken, NJ: John Wiley & Sons.

14

New Desalination Processes

14.1 Introduction

This chapter introduces the reader to new desalination process. Traditional processes were reviewed in previous chapters and five new options for desalination developed by the authors are presented here. These include: a simple evaporation (still) unit, a heat conduit evaporator, a geothermal evaporator, a freeze desalination process, and a mangrove process. Descriptions of each of the five processes are included in the material to follow, along with legal considerations and future prospects for new desalination technologies. An application section provides six illustrative examples related to the general subject of new water desalination options.

14.2 The Theodore Simple Still (THESS)

The Theodore Simple Still (Ganiaris et al. 2018). This first new water desalination process proposed by GADUTH (GAniaris, DUpont, and THeodore) Consulting Engineers is one that would be suitable for domestic applications. The main requirement is that relatively inexpensive energy be available for domestic households. The only equipment of a unit operations (Theodore 2016; Theodore and Rici 2010) nature would be an efficient vacuum pump or vapor compressor (or the equivalent). As indicated in Figure 14.1, feed water (1) is introduced to a ministill (2) where vapors in the form of steam (3) are suctioned off by a vacuum pump or vapor compressor (or the equivalent) (4). When a vacuum pump is employed, the steam is directed to a canister (5) where condensation occurs due to radiation losses in the line and clean water is available (6). When the vacuum pump (4) is activated, a proportional controller opens the valve (7) to allow the feed water to enter the ministill. This in turn proportionally opens a valve (8) allowing part of the more concentrated brine solution to leave the still. The control valves can be set during operation for 50% of the feed water to be evaporated and the remaining to be discarded as a 50% more concentrated salt solution. A slightly modified process is employed

Introduction to Desalination: Principles, Processes, and Calculations, First Edition. Louis Theodore and R. Ryan Dupont.

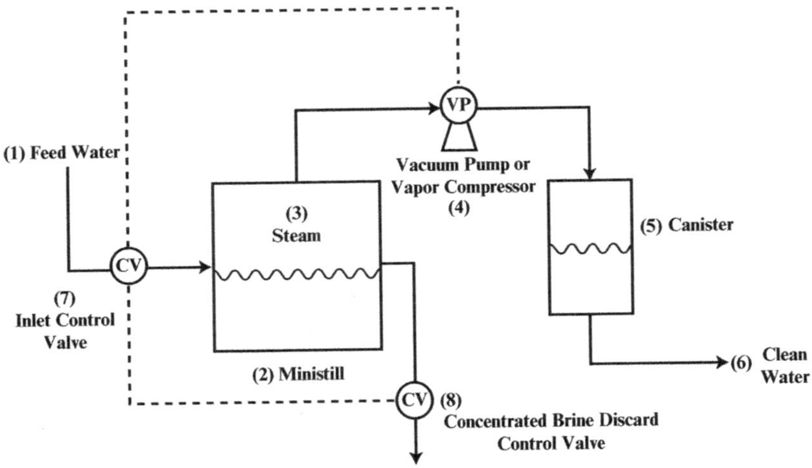

Figure 14.1 The Theodore Simple Still.

when a vapor compressor replaces the vacuum pump. Here, the vapor compressor takes in water vapor from the evaporator and compresses it to both a higher pressure and temperature (Theodore 2016). This vapor passes through the coils in the evaporator and heats the seawater while the vapor is condensing into fresh water. This heat exchanger can salvage heat from both the discharged concentrated brine and fresh water for re-use in the evaporator, subject to economic considerations.

14.3 The Heat Conduit Evaporation (HCE) Process

The HCE process (Theodore 2017) involves both geothermal energy and the utilization of a heat pipe. Both geothermal energy and heat pipes are briefly described before proceeding to a description of the HCE processes.

Geothermal energy generally refers to the utilization of energy from the stored heat in the Earth's crust. Geothermal energy (also referred to by some as *geothermics*) is virtually limitless and is based on the fact that the Earth is hotter the deeper one drills below the surface. The Earth's core lies almost 4,000 miles beneath the Earth's surface. The double-layered core is made up of very hot molten iron surrounding a solid iron center. Estimates of the temperature of the core range from 5,000 to 11,000°F! Surrounding the Earth's core is the mantle, which is partly rock and partly magma. The mantle is about 1,800 miles thick. The outermost layer of the Earth is the insulating crust; it is not one continuous sheet of rock but is broken into pieces called *plates*.

The magma can reach the surface and form volcanoes, but most remains underground where it can underlie regions as large as spacious mountain ranges. The magma can take from 1,000 to 1,000,000 years to cool as its heat is transferred to surrounding rocks by conduction. In areas where there is underground water, the magma can fill rock fractures and porous rocks. The water becomes heated and can circulate back to the surface to create hot springs and mud pots, or it can become trapped underground, forming deep geothermal reservoirs. The Earth's geothermal resources are theoretically more than adequate to supply humanity's energy needs, but only a very small fraction may be profitably exploited at this time. Unfortunately, drilling and exploration for deep resources is very expensive.

Seismically active hotspots are not the only places where geothermal energy can be found. There is a steady supply of milder heat, useful for direct heating purposes, virtually in any location on Earth at depths of anywhere from 10 to a few hundred feet below the surface. In addition, there is a vast amount of heat energy available from dry rock formations deeper below the surface (4–10 km). Some of the emerging technologies may be able to capture this energy on a much larger scale than some of the conventional technologies.

The areas with the highest underground temperatures are in regions with active or geologically young volcanoes. These "hot spots" occur at plate boundaries or at places where the crust is thin enough to allow the heat to pass. The Pacific Rim, often called the "Ring of Fire" for its many volcanoes, has many hot spots in the United States, including those in Alaska, California, and Oregon. In addition, Nevada has hundreds of hot spots, covering much of the northern part of the state. A similar "ring" extends through southern Europe and across the middle of Asia connecting with some of the South Sea Islands.

The bulk of thermal energy stored on Earth is located within its core. Some of this energy is transmitted from the interior by means of conduction through the Earth. The average temperature difference driving force is approximately 10°C/km. The average rate of flow of heat to the surface has been estimated to be 0.063 W/m^2. Since the Earth's surface is 510×10^{12} m^2, the total heat flow amounts to 32×10^{12} W. Only 1 percent of this total can be attributed to heat transfer by convection (Harder 1982). The heat flow rate to the surface by conduction is approximately 30–40 terawatts (TW). This energy is replenished by radioactive decay at a rate of approximately 30 TW. This nuclear process has been going on since the Earth's formation and will effectively continue for many millennia. Harder (1982) has provided approximate temperature data at various depths. This information is provided in Table 14.1. Understandably, future development potential is highest for hot water/hot temperature resources.

A heat pipe (Kern 1950) may be circular, square, or rectangular, and is a relatively new heat-transfer device that combines the principles of both thermal conductivity and phase transition to efficiently manage the transfer of heat between two interfaces (Figure 14.2). When heat is applied to the heat pipe, the working liquid in the wick (chosen for compatibility with both the heat pipe material of construction and the temperatures of the hot and cold surfaces)

Table 14.1 Earth Temperatures at Various Depths.

Depth	Temperature, F°
Surface	50
1,000 ft (305 m)	65
1 mile (1.61 km)	129
1.4 miles (2.25 km)	161
2.05 miles (3.29 km)	212
10 miles (16.1 km)	842

heats and evaporates (high temperature applications almost always use liquid metals such as sodium and potassium). As the evaporating fluid fills the heat pipe's hollow center, it diffuses axially along its length. Condensation of the downstream vapor occurs wherever the temperature is even slightly below that of the evaporation area. As it condenses, the vapor gives up the heat it acquired during evaporation and establishes a reverse vapor flow pattern in the pipe. The condensed liquid flows, through capillary action within the wick, back to the heat source, completing the heat transfer cycle. Thus, the heat pipe thermal cycle operates in the following manner (refer to Figure 14.2).

1. Working fluid evaporates to vapor, absorbing thermal energy.
2. Vapor migrates along the cavity to the lower temperature end.
3. Vapor condenses back to the fluid releasing thermal energy in the cooler end and is absorbed by the wick.
4. The working fluid flows back to the higher temperature end.

The unit process itself consists of three operating sections: the evaporator section, the adiabatic section, and the condenser section.

Heat pipes have been employed within computer systems to transfer heat from CPU's to heat sinks so the heat can be rejected to the environment. Heat pipes are also used in HVAC systems to recover heat by transfer from an exit stream to a

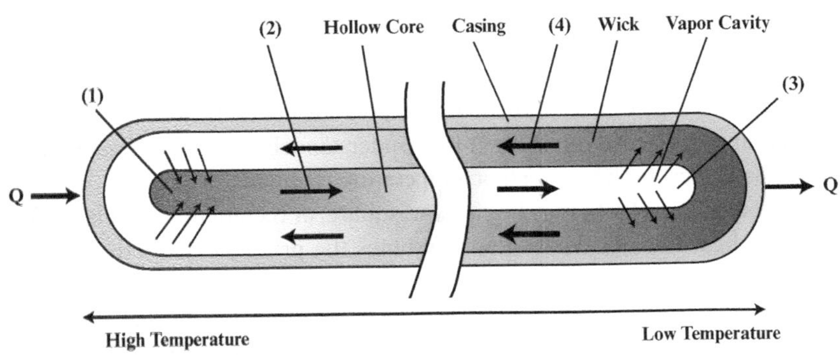

Figure 14.2 Heat Pipe Principles.

fresh air stream in the winter and to cool an entering air stream in the summer. For example, a heat pipe buried deep in the ground transports energy from below the Earth's surface to the ground where the accumulation of ice and snow can be reduced/eliminated. Thus, heat pipes are capable of transferring heat from a high temperature source to a cooler surface/location. Additional information is available in the literature (Kern 1950). Calculational details regarding heat pipes are provided by Hagen (1999). Information on the pipe thermal resistances, the heat source, the cooling (sink) source, evaporator and condenser pipes, and for the wick on both the evaporator and condenser sections are also provided.

The HCE process (Theodore 2017) combines the principles associated with both geothermal energy and heat pipes. This appears to be a promising prospect for those locations on Earth where volcanic (or the equivalent) activity can provide temperatures in excess of 212°F at reasonable depths below the Earth's upper surface. The discharge of heat from the top section of conduit can be transferred to seawater that will produce steam that can be recovered as clean water. The first (original) flow diagram developed by the authors is provided in Figure 14.3.

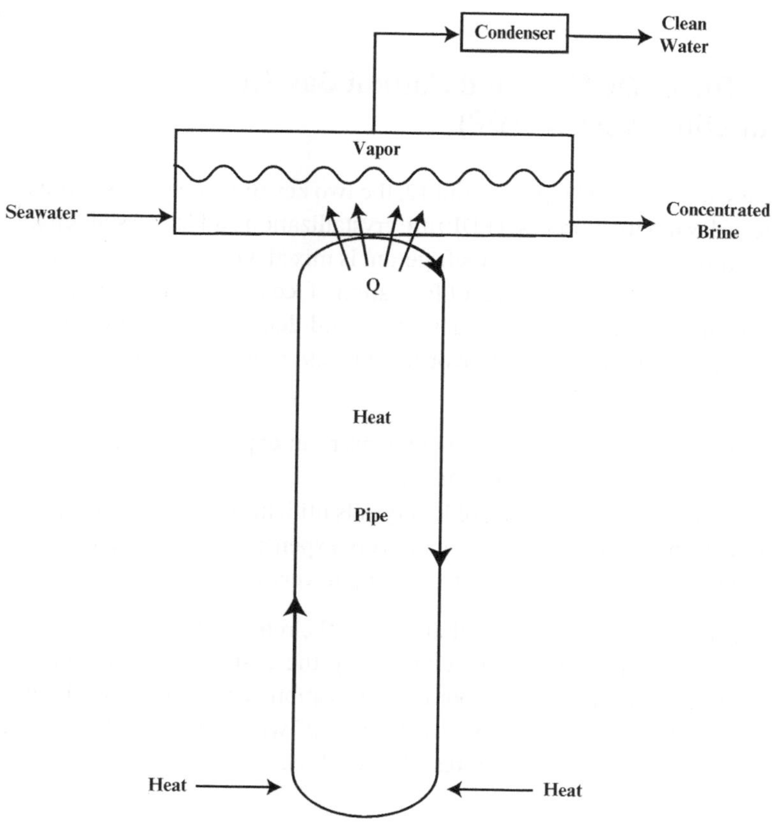

Figure 14.3 The Heat Conduit Evaporator Process.

14.4 The Geothermal Evaporator (GEO) Process

The GEO process (Theodore 2018) involves a very simple extension of the aforementioned HCE process. Refer, once again, to Figure 14.3. Once a well has been dug, seawater is fed directly into the Earth's cavity (without the presence of a heat pipe) where steam is generated at greater depths due to the high underground temperatures. Additional horizontal cavities at various depths can be included to increase the contact area for transferring heat from the hot underground sources of the downward flowing seawater. The steam output can be recovered by any of the means described earlier. The hot concentrated brine Solution that remains at the bottom of the well is withdrawn through a separate conduit attached to the bottom of the well that discharges at ground-level. The energy content of this hot brine can be used to either ensure that the rising steam does not condense or heat the seawater downflow, or both. Alternatively, the feed seawater can empty into a dome that would provide a depository for the salt. If needed, the deposited salt could be removed with (polluted) hot water or any other suitable means.

14.5 The GADUTH Liquid Natural Gas Freeze Desalination Process (FDP)

As discussed earlier in Chapters 9 and 12, the two key unit operations involved in a freeze desalination process (FDP) are crystallization and ice crystal separation. In a direct contact process, a refrigerant is mixed with sea water in a crystallization zone with the subsequent formation of ice crystals. The crystals are then separated. This concept was developed and demonstrated 50 years ago; however, the economics did not favor this process at the time. The two main problems with the FDP were:

1. The cost of refrigeration in the condenser, an expensive form of energy required for ice crystal formation, and
2. The separation and washing of ice crystals utilizing centrifuges or equivalent is technically feasible but again, very expensive, while a water "wash column" is extremely difficult to scale up to very large sizes.

GADUTH has proposed a process that utilizes the refrigeration potential from liquefied natural gas (LNG), thus eliminating the cost of refrigeration. The LNG has to be re-gasified at the point of utilization. The process would also replace the separation step and/or "wash column" with refrigerant tanks. The proposed process is shown in Figures 14.4 and 14.5.

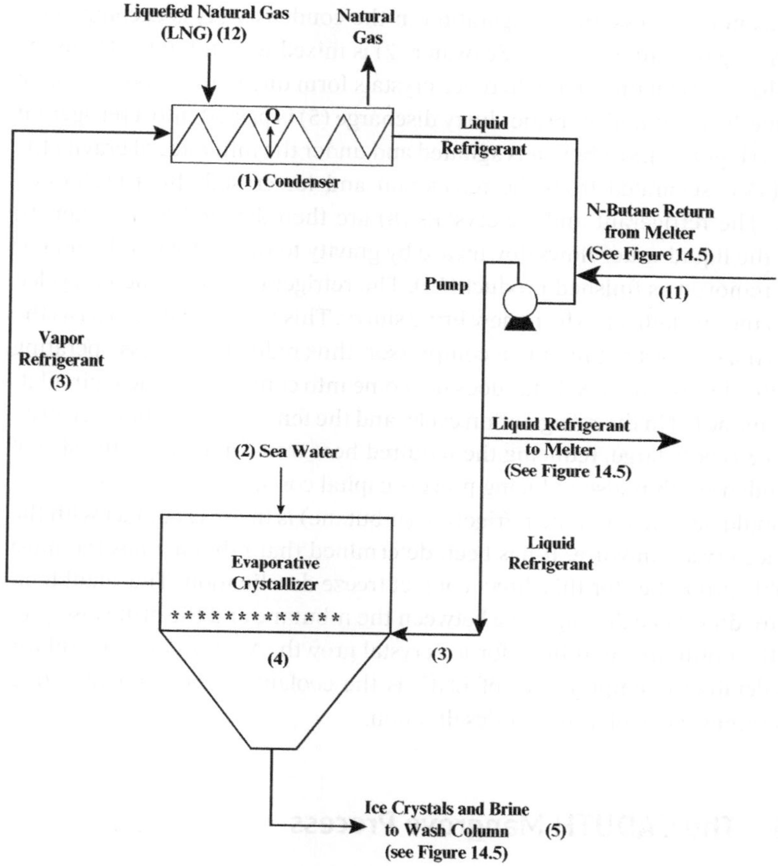

Figure 14.4 The GADUTH Freeze Desalination Process Refrigeration Cycle.

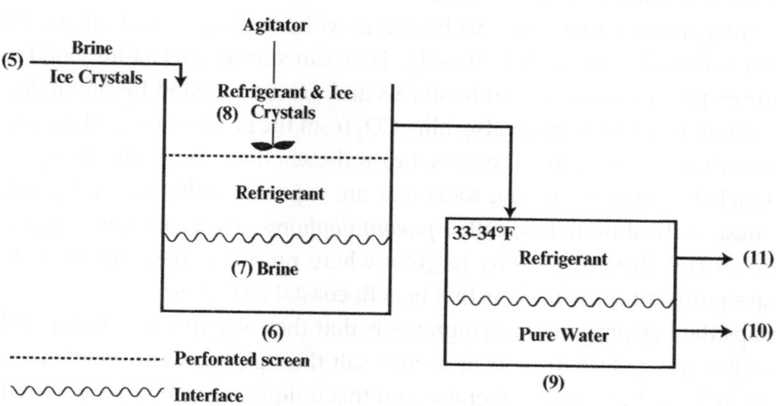

Figure 14.5 The GADUTH Freeze Desalination Process Refrigerant Tanks.

In this new process, the refrigeration in the condenser (1) is accomplished with the regasification of LNG. Seawater (2) is mixed with n-butane (3) in the crystallizer (4) (Figure 14.4) where ice crystals form due to the vaporization of n-butane. The crystallizer brine slurry discharge (5) is passed into a refrigerant tank (6) (Figure 14.5) where it is agitated and under the influence of gravity the brine (7) is separated from the refrigerant and ice crystals by a perforated screen. The refrigerant and ice crystals (8) are then directed to a melter (9) where the liquid water flows downward by gravity to the bottom of the melter and is removed as finished product (10). The refrigerant (11) is then recycled back to the crystallizer to form new brine slurry. This process using LNG as the liquid coolant does not require a compressor, thus reducing process operating costs. In addition, the LNG (12) does not come into contact with the recirculating n-butane (3) in the refrigeration cycle, and the temperature difference driving force is very large, reducing the required heat transfer area, i.e. the size of the condenser, thus also reducing process capital costs.

It should be noted that the refrigerant (n-butane) is in *direct* contact with the brine/ice crystal mixture. It has been determined that n-butane has the most desirable properties for this direct contact freeze desalination. The small temperature difference driving force between the n-butane and the brine also provides the optimum conditions for ice crystal growth. A 1972 patent (Ganiaris 1972) details the employment of LNG as the coolant in the condenser of a refrigeration cycle for seawater desalination.

14.6 The GADUTH Mangrove Process

Mangrove is a common name applied to several kinds of flowering plants that are members of three different families. From a desalination perspective, mangroves are a group of trees and shrubs that grow in shallow coastal saline and brackish waters that are usually muddy. They can survive in a salinity of 9%. Mangroves protect shorelines from storms and prevent erosion by stabilizing sediments and are excellent at adsorbing CO_2 from the atmosphere and storing carbon (as leaves, roots, and branches die) at the sea floor. Typically, they produce tangled masses of arching roots that are exposed to the air during low tide. These vertical branches, called *pneumalophores*, act as aerating organs. Mangroves thus have the ability to grow where no other trees can, thereby making significant contributions that benefit coastal ecologies.

Perhaps the key property of mangroves is that they absorb saline water and either reject salt with their roots or excrete salt through glands of their leaves. There is little in the technical literature on this unique separation process and the mechanism has not been investigated to any great extent for the purpose of accomplishing desalination. The GADUTH engineers are currently investigating the commercialization of this potential desalination process (Ganiaris 2018).

14.7 Legal Considerations

This section is concerned with, as can be noted from the title, legal considerations associated with the development and deployment of new technologies. This is a topic that inevitably arises when discussing new process schemes, and includes considerations of intellectual property protection, contract law, and patents. As one might suppose, it is not possible to address all topics directly or indirectly related to legal considerations. However, additional details may be obtained from the U.S. Patent and Trademark Office (www.uspto.gov).

Technology affects almost every area of human activity in one way or another. Therefore, one can expect legal relations between people will have to be taken into account. Even though unique developments in technology will continue, those involved with any field of engineering and science plus those involved with law should consider how technology and law might interact.

It is incumbent upon those engaged in any area of technological development to acquire a basic understanding of patent law, and legal considerations in general, because the patent portfolio of a company, particularly one focused on research and development, may represent its most valuable asset(s). Certain activities, such as premature sale or public disclosure, can jeopardize one's right to obtain a patent; the reader should note that patents are creatures of the national law of the issuing country and are enforceable only in that country. Thus, a U.S. patent is enforceable only in the United States. To protect one's invention in foreign countries, one must apply in the countries in which protection is sought. In addition, one can obtain a general idea of the development/progress of a new technological field by monitoring the number of patents issued in that field.

As with any invention, the qualities that make technology-related inventions patentable are novelty, nonobviousness, and utility. While many unique aspects of technology are already known, more are being discovered as are new ways of exploiting known properties, all of which can lead to patentable inventions.

Another area of concern is government regulations. Environmental regulations, in particular, have become an important component of new technological development. In fact, for many industrial sectors, it is the increasingly stringent clean air and water regulations that drive new technological development. This is particularly true, for example, in the automotive industry and in the heavy industries where emissions are released into the atmosphere or water. What effects will new technology and products have on human health and the environment? This must be determined as part of a new process, technology, and product development. Suppose, for example, new products are developed, that can be ingested or inhaled, and that can act upon the interior organs of the human body or interact within natural ecosystems. Also, suppose that such products are released into the atmosphere. They might be carried by air within a community, state, or across state lines, and may enter water bodies through dry or wet deposition. State and federal environmental regulatory agencies would understandably take great interest in the human health and environmental impact of such a release from these new products created from new processes.

An important area of law with which engineers must be concerned is *intellectual property law*. Technological development is all about ideas. Ideas have commercial value only if they can be protected by excluding others from exploiting those ideas. Typically, the way to protect ideas is through intellectual property rights such as *patents, trademarks, copyrights, trade secrets*, and *financial secrets*.

Patents can be used to protect useful inventions, ornamental designs, and even botanical plants. The patent allows the owner of the patent the right to prevent anyone else from making, using, or selling the "invention" covered by the claims of the patent. Trademarks are distinctive marks associated with a product or service (these are usually referred to as service marks) which the owners of the mark can use exclusively to identify themselves as the source of the product or service. Copyrights protect the expression of an idea, rather than the idea itself, and are typically used to protect literary works (such as this book) plus visual and performing arts, such as photographs, painting and drawings, sculptures, movies, songs, etc. Trade secret laws protect technical or business information that a company uses to gain a competitive business advantage by virtue of the secret being unknown to others. Customer or client lists, secret formulations, or methods of manufacture are typical business and trade secrets.

Another legal area that is relevant to engineers is *contract law*. Any time two or more parties agree upon something, the principles of contract law come into play. The essential components of contract law are:

1. Parties must be competent to enter into a contractual agreement and subject matter, i.e. what the contract is about.
2. Legal considerations (the inducement to contract such as the promise(s)) or payment exchanged or some other benefit or loss or responsibility incurred by the parties.
3. Mutuality of agreement.
4. Mutuality of contract.

While verbal contracts can be legally binding in the event of a dispute, it may be difficult to establish in court who said what. It is far better to memorialize the agreement in the form of a written contract. Who are the entities engaged in the contract? Typically, these are business entities. Also, one must also consider whether there may be some peripheral issues, or corporate or partnership law obligations that apply.

There is a branch of law that can retrospectively address certain situations in which property or people are harmed. That is *tort* law, a topic that continues to receive significant attention by politicians. A tort is a civil wrong, other than a break of contract, for which the law provides a remedy. One can recover damages under tort law if a legal duty has been breached that causes foreseeable harm. These duties are created by law other than duties created by criminal law. governmental regulations, or those agreed to under a contract.

Finally, in order to encourage new discoveries for the benefit of society/mankind, the U.S. Constitution provides for *patents*. These are limited monopolies provided in exchange for the public disclosure of new products and inventions. Patents are an integral part of a free enterprise system. The process discourages secret behavior by rewarding the aforementioned "monopoly" for prompt and adequate disclosure. The U.S. patent system is responsible for much of the growth in industry because it encourages research upon which the growth is based. Attaining a patent is a procedure requiring skilled and experienced guidance. The patent must be fully disclosed, and the essentials must be covered by the claims.

Of all the intellectual property rights, the most pertinent for products and inventions developed are patents. A patent can protect, for example, a composition of matter, an article of manufacture, or a method of doing something. Patent rights are private property rights. *Infringement* of a patent is a civil offense, not criminal. The patent owner must come to his or her own defense through litigation, if necessary. And, this is often a very expensive undertaking. Patent lawsuits costing more than a million dollars are not unusual. But at stake can be exclusive rights to technology worth several orders of magnitude more.

Patents pertaining to any new product, process, equipment, use, or application should be reviewed by a patent attorney. The product or inventions of concern may be involved in other patents. Patents available for purchase and lease as well as participation in patent pools should be reviewed by a legal expert.

As with any invention, the qualities noted earlier that make technology-related inventions patentable are *novelty, nonobviousness*, and *utility*. While many "processes" of technology are already known, more are being discovered as are new ways of exploiting existing technologies, all of which can lead to patentable inventions. It is fair to say that a patent is essentially a contract between an inventor and the public. By full disclosure of the invention to the public, the patentee is given exclusive rights to control the use and practice of his/her product or invention. A patent gives the holder the power to prevent others from using or practicing the invention for a period of years from the date of being granted such a patent. In contrast, *trade secrets* receive protection as long as the information is not public knowledge.

As noted earlier, a patent may be obtained on any new and useful process, method of manufacture, composition of matter, etc., provided it has not been known or used by others before the patentee made the invention or discovery. The invention must not have been described in a printed publication or been in public use or on sale for more than 1 year prior to the patent application. A patentable item must result from the use of creative ability above and beyond that which would be expected of a person working in a particular field. A patentable item cannot be something requiring purely mechanical skill. Furthermore, a patent will not be granted for a change in a previously known item or process unless the change involves something entirely new.

A patent application consists of the following:

1. A petition is directed to the Commissioner of Patents requesting the grant of a patent.
2. An oath is sworn to before a notary public or other designated officer of the validity of the patent request.
3. Specifications and claims are presented in which the claims to be patented are indicated along with detailed specifications, including drawings (if applicable), and other pertinent information defining the paten request.
4. Payment of the application filing fee.

The application is then examined and after a period of time, official action on the claim is taken.

14.8 Future Prospects

The new processes proposed by GADUTH offer the technical community the opportunity to commercialize these ideas. The sequence in the development of a process from an idea to a plant will vary with the magnitude and nature of each of the GADUTH projects. Each process should be studied so that any unique factors are analyzed. As the projects develop, information should become more complete for calculating the return on investment, which is the key to commercialization.

The following suggested sequence pertains to a new plant and is subject to modification to suit each specific application, i.e. the logical evolution of a process may proceed via the following stages:

1. The idea
2. Process development
3. Preliminary design
4. Estimating
5. Final design

It is important to evaluate the idea at an early stage so that the probable success can be determined and if further work is justified. This step can include a search of the literature for pertinent authoritative references.

Seawater conversion is one of the most difficult problems ever proposed for chemical engineering process analysis. This is mainly due to process complexities and the unusual economic requirements. However, the authors believe that the market for producing potable water from seawater is not only near unlimited but also astronomical from a financial perspective. Domestic clean water ministills in arid regions short on pure water supplies may soon be as common as filtered water systems in many domestic homes in the United States.

14.9 Illustrative Examples

Six illustrative examples complete the presentation of new desalination technologies highlighted in this chapter.

14.9.1 Illustrative Example 1

Discuss why there is often a need for collecting process data via pilot-plants for developing new desalination processes.

Solution. The need for pilot-plant operation to test the design of new desalination processes is discussed under the various unit operations incorporated into new desalination process development. Aside from the considerations of separation design reliability, it is often necessary to test the integrated operation of all units that comprise the new process. If pilot plant operation is needed for that purpose, all units must be included in the pilot-plant whether or not testing their individual performance is required. Interaction among process flow streams and recycle lines are particularly important to understand from pilot-plant data to ensure the reliability and functionality of new process designs.

14.9.2 Illustrative Example 2

Discuss the role that the number of processing steps plays in the technical and economic feasibility of new desalination process designs.

Solution. An important variable to consider in every newly developed process is the number of processing steps required to accomplish a desired outcome, in the case of desalination systems, the desired separation. The number of steps is not an absolute determinant of the feasibility or cost-effectiveness of a process, and there are many instances in which it is more economical to install a desalination sequence incorporating two or more steps even if it would be possible to accomplish the separation with a single-step process. However, in the majority of cases, a single-step process will prove more economical than a multistep process in terms of both capital and life-cycle costs due to its inherent simpler configuration. The KISS (Keep It Simple Stupid) principle generally applies to process design considerations as it does in many other aspects of life.

14.9.3 Illustrative Example 3

With regard to legal considerations, briefly describe the concept of *interference*.

Solution. A situation can arise in which two or more independent patent applications, covering essentially the same invention, are on file in the U.S. Patent Office. Although this rarely occurs, a procedure called *interference* is instituted to determine who is entitled to the patent. For example, an interference may also be instituted between a pending application and granted patent. Generally, interferences are decided on the basis of priority. The patent is granted to the applicant who was first to conceive the idea for the invention.

Because of the role of priority in any type of interference process, it is very important for an inventor to maintain complete records. A written description and sketches should be prepared by an inventor as soon as possible after the conception of an idea that might eventually be patented. This material should preferably be disclosed to one or more witnesses who should indicate in writing that they understand the purpose, method, and structure of the invention. The disclosure should be signed and dated by the inventor and the witnesses. Additional details are available through the U.S. Patent and Trademark Office (www.uspto.gov).

14.9.4 Illustrative Example 4

A proposed new desalination plant treats sea water with a salt concentration of 35,000 mg/L at a volumetric flow rate, Qin, of 1.75 m^3/s. The recovery is 75% and the process salt rejection rate is 99.6%. Determine if the concentration of salt in the discharge brine is below the regulatory requirement of a maximum 15% salt content.

Solution. Based on a recovery of 75%, the potable water discharge flow rate, Q_P, is:

$$Q_P = (0.75)\ Q_{in} = (0.75)(1.75\ \text{m}^3/s) = 1.31\ \text{m}^3/s$$

The brine discharge, Q_B, is

$$Q_B = Q_{in} - Q_p = 1.75 - 1.31 = 0.44 \text{ m}^3/s$$

Since the salt rejection is 99.6%, the concentration of salt in the potable water stream, C_P, is:

$$C_P = (1 - 0.996)(35{,}000 \text{ mg/L}) = (0.004)(35{,}000 \text{ mg/L}) \\ = 140 \text{ mg/L}$$

Carrying out a salt mass balance and solving yields a value for the brine salt concentration, CB:

$$(1.75 \text{ m}^3/\text{s})(35{,}000 \text{ mg/L}) = (1.31 \text{ m}^3/\text{s})(140 \text{ mg/L}) \\ + (0.44 \text{ m}^3/\text{s})(C_B)$$

$$C_B = \left[(61{,}250 \text{ m}^3\text{-mg/L-s}) - (183.4 \text{ m}^3\text{-mg/L-s})\right] / (0.44 \text{ m}^3/\text{s}) \\ = 138{,}788 \text{ mg/L}$$

This brine salt concentration is ≈ 140,000 mg/L ≈ 14%, so the brine discharge satisfies the regulatory limit of maximum 15% concentration.

14.9.5 Illustrative Example 5

Plans are underway to construct and operate a commercial desalination plant. The company is still undecided as to whether employ an RO unit or a heat conduit evaporator (HCE) unit to accomplish the separation. The RO unit is less expensive to purchase and operate than an HCE unit. However, the projected income from the HCE unit is higher since it can handle a larger quantity and different sea water concentrations. Based on economic and financial data provided here, select the process that will yield the higher annual profit. Calculations should be based on an interest rate, *i*, of 12% and a lifetime, n, of 20 years for both processes.

Solution. Calculate the capital recovery factor (CRF) to generate annual costs for the capital and installation costs of the units using Equation 14.6 as shown in Table 14.2:

$$\text{CRF} = \frac{i}{1-(1+i)^{-n}} = \frac{0.12}{1-(1.12)^{-20}} = \frac{0.12}{1-0.104} = 0.134 \qquad (14.1)$$

Table 14.2 Cost and Income Data for Desalination Options for Illustrative Example 5.

Capital and income	RO unit	HCE unit
Capital ($)	2,625,000	2,975,000
Installation ($)	1,575,000	1,700,000
Operation ($/yr)	400,000	550,000
Maintenance ($/yr)	650,000	775,000
Income ($/yr)	2,000,000	2,500,000

Using this CRF, the annualized cost for capital and installation for the two units are:

For RO; Annualized Cost for Capital + Installation is:

$$\begin{aligned}(\text{Capital} + \text{Installation})(\text{CRF}) &= (\$2{,}625{,}000 + \$1{,}575{,}000)(0.134)\\ &= \$562{,}800/\text{yr}\end{aligned}$$

For EV; Annualized Cost for Capital + Installation is:

$$\begin{aligned}(\text{Capital} + \text{Installation})(\text{CRF}) &= (\$2{,}975{,}000 + \$1{,}700{,}000)(0.134)\\ &= \$626{,}450/\text{yr}\end{aligned}$$

Table 14.3 summarizes the total annual costs for the RO and EV units, and provides the annual profit based on the total annual costs minus the annual income. Based on results shown in Table 14.3, the HCE unit provides a higher annual income, and therefore should be the unit of choice for the company's new desalination plant.

Table 14.3 Comparison of Total Annual Cost and Net Annual Income Data for Desalination Options for Illustrative Example 5.

Cost and income	RO unit	HCE unit
Total installed annual cost ($)	562,800	626,450
Operation ($)	400,000	550,000
Maintenance ($/yr)	650,000	775,000
Total annual cost ($/yr)	1,612,800	1,951,450
Income ($/yr)	2,000,000	2,500,000
Net annual profit ($/yr)	387,200	548,550

14.9.6 Illustrative Example 6

A large, deep cavern (formed from a salt dome) is located near the site of a proposed desalination plant and has been suggested as a long-term disposal site for the plant's brine reject flow. A risk assessment was conducted during the preliminary study and the results indicate that there was a greater than 99% probability that no hazardous or toxic material would "meander" beyond the cavern walls during the next 25 years. The company preparing the permit application for the proposed desalination plant has provided the following data and information:

Approximate total volume of cavern = 0.08 mi^3.
Approximate volume of cavern available for waste depository = 75% of total volume.
Proposed maximum brine feed rate to the cavern = 1,900,000 lb/day.
Feed rate schedule = 7 days/week.
Average bulk density of brine = 72 lb/ft^3.

Calculate the volume of the cavern in units of mi^3 and ft^3 that would be available for brine disposal and estimate the minimum amount of time it will take to fill this cavern with desalination plant brine water. Comment on this option as a long-term solution for plant brine disposal.

Solution. The volume of the cavern, V, in mi^3 available for desalination plant brine disposal is:

$$V = (0.75)(0.08\ \text{mi}^3) = 0.06\ \text{mi}^3$$

This volume can be converted to ft^3 as:

$$V = (0.06\ \text{mi}^3)(5{,}280\ \text{ft}/\text{mi})^3 = 8.83\ x 10^9\ \text{ft}^3$$

The daily volume rate of plant brine to be deposited within the cavern in ft^3/day, QB, is:

$$Q_B = (1{,}900{,}000\ \text{lb}/\text{d})/(72\ \text{lb}/\text{ft}^3) = 26{,}389\ \text{ft}^3/\text{d}$$

The brine disposal volumetric rate can now be converted to ft^3/year as:

$$Q_B = (26{,}389\ \text{ft}^3/\text{d})(7\ \text{d}/\text{wk})(52\ \text{wk}/\text{yr}) = 9{,}605{,}596\ \text{ft}^3/\text{yr}$$

The time it will take to fill the cavern with the brine waste is therefore:

$$t = V/Q_B = \left(8.83\,x10^9\ \mathrm{ft}^3\right)/\left(9.610^6\ \mathrm{ft}^3/\mathrm{yr}\right) = 920\ \mathrm{yr} \tag{14.2}$$

This cavern obviously has the long-term capacity to act as a disposal site for the proposed water desalination plant. It should be noted that deep-well injection of this sort is an ultimate disposal method that transfers wastes far underground and away from freshwater sources. Like landfarming this disposal process has been used for many years by the petroleum industry. It is also used to dispose of saltwater in oil fields. When the method first came into use, the injected brine would often eventually contaminate groundwater and freshwater sands because the sites were often poorly chosen. The process has since been improved, and laws such as the Safe Drinking Water Act of 1974 ensure that sites for potential disposal wells are better surveyed and evaluated than have been in the past.

Many factors are considered in the selection of a deep-well injection site. The rock formation surrounding the disposal zone must be strong but permeable enough to absorb the wastes, and the site must be far enough from drinking water sources to prevent contamination. Once a site is selected, it must be tested by drilling a pilot well. The performance data from the pilot well, besides testing permeability and water quality, also aid in the design of the final well and in determining the proper injection rate for sustainable operation.

14.10 Problems

1. Discuss the limitations associated with extraction processes.
2. Qualitatively discuss the freeze crystallization process.
3. Briefly describe copyrights and trademarks.
4. Regarding problems associated with brine water management and disposal, highlight the difference between a "direct discharger" and an "indirect discharger."
5. A small evaporator is fed 1,000 kg/min of sea water containing 0.005 mass fraction salt. The overhead flow rate, D, of pure water is 400 kg/min. Calculate the rate of flow, W, and the salt concentration, x_W, of the brine wash.
6. A tank contains 1,000 kg of brackish water. It is desired to crystallize 400 kg of ice, D, from this brackish water. Calculate the resulting brine volume, W, and concentration, x_W. Assume a brackish water salt concentration of 0.5% and a pure water salt concentration of 0.0%
7. Refer to Illustrative Examples 4 and 6 earlier. For the desalination plant described in Illustrative Example 4, determine the plant production capacity,

Q_P, and brine flow, Q_B, required for disposal in units of ft^3/d and MGD. Asses the suitability of the brine disposal cavern described in Illustrative Example 6 in terms of the time to fill the cavern with the brine flow generated from this plant.

References

Ganiaris, N. (1972). *Desalination Process*. U.S. Patent # 3,675,436. July 11.

Ganiaris, N. (2018). © *The GADUTH Mangrove Process*. New York, NY.

Ganiaris, N., Dupont, R.R., and Theodore, L. (2018). © *The Theodore Simple Still*. East Williston, NY.

Hagen, K. (1999). *Heat Transfer with Applications*. Upper Saddle River, NJ: Prentice-Hall.

Harder, E. (1982). *Fundamentals of Energy Production*. Hoboken, NJ: John Wiley & Sons.

Kern, D. (1950). *Process Heat Transfer*. New York City, NY: McGraw-Hill.

Theodore, L. (2016). *Chemical Engineering: The Essential Reference*. New York City, NY: McGraw-Hill.

Theodore, L. (2017). Patent No. 63/184,945, 5/16/2021. *The Heat Conduit Evaporator*. East Williston, NY.

Theodore, L. (2018). © *The Geothermal Evaporator*. East Williston, NY.

Theodore, L. and Rici, F. (2010). *Mass Transfer Operations for the Practicing Engineer*. Hoboken, NJ: John Wiley & Sons.

15

Non-Desalination Processes

15.1 Introduction

Here is the reality as it relates to desalination: potable water demands continue to increase as populations around the world keep growing as "fresh water" reserves continue to decline due to increased consumption and pollution. In addition, energy demands are also increasing due to global industrialization, with natural gas emerging as the primary source of energy for most of the world. Developing economically attractive and viable non-desalination systems may soon become the approach of choice for producing potable water.

Developing new sources of potable water other than from the traditional sources and the new desalination options that were described in Chapters 13 and 14 is addressed in this chapter. In the near term, the main author believes that the preferred approach for developing new water sources will not involve seawater, deep wells, saline aquifers, and/or wastewater, but rather employing fossil fuels as the primary source of additional potable water. This could hopefully be used in disaster situations, in response to water crises, or poverty-stricken areas and developing countries. The proposed technologies rely on other sources to produce large amounts of potable water. Although desalination has been employed over the years to produce potable water, and these methods are well documented, little has been done in commercially implementing these non-desalination processes.

This chapter introduces the reader to non-desalination process. As noted, processes were reviewed in Chapter 14, and four new options for desalination developed in part by the authors were presented here. Details of four non-desalination topics/processes in the material to follow includes: water obtained from fossil fuels (WOFF), water harvested from other contaminated flow streams, and other sources. An application section provides six illustrative examples related to the general subject of non-desalination options for potable water production, with most of these illustrative examples keying on the WOFF process.

Introduction to Desalination: Principles, Processes, and Calculations, First Edition. Louis Theodore and R. Ryan Dupont.

15.2 Water Obtained from Fossil Fuels (WOFF)

GADUTH Engineers have proposed a process that generates Water that is Obtained from Fossil Fuels (WOFF) following any energy conversion device. The WOFF process (Theodore 2021) involves the reaction of a fossil fuel or any hydrogen-bearing compound that produces a product at an elevated temperature that contains water. For a combustion reaction, a flue gas at elevated temperatures is normally produced. This flue gas may be subsequently condensed by any suitable traditional means, including decreasing the temperature and increasing the pressure, to condense the water vapor present. The water vapor arises primarily from the combustion of the hydrogen in the fossil fuel; additional water may be present due to water in the fossil fuel and water vapor in the air. In effect, potable water can be recovered from flue gas discharges.

Combustion flue gas emissions vary significantly with the type of fossil fuel fired. These emissions can significantly impact the quality of any water produced and condensed following combustion. Problems arise if coal is the fossil fuel source; the problems with oil are not nearly as significant. However, environmental emission problems are not a concern when employing natural gas as the fossil fuel; the USEPA, USDOE, and EPRI have reported only trace amounts of nitrogen oxides, CO, VOC, sulfur oxides, and particulate matter in flue gases from natural gas combustion. In addition, only trace amounts of these trace gases ultimately are absorbed by the condensed water, which may account for the fact that no contaminants in water have been recorded when combusting natural gas. One may thus conclude that the flue gas components on combusting natural gas are CO_2, H_2O, and N_2.

Although many fossil fuels and related compounds may be employed in the WOFF process, the combustion of natural gas, consisting primarily of methane with stoichiometric. air, would be the most attractive choice from a process and economic perspective. (See also Table 15.1.) One can show that for this WOFF case, the water present (approximately 20% by volume) will condense, as relatively pure potable water, when the temperature is cooled to 138.6°F; the fossil fuel propane scenario produces a condensation temperature of 130.8°F. These represent the temperatures at which water will start to be produced. Furthermore, approximately 50% and 75%, of the water will condense at 112°F and 90°F, respectively. These calculations are based on the combustion of natural gas (95% CH_4, 5% C_3H_8) producing a partial pressure of H_20 in the 0.155 to 0.190 atm range with a corresponding saturated temperature of approximately 135°F. Similar calculations have been performed for other classes of fossil fuels.

What do the calculations indicate, i.e. what conclusions can be drawn? If the WOFF flue gas produced following the combustion of natural gas is cooled

Table 15.1 Physical and Chemical Properties of Methane.

Property	Value
Molecular weight, MW	16.04
Boiling point at 1.0 atm	–263.2°F
Melting point at 1.0 atm	–296.5°F
Heat capacity	0.528 Btu/lb-°F at 60°F, 1 atm
Critical temperature, Tc	–116.5°F
Critical pressure, Pc	45.8 atm
Density	0.0448 lb/ft^3 at 32°F, 1 atm
Density at boiling point	0.1124 lb/ft^3
Density (of liquid) at boiling point	26.47 lb/ ft^3
Heat of vaporization at boiling point	248.4 Btu/lb
Higher heating value, HHV	23,890 Btu/lb (978 Btu/ft^3) at 77°F, 1 atm

below approximately 135°F, H_20 will start to condense. Obviously, more water will be condensed as the temperature is decreased. As noted, 50% the water present will condense if the temperature is reduced to approximately 112°F and approximately 75% of the water will condense at approximately 90°F. For the former case, the author's home heating system would annually produce approximately 2,500 lb of H_20 if the flue gas condensation process was operated at 112°F.

GADUTH have shown that the WOFF process can be enhanced by a variety of methods including the following four.

1. Employ stoichiometric oxygen (Equation 15.2) as opposed to stoichiometric air (Equation 15.1). This eliminates the nitrogen in the flue gas producing a higher concentration, i.e. higher partial pressure of water vapor in the flue gas.

$$CH_4 + 2\,O_2 + 7.53\,N_2 \rightarrow \underset{\text{(flue gas)}}{CO_2 + 2\,H_2O + 7.53\,N_2} \tag{15.1}$$

Concentration of water vapor = 2/(1 + 2 + 7.53) = 2/(10.53) = 0.19 = 19%

$$CH_4 + 2\,O_2 \rightarrow \underset{\text{(flue gas)}}{CO_2 + 2\,H_2O} \tag{15.2}$$

Concentration of water vapor = 2/(1 + 2) = 2/(3) = 0.67 = 67%

2. The WOFF can be enhanced by producing elemental carbon (carbon black/graphite) via a catalytic process, e.g. the Thermal Black process (Equation 15.3) which converts methane to amorphous carbon and hydrogen. This requires less air (and thus less nitrogen) and produces no carbon dioxide, thus increasing the concentration of water vapor in the flue gas. Although half the energy normally generated is lost, it can be recovered if the carbon is combusted to form carbon dioxide.

$$CH_4 + O_2 + 3.77\ N_2 \rightarrow C\downarrow + \underset{\text{(flue gas)}}{2\ H_2O + 3.77\ N_2} \tag{15.3}$$

   Concentration of water vapor = 2/(2 + 3.77) = 2/(5.77) = 0.35 = 35%

3. The WOFF process can be enhanced by producing carbon monoxide rather than carbon dioxide via the incomplete combustion of methane. This will marginally increase the concentration of water vapor (Equation 15.4).

$$CH_4 + 1.5\ O_2 + 5.65\ N_2 \rightarrow \underset{\text{(flue gas)}}{CO + 2\ H_2O + 5.65\ N_2} \tag{15.4}$$

   Concentration of water vapor = 2/(1 + 2 + 5.65) = 2/(8.65) = 0.23 = 23%

4. The WOFF process can be enhanced by recovering carbon dioxide from the flue gas by any traditional capture method (e.g. scrubbing with magnesium oxide). The elimination of carbon dioxide in the flue gas will again increase the concentration of water vapor as shown in Equation 15.5.

$$CH_4 + 2\ O_2 + 7.53\ N_2 \rightarrow \underset{\text{(flue gas)}}{CO_2 + 2\ H_2O + 7.53\ N_2} \tag{15.5}$$

   Concentration of water vapor = 2/(2 + 7.53) = 2/(9.53) = 0.21 = 21%

5. The WOFF process can be enhanced by increasing the pressure of the system. This step will increase the quantity of water condensed.

Based on the WOFF calculations, one can apply the WOFF process to not only a power plant but also to small domestic (or commercial) units for generating water that is either potable or water that can be further easily treated by traditional methods to ensure that it is potable. It has been shown that the combustion of natural gas (a clean fossil fuel) produces only trace quantities of undesirable gases and of which only trace quantities will be absorbed by the condensed water. Preliminary economic calculations indicate that the cost for producing byproduct water from a WOFF energy combustion process employing natural gas are favorable.

GADUTH's proposed patent-pending claims regarding the WOFF process are provided shortly. It should be noted that all GADUTH's present WOFF activities have centered on natural gas as the fuel source.

1. A process designed to recover water from the combustion of any fossil fuel, including:
   a. Natural gas
   b. Oil
   c. Coal
   d. Shale
   e. Tar sands

   Any combustion equipment or any energy conversion device, including utility boilers, domestic boilers, diesel engines, thermal and catalytic reactors, etc., may be employed.
2. With reference to Claim 1, any process to remove nitrogen from the flue gas produced in the combustion process (e.g. selective adsorption).
3. With reference to Claim 1, any process to remove carbon dioxide from the flue gas produced in the combustion process.
4. With reference to Claim 1, any process that converts the carbon in the fossil fuel to either elemental carbon and/or carbon monoxide.
5. Enhance the condensation of water vapor in the flue gas by increasing the pressure of the condensation system.
6. Enhance the condensation of water by both increasing the pressure and decreasing the temperature of the condensation process.
7. Employ nitrogen depleted air, including operations involving membranes.
8. Employing natural gas (or a fossil fuel) containing water vapor.
9. Humidifying the combustion air prior to combustion.
10. Employ a two (or multi) stage heat exchanger (condenser) to enhance acid gas removal.
11. Employ the energy (heat of combustion) to cool via refrigeration and/or pressurize the flue gas.
12. Any combination of the above.

15.3 Water Harvested from Contaminated Flow Streams

Vacuum systems can be employed to "extract" water from not only saline water, but water of any quality. Decreasing the pressure above a water body will effectively evaporate pure water from most contaminated streams. There

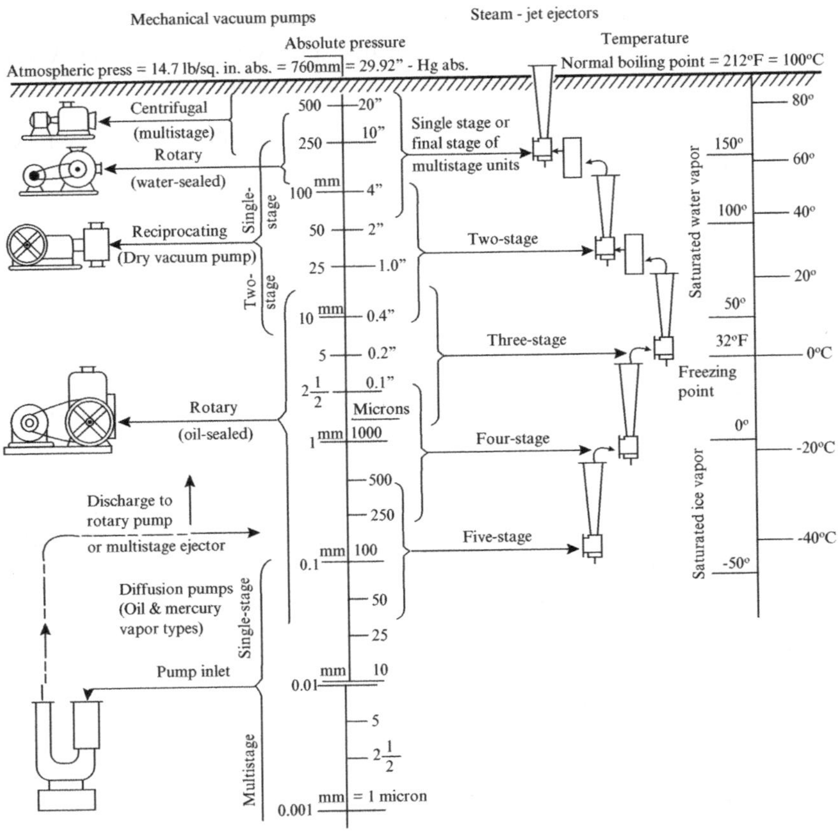

Figure 15.1 Vacuum Levels Attainable with Various Types of Equipment (Green and Southard 2019).

is various vacuum producing equipment to accomplish this operation. A brief write-up on vacuum equipment follows.

The attainment of various levels of vacuum is related to available equipment as presented in Figure 15.1. The equipment in Figure 15.1 does not include diffusion pumps. In a diffusion pump, a liquid of low absolute vapor pressure is boiled in a reservoir. The vapor is ejected at high velocity in a downward direction through multiple jets and is condensed on the walls, which are cooled by surrounding coils. Molecules of the gas being pumped enter the vapor stream and are driven downward by collisions with the vapor molecules. The gas molecules are removed through the discharge line by a pump such as a rotary oil-sealed unit. These diffusion pumps operate at very low pressures. The ultimate vacuum attainable depends somewhat upon the vapor pressure of the pump liquid at the temperature of the condensing surfaces. By providing a cold trap between the diffusion pump and the region being evacuated, pressures as low as 10^{-7} mmHg absolute are achievable. Liquids used for diffusion pumps are mercury and oils of low vapor pressure.

The unit of choice in most applications, however, is often an ejector. Ejectors are easy to operate and require little maintenance. Installation costs are low. Since they have no moving parts, they have long life, sustained efficiency, and low maintenance costs. Ejectors are suitable for handling practically any type of gas or vapor. They are also suitable for handling wet or dry mixtures or gases containing sticky or solid matter such as chaff or dust. Ejectors are available in many materials of construction to suit process requirements. If the gases or vapors are not corrosive, the diffuser is usually constructed of cast iron and the steam nozzle of stainless steel. For more corrosive gases and vapors, many combinations of materials such as bronze, various stainless steel alloys and other corrosion-resistant metals, carbon, and glass can be used (Green and Southard 2019).

One can conclude from this discussion that any liquid source of water (not only seawater) when exposed to pressures below atmospheric conditions will be transferred from its original liquid state to the vapor state. The water vaporized would be essentially pure potable water. GADUTH Engineers are currently investigating this evaporative process. For example, if polluted water is exposed to a vacuum, pure potable water will be separated from its original source, expanding the possible sources of potable water production beyond saline or brackish waters alone.

15.4 Other Sources

There are other sources of water including:

1. The atmosphere
2. Inorganic chemicals
3. Salination

Each are briefly discussed here.

15.4.1 The Atmosphere

Simply put, there is water in the atmosphere. What remains is to develop a viable, economic process to "extract" the water (humidity) from the air. Early efforts to collect water from the atmosphere focused on fog/dew collection that could coalescence small water droplets, generating big droplets that could be recovered by airflow or gravity (Zhou et al. 2020), with the primary objective to accelerate the growth of water droplets adhering to the surface of collectors and decrease the critical size of water droplets falling from the surface.

However, the dependency on saturated local humidity (relative humidity, RH = 100%) significantly hinders the practical application of fog/dew collection in arid regions where the water is needed the most.

Zhou et al. (2020) provide an excellent review of Advanced Water Harvesting (AWH) which is a newly developing technology primarily focused on moisture harvesting in the gaseous form. The objective of AWH is to liquefy moisture, generating collectible liquid water, not using the conventional, energy intensive (~2450 kJ/kg at 20°C) method of cooling the ambient air below its dew point and collecting the condensate, but using innovative materials for tailoring the interaction with water molecules in the atmosphere. The ideal AWH system might utilize hydrogels with high saturated water to enhance water uptake, along with environmentally responsive polymers that can directly release water in the liquid form. The ideal harvesting materials should have a porous structure for quick vapor diffusion during moisture capture, appropriate interaction with water for easy release, and a stable structure for long-term cycling stability. Trade-offs between performance and material stability would be expected, and extensive optimization of these newly conceived AWH systems is needed in order for this innovative water harvesting approach to become an economic reality to alleviate looming water scarcity problems across the globe.

15.4.2 Inorganic Chemicals

React an acid with a base and what is the product? A salt plus water (Equation 15.6).

$$\text{Acid} + \text{Base} \rightarrow \text{Salt} + H_2O \tag{15.6}$$

This is, of course, a source of water. Any inorganic reaction that produces water is an eligible candidate as a source of water, as long as the resulting salt has a low enough solubility that the salt concentration is below 500 mg/L TDS. Above this TDS level the water is not suitable for potable use but may be useful for some other beneficial purpose, i.e. irrigation, stock watering, etc.

15.4.3 Salinization

While desalination will produce potable water, a by-product of the process is salinization. The process results in a "salt pickup" in the discharge brine stream. This brine has traditionally been viewed as an environmentally unfriendly and useless by-product. However, this discharge stream contains two valuable chemicals, salts *and* water. Once again, this represents another source of water that may be available for recovery.

Evaluation of the desirability and cost effectiveness of additional processing of desalination system discharge brine has been carried out by the Bureau of Reclamation and its contractors over several decades and has resulted in a range of systems considered to be Zero Liquid Discharge (ZLD), systems that include: a Brine Concentrator and Recovery System (BCRS) for salt and water recovery from RO brine (University of Texas, El Paso 2002); a ZLD process with enhanced freshwater yield and production of salable sodium chloride (NaCl), magnesium hydroxide ($Mg(OH)_2$), and bromine (Br_2) from the SWRO reject using electrodialysis (ED) to reduce the salinity of the reject stream so that the salt-depleted reject stream can be recycled to improve freshwater yield (University of South Carolina 2006); and a Zero Discharge Desalination (ZDD) process capable of extracting up to 98% of the water from brackish groundwater that has high levels of calcium sulfate, silica, and other dissolved constituents by combining reverse osmosis or nanofiltration, and a secondary process called electrodialysis metatheses to recover salts and pure water from an RO reject stream (Cappelle et al. 2014).

Widespread interest in increasing desalination water recovery efficiency and recovery of valuable minerals from brine reject water emphasizes the desire to use brine reject water as an additional resource rather than a waste product. Al Bazedia et al. (2013) indicate that using nanofiltration (NF) as pretreatment and adopting salt recovery schemes, especially membrane crystallization, provide higher water recovery and produce valuable products in both a technically feasible and economically viable manner. Morillo et al. (2014) reviewed a range of options for both increasing desalination water recovery, through multiple step processing of reject brine, and options for commercial salt recovery from produced brine streams. Finally, Hassan (2014) has shown that ZLD techniques for desalination brine treatment in Egypt will be cost effective for both yielding additional water and valuable salt recovery.

15.5 Illustrative Examples

Six illustrative examples complete the presentation of non-desalination technologies highlighted in this chapter.

15.5.1 Illustrative Example 1

Discuss some environmental issues associated with natural gas utilization for energy.

Solution. All fossil fuels, i.e. coal, petroleum, and natural gas, release pollutants into the atmosphere when burned. However, of these primary fuels, natural gas is the most environmentally friendly fossil fuel. Combusting natural gas produces less oxides of sulfur, carbon, and nitrogen than other fossil fuels. Natural gas also emits little to no ash (particulate) into the environment.

The main sources of environmental concern related to natural gas may be placed in these four categories:

1. Pipeline leakage
2. Liquefied natural gas transport risks
3. Combustion: carbon emissions, SOx, NOx
4. Health and safety

Items 1 and 2 are generally not considered major problems. Combustion issues have yet to be resolved, but release of CO_2 is preferred over direct methane injection due to methane's increased global warming potential (approximately 20 fold) compared to CO_2. Health and safety remain a concern for any industrial process (Theodore and Dupont 2012).

15.5.2 Illustrative Example 2

You have been requested to provide comments on the merits of traditional desalination processes compared to non-desalination processed for the production of potable water. Provide a response.

Solution. Experience and past history are the keys to the merits associated with traditional desalination processes. Non-desalination processes, for the most part, suffer primarily from the lack of performance data and unknown process performance information. On the positive site, new potential sources of potable water may be realized with successful non-desalination processes.

15.5.3 Illustrative Example 3

Natural gas is at the heart of the WOFF process. Provide some background material on this fossil fuel (Skipka and Theodore 2014).

Solution. Natural gas is a nonrenewable fossil fuel, because most scientists and engineers believe that natural gas (as with oil) was formed from the remains of tiny sea animals and plants that died 300 to 400 million years ago. When these sea animals and plants died, they sank to the bottom of the oceans where they were buried by layers of sediment that turned into rock. Over the years, these layers increased in height to several thousands of feet, subjecting the remains to high pressure, which, when combined with the high temperature of the Earth, converted this matter into not only oil, but also natural gas. Over time, high concentrations of natural gas became trapped in the layers of rock.

As with other fossil fuels, natural gas was formed in the Earth's crust, possessing the ability to rise toward the Earth's surface. It built up in pockets when its progress was blocked by an upper layer of solid impervious rock. This process has been described as being analogous to collecting water with a sponge. These gas reservoirs can be reached by drilling wells into these pockets. It should be mentioned that natural gas is produced on all continents except Antarctica.

Natural gas consists of a mixture of different gases. The main component is methane (CH_4) which is odorless, colorless, and tasteless. Companies add a chemical odorant called mercaptan (with an odor like rotten eggs) so that any escaping gas can be detected. This process is employed as a safety measure. Other components of natural gas include ethane, propane, butane, and pentane. Other gases include carbon dioxide, nitrogen, and helium, all of which reduce the heating value of the gas mixture. The density of natural gas at ambient conditions is approximately 0.60 lb/ft^3, compared to that of air at 0.75 lb/ft^3. The heating value is approximately 1,000 Btu/scf. The high hydrogen content of natural gas, compared to oil or coal, results in more water vapor being produced in the products of combustion. This results in a correspondingly lower thermal efficiency but also reduced CO_2 emissions. The important physical and chemical properties of CH_4, in engineering units are provided in Table 15.1.

15.5.4 Illustrative Example 4

The composition and combustion properties of a gas mixture are given in Table 15.2. Determine the gross heating value (*HHV*) of this gas (Theodore et al. 2007).

Solution. The *HHV* represents the enthalpy change or heat released when a gas is stoichiometrically combusted at 60°F, with the final (flue) products at 60°F and any water present in the liquid state. Stoichiometric combustion requires that no oxygen be present in the flue gas following combustion of the hydrocarbons.

Table 15.2 Composition and Combustion Properties of a Gas Mixture for Illustrative Example 15.4.

Component	Mole fraction, x_i	Gross heating value, Btu/scf
N_2	0.0515	0
CH_4	0.8111	1,013
C_2H_6	0.0967	1,792
C_3H_8	0.0351	2,590
C_4H_{10}	0.0056	3,370
Σ	1.0000	

Using values from the table, the gross heating value of the gas mixture in Btu/scf is given by Equation 15.7 as:

$$HHV = \sum x_i HHV_i \tag{15.7}$$

where HHV_i = the gross heating value of individual gases. Substituting and solving yields:

$$HHV = (0.05155)(0) + (0.811l)(l{,}013) + (0.0967)(1{,}792) + (0.035l)(2{,}590) + (0.0056)(3{,}370)$$

$$HHV_i = 1{,}105 \text{ Btu/scf}$$

The gas described in this example is typical of natural gas. The net heating value (*NHV*) is similar to *HHV* except this is the energy value of the mixture with the water in the vapor state, accounting for the heat of vaporization of water. The *NHV* is also known as the *lower heating value*.

15.5.5 Illustrative Example 5

Assume perfect combustion of a contaminated fuel oil sample with stoichiometric air. The gravimetric weight percentage of a sample of this fuel oil is given as: 88.52% carbon, 10.87% hydrogen, 0.4% sulfur, 0.1% nitrogen, 0.06% oxygen, and 0.05% ash. Calculate the gravimetric analysis of the flue gas and the total volume of flue gas (at 500°F and 1 atm) per pound of oil burned.

Solution. Convert the gravimetric percent of each component to lbmol selecting 100 lb of fuel as a basis.

C 88.52/12 = 7.38 lbmol
H 10.87/1 = 10.87 lbmol
S 0.40/32 = 0.0125 lbmol
O 0.06/16 = 0.00375 lbmol
N 0.10/14 = 0.00714 lbmol

Since the molecular weight of the ash is unknown and its content is small, it is neglected in the calculations. Write the combustion reaction for each component in the fuel:

$$C + O_2 \rightarrow CO_2$$

$$2\,H + O_2 \rightarrow H_2O$$

$$S + O_2 \rightarrow SO_2$$

Determine the amount of oxygen required for each component. Use the previous results to complete Table 15.3:

$$\text{Amount of } N_2 = (\text{Amount of } O_2)(0.79 \text{ lbmol } N_2)/(0.21 \text{ lbmol } O_2)$$
$$= (10.1)(3.76) = 38.0 \text{ lbmol}$$

Determine the gravimetric composition (percentage) of the flue gas using Table 15.4.

Calculate the total lbmol of flue gas produced per pound of oil burned:

$$\text{lbmol of flue gas} = \text{total lbmol/100 lb fuel} = (50.85 \text{ lbmol})/100 \text{ lb fuel}$$
$$= 0.5085 \text{ lbmol/lb oil}$$

Table 15.3 Composition and Combustion Properties of the Fuel Oil for Illustrative Example 15.5.

Component	lb in fuel	lbmol O_2 required
C	7.38	7.38
H	10.87	2.72
S	0.0125	0.0125
O	0.00375	-0.00188
Σ		10.1

Note: Oxygen in the fuel reduces air requirements. Therefore, the amount of oxygen in the fuel has been subtracted from the amount of O_2 required. The amount of nitrogen is calculated from the oxygen requirement.

Table 15.4 Gravimetric Composition of the Flue Gas for Illustrative Example 15.5.

Component	lbmol	MW	lb	Wt%
CO_2	7.38	44	324.7	21.8
H_2O	5.44	18	97.9	6.6
SO_2	0.0125	64	0.08	0.005
O_2	0.0	32	0	0
N_2	38.0	28	1,064	71.6
Σ	50.85		1,486.7	100

Table 15.5 Composition of the Dry Flue Gas from Illustrative Example 15.6.

Component	lbmol
CO_2	7.38
SO_2	0.0125
O_2	0.0
N_2	38.0
Σ	45.39

Remember that 100 lb of fuel was used as a basis. Calculate the total volume of flue gas at 500°F and 1 atm using the ideal gas law (Equation 15.8) as follows:

$$V = \frac{nRT}{V} = \frac{(\text{lbmol flue gas / lb oil})\left(\begin{array}{l}0.7302\text{ft}^3\\ \text{-atm/lbmol-°R}\end{array}\right)(500+460°\text{R})}{1.0\text{ atm}}$$
$$= 356.1\ \text{ft}^3/\text{lb oil} \qquad (15.8)$$

15.5.6 Illustrative Example 6

Calculate the volume percent of CO_2 in the *dry* flue gas from Illustrative Example 5.

Solution. For the dry flue gas from Illustrative Example 5, the composition is as shown in Table 15.5.

The volume percentage of CO_2 in the dry flue gas is:

$$\%CO_2 = \frac{\text{lbmol } CO_2}{\text{total lbmol dry flue gas}}(100) = \frac{7.38}{45.39}(100) = 16.26\%$$

15.6 Problems

1. Discuss environmental management approaches that should be employed when combusting fossil fuels.
2. Discuss the three major types of boilers used for natural gas combustion in commercial, industrial, and utility applications.
3. Provide a balanced chemical combustion reaction for the natural gas given in Illustrative Example 4.

Table 15.6 Composition and Combustion Properties of a Gas Mixture for Problem 6.

Component	Mole Fraction, x_i
N_2	0.04
CH_4	0.85
C_2H_6	0.007
C_3H_8	0.03
C_4H_{10}	0.01
Σ	1.0000

4. Determine the total cubic feet of combustion products per scf of natural gas burned based on results of Problem 3.
5. Why is energy the "big enemy" of desalination processes?
6. The composition and combustion properties of a natural gas are provided in Table 15.6. Determine the gross heating value (HHV) of this natural gas employing the HHV values provided in Table 15.2.
7. A waste material is planned for disposal via combustion and the engineering firm designing the combustion unit is considering the WOFF process for water recovery. Assume complete combustion with stoichiometric air, and the gravimetric weight percentage of a sample of this waste is: 88% carbon, 10% hydrogen, 0.1% sulfur, 0.5% nitrogen, 1.35% oxygen, and 0.05% ash. Calculate the gravimetric analysis of the flue gas, the total volume of flue gas (at 500°F and 1 atm) per pound of waste burned, determine the concentration of water vapor in the flue gas potentially recoverable, and comment on these results compared to the use of natural gas for water recovery.

References

Al Bazedia, G., Ettouney, R.S., Tewfika, S.R., Soroura, M.H., and El-Rifai, M.H. (2013). Salt recovery from brine generated by large-scale seawater desalination plants. *Desalination and Water Treatment* 52 (25–27): 4689–4697. 10.1080/19443994.2013.810381.

Cappelle, M., Davis, T.A., and Gilbert, E. (2014). *Demonstration of Zero Discharge Desalination (ZDD)*. Desalination Research and Development Program Report No. 165, submitted to the Bureau of Reclamation. Denver, CO: Water Treatment Engineering and Research Group. https://www.usbr.gov/research/dwpr/reportpdfs/report165.pdf.

Green, D. and Southard, M. (2019). *Perry's Chemical Engineers' Handbook*, 9th Edition. New York, NY: McGraw-Hill.

Hassan, R.M.A. (2014). Using ZLD technique for safe and economic disposal of desalination plant brines in Egypt. *Journal of the Egyptian Academy of Environmental Development* 15 (2): 67–81.

Morillo, J., Usero, J., Rosado, D., El Bakouri, H., Riaza, A., and Bernaola, F.J.B. (2014). Comparative study of brine management Technologies for desalination plants. *Desalination* 336 (2014): 32–49.

Skipka, K. and Theodore, L. (2014). *Energy Resources: Availability, Management, and Environmental Impacts*. Boca Raton, FL: CRC Press/Taylor & Francis Group.

Theodore, L. (2021). Provisional Patent No. 63/145,701, 2/4/2021. *Water that Is Obtained from Fossil Fuels (WOFF)*. East Williston, NY.

Theodore, L. and Dupont, R.R. (2012). *Environmental Health and Hazard Risk Assessment: Principles and Calculations*. Boca Raton, FL: CRC Press/Taylor Francis Group.

Theodore, L., Ricci, F., and Van Vliet, T. (2007). *Thermodynamics for the Practicing Engineer*. Hoboken, NJ: John Wiley & Sons.

University of South Carolina (2006). *Zero Discharge Seawater Desalination: Integrating the Production of Freshwater, Salt, Magnesium, and Bromine*. Desalination Research and Development Program Report No. 111, submitted to the Bureau of Reclamation. Denver, CO: Water Treatment Engineering and Research Group. https://www.usbr.gov/research/dwpr/reportpdfs/report111.pdf.

University of Texas, El Paso (2002). *Zero Discharge Waste Brine Management for Desalination Plants*. Desalination Research and Development Program Report No. 89, submitted to the Bureau of Reclamation. Denver, CO: Water Treatment Engineering and Research Group. https://www.usbr.gov/research/dwpr/reportpdfs/report089.pdf.

Zhou, X., Lu, H., Zhao, F., and Yu, G. (2020). Atmospheric water harvesting: A review of material and structural designs. *ACS Materials Lett* 2 (7): 671–684. https://doi.org/10.1021/acsmaterialslett.0c00130.

Part III

Select Related Topics

Chapters in this Part include the following:

16

Water Conservation

16.1 Introduction

Since ancient times water has been considered the most suitable medium to clean, disperse, transport, and dispose of waste from human activities. The use of this resource for culinary, agricultural, industrial, and recreational needs puts an increasing demand on its management for multiple beneficial purposes. With the advent of industrialization and increasing population, the range of beneficial uses for water has expanded as has the demand for higher quality water. Over time, water requirements have emerged not only for drinking, but also for personal hygiene, fisheries, agriculture (irrigation and livestock supply), navigation for transport of goods, industrial production, cooling in power plants, hydropower generation, and recreational activities such as swimming or fishing. Fortunately, the largest demands for water quantity, such as for agricultural irrigation and industrial cooling, require the least stringent level of water quality. However, drinking water supplies generally require the most stringent water quality levels to ensure the health and safety of the public.

Water conservation is a cost effective alternative to the development of new water supplies, and refer to the actions that must be instituted to reduce the usage of potable water by directly reducing water use that demands potable water level quality, as well as to substitute the use of potable water for alternative water supplies for those beneficial uses not requiring potable water level treatment. This latter option would include the use of untreated surface water from reservoirs or direct use of reclaimed wastewater for municipal irrigation purposes (secondary irrigation water use) as is becoming more and more common across the Western U.S. It should be noted that water conservation steps are not static but subject to modification with changing circumstances, i.e. during times of drought, natural disasters, etc.

Societies worldwide identify water issues as a critical concern. Water is essential not only for domestic and agricultural purposes but is also essential in a range of industrial processes and support functions, facility operations, and as a feed ingredient for many products. Water use is increasing much faster

Introduction to Desalination: Principles, Processes, and Calculations, First Edition. Louis Theodore and R. Ryan Dupont.

than population growth as individual standards of living are improved across the globe. Regional and seasonal differences in water availability can lead to shortages and water-related conflicts, and the Water Footprint Network (2021) reported on a study that projects, that if present trends continue, over 9.7 billion people (52% of the world's population) could face water stress by 2050.

All sectors of the economy are recognizing the role they play in addressing water resource challenges, and industries, municipalities, and the agricultural sector are responding to looming water shortages by engaging in efforts to measure, report, and reduce their water use. A key reason why industries and agriculture choose to use water more efficiently is to improve their relationship with the communities that surround their facilities and fields. Water use is of paramount importance to a businesses' relationship with its surrounding community. Businesses that use less water in drought prone areas leave more water available for agricultural and residential uses, and more efficient agricultural water use can lead to less water stress on the community as a whole. Some industrial and agricultural businesses even risk being forced to withdraw from regions where water is a scarce resource, a problem that is likely to recur as water scarcity increases in many regions around the globe. Today, communities around industrial facilities and large agricultural water users show much greater concern for the amount of water a business uses than the amount of energy it uses.

This chapter focuses on a review of a range of water conservation approaches and ideas in current practice and/or potentially viable for the reduction of water use across all sectors of society. Topics to be reviewed include water usage; water waste; and conservation efforts in power generation, by agriculture, in municipal and domestic applications, and by industry. An application section provides six illustrative examples of concepts and implementation of water conservation practices throughout society.

16.2 Quantifying Water Use

Water use is usually defined and measured in terms of *withdrawal* or *consumption*, i.e. that which is taken and that which is used up. Withdrawal refers to water extracted from surface or groundwater sources, with consumption being that part of a withdrawal that is ultimately used and removed from the immediate water environment whether by evaporation, transpiration, incorporation into crops or a product, or other consumption. Conversely, *return flow* is the portion of a withdrawal that is actually not consumed, but instead is returned to a surface or groundwater source from a point of use and becomes available for further use (Equation 16.1).

$$\text{Water Withdrawal} = \text{Consumption} + \text{Return Flow} \tag{16.1}$$

Water use can also be divided into *offstream* and *instream* uses. Offstream water use involves the withdrawal or diversion of water from a surface or groundwater source for:

1. Municipal and residential use
2. Industrial use
3. Agricultural use
4. Energy development use

Instream water uses are those which do not require a diversion or withdrawal from surface or groundwater sources, such as:

1. Water quality and habitat improvement
2. Recreation
3. Navigation
4. Fish propagation
5. Hydroelectric power production

The distribution of total global water use across all sectors was summarized by Perkins (2018) as shown in Figure 16.1, where the major water use across the globe is for power generation, followed by irrigation for food production. Potable water, industrial and livestock watering/aquacultural are minor use categories compared to power generation and irrigation demands.

Water users can be divided into two basic groups: system users (such as residential users, industries, and farmers) and system operators (such as municipalities, state and local governments, and privately owned suppliers). These users can choose from among many different water use efficiency practices, which fall into two categories:

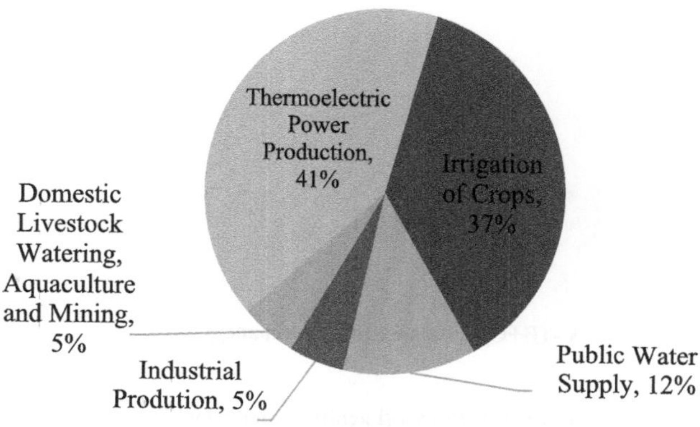

Figure 16.1 Global Water Withdrawals by Sector. Adapted from Perkins (2018).

1. Engineering practices: practices based on modifications in plumbing, fixtures, or water supply operating procedures
2. Behavioral practices: practices based on changing water use habits

One way to approach understanding water conservation options and determine the optimal practices appropriate for a given water user is to carry out a water balance. A typical water balance calculation is provided in Figure 16.2. When creating a water balance, all water use should be noted, and water flows from the source through all operations including indoor and outdoor uses, evaporative losses, input to products, and wastewater discharges should be tracked. Common end uses of water include processing and auxiliary processing, heating and cooling, indoor domestic use, and landscape irrigation. A water balance should capture all of the flows into and out of a system and should verify that there is a balance between flows in and flows out.

The process of putting together a water balance helps to capture a baseline of current water usage and help identify sources of leaks and any potential compliance concerns. Conducting water balances on a regular basis aids in identifying ways to reduce water use and discover hidden water-savings dividends.

To complete a water balance, data should be collected on the volume of water flows for each relevant use being considered. Sources of data can include:

1. Water and sewer bills from the previous 1–3 years
2. Numbers, sizes, and locations of water meters
3. Sources of incoming water
4. Diagrams of plumbing and water pipes, as well as irrigation schematics

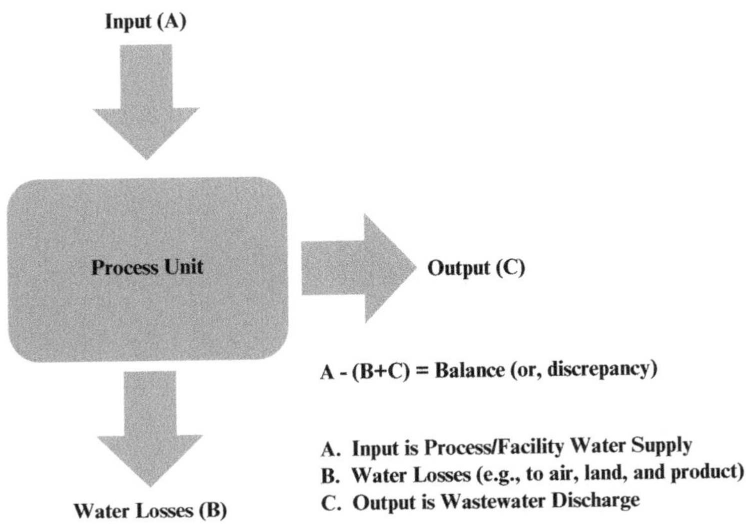

Figure 16.2 General Water Balance Approach.

5. List of water consuming activities
6. If applicable, information about unsewered wastewater (wastewater not discharged to municipal sewer systems, such as through separate permitted discharges)

Benefits of compiling a water balance include:

1. Ability to identify the best targets for reducing water use
2. Establishment of baseline water use data, from which improvements in efficiency can be benchmarked
3. Creation of a tool with which to compare different water-efficiency improvement strategies

16.3 Water Waste

Water is one of the most critical resources in the world and is necessary for all types of applications. Water is required in substantial quantities to create goods from food product to silicon chips. The term "water waste" refers to use of water beyond the point that optimizes value to the user. This includes, for example, non-value-added water use throughout a facility's operations and support processes, and from the supply chain through production, product use, and waste disposal. Water waste can cause harm to health and well-being of people and ecosystems by diverting water from other needs and beneficial uses. Reducing water waste can help ensure that present and future generations have access to a sustainable water supply.

Water use efficiency can be improved by reducing water waste, not only by reducing water intake but also by finding creative ways to use water more efficiently. Water use efficiency refers to the minimum amount of water needed to perform a task. (Note that water-efficiency improvements do not necessarily mean that total water use is decreased since increases in production may offset water efficiency gains for some uses.) Examples of water waste include:

1. Loss of water through material failures, such as leaky hose nozzles
2. Discarding water that could be reused, such as treated rinse water
3. Wastewater discharges
4. Water used by high-flow appliances instead of more efficient alternatives
5. Use of water in excess of necessity, such as the use of too much water for facility or parts cleaning

Understanding the many places where water waste occurs is an important first step for identifying areas for improvement. The following is a list of water saving strategies that are applicable to a wide range of water uses.

1. Adjust water flow
2. Install water-saving devices
3. Change to more water-efficient equipment
4. Reuse or recycle water
5. Shift to a low-water or waterless process

There are three key questions to ask to identify water efficiency opportunities:

1. Can water losses, e.g. leaks, be stopped or prevented?
2. Can water use be reduced by changing equipment, plumbing, processes, and/or behaviors?
3. Can water be recycled or reused, i.e. recirculating water within a process, reusing process water for another application, capturing rainwater for irrigation, etc., for another purpose?

The following sections address opportunities for reducing water waste and increasing water use efficiency via conservation measures applicable for power generation, agriculture, and domestic and industrial uses.

16.4 Water Conservation in the Power Generation Sector

As indicated in Figure 16.1, power generation accounts for greater than 40% of total global water withdrawals. Water for thermoelectric power is used in the process of generating electricity from fossil fuels, nuclear energy, or geothermal energy with steam-driven turbine generators. Most water withdrawn for thermoelectric power production is used for condenser and reactor cooling. *Once-through cooling* refers to cooling systems in which water is circulated through heat exchangers, and then returned to the source. *Recirculating cooling* or *closed-cycle cooling* refers to cooling systems in which water is circulated through heat exchangers, cooled using ponds or towers, and then recirculated. Subsequent water withdrawals for a recirculating system are used to replace water lost to evaporation, blowdown, drift, and leakage.

Thermoelectric power cooling water sources include fresh and saline water from both surface water and groundwater sources, with more than 99% of the water used for thermoelectric power production coming from self-supplied surface water and less than 0.2 percent from public supplies. Reclaimed wastewater is a supplemental source of water for thermoelectric power, especially in areas where additional water sources are needed for plant operations (Diehl et al. 2013).

Thermoelectric power generation is the largest single water withdrawal category across the globe. In contrast, thermoelectric consumption, the water evaporated or incorporated into byproducts as a result of the production of electricity from heat, has been estimated to be about 2% of total U.S. water consumption, though locally it can be greater (Solley et al. 1998). Evaporation from cooling systems accounts for most of the water consumption at most thermoelectric plants (U.S. Department of Energy 2010). Thermoelectric water consumption is a critical water use as thermoelectric plants cannot operate without consuming water. At the same time, water consumed in thermoelectric power generation is unavailable for other uses. The Mid-Atlantic, South Atlantic Gulf, Ohio, and Great Lakes water resource regions use the largest amounts of water for thermoelectric production. The eastern United States uses about five times more water than the West to produce about twice as much thermoelectric power (U.S. EPA 1990).

While water conservation strategies are plant specific, and vary greatly from one plant to the next based on the availability of water resources, location, climate, etc., a few of the most common water-saving approaches for thermal power plants include:

1. *Using alternative water sources* including treated sewage, high salinity water (seawater, brackish water, some groundwater), water from mining operations, produced water from oil and gas wells, agricultural runoff, and storm water. In some locations, the lower cost of degraded water can more than make up for the additional costs of pretreatment for cooling needs and wastewater treatment prior to discharge.
2. *Using recycled water* that has been used for other processes in the plant. Recycling water for cooling purposes typically calls for some form of treatment to make the stream safe for high-heat applications, to prevent scaling, or to otherwise meet the specifications of cooling equipment.
3. *Switching to a dry or hybrid cooling process*. Dry cooling systems, also known as air condensers use air circulation. Dry cooling systems accomplish this by using fans and heat exchangers, and therefore consume little to no water. Dry cooling systems have higher upfront and higher operating costs, are less efficient and reliable at warmer temperatures, and demand a larger footprint than traditional wet cooling methods. These dry systems are attractive, however, in drought-prone areas, or in locations where raw water costs are high or expected to increase in the future. Hybrid systems combine wet cooling with dry cooling systems to overcome some of these drawbacks.
4. *Switching from once-through to closed-cycle cooling*. As indicated earlier, Closed-cycle cooling systems use significantly less water than once-through systems, making them attractive options where water costs are high or where water is scarce. However, closed-cycle cooling systems can demand significant capital investment, and are often less efficient in terms of energy consumption and emissions than conventional cooling systems.

While these water conservation measures can be implemented in all thermal energy production systems, the largest water withdrawal and water consumption reductions related to energy production can be realized through the implementation of energy production systems that are not based on thermal power generation, i.e. alternative energy sources including wind and solar electricity production. At the time this chapter was being prepared, the western U.S. was in a significant and extended drought. In response to this in the State of Utah, Stafford (2021) encouraged the aggressive adoption of wind and solar energy production in the state to address current and future water scarcity problems. Wind and solar energy production require essentially no water while coal and natural gas consume approximately 690 and 260 gal water/MWh produced, respectively. In Utah's case, 61% of its 39.4 TWh 2020 electricity consumption comes from coal fired power plants, resulting in an annual water consumption of 16.6 billion gal of water for electricity generation. Conversion to low-water demanding power production methods could save billions of gallons of water in Utah alone that could be used for other beneficial uses.

16.5 Water Conservation in the Agricultural Sector

Agricultural water use can be divided between crop irrigation and livestock watering. Irrigation includes all water applied to farm or horticultural crops; livestock watering incorporates water used for livestock, dairies, feedlots, fish farms, and other farm needs. As indicated in Figure 16.1, irrigation of crops represents the largest water use in the agricultural sector, amounting to 37% of total global water withdrawals.

Not only can the loss of water from irrigation conveyance systems be significant, but the percentage of consumptive water use by irrigated agriculture is high as well due to evapotranspiration, amounting to approximately 56% of total water withdrawals by the sector (U.S. EPA 1990). Irrigation causes the movement of pollutants from land into surface or groundwater. This pollutant movement is affected by:

1. The fate of both applied irrigation water and precipitation
2. The physical, chemical, and biological characteristics of the irrigated land
3. The type of irrigation system used
4. The crop being irrigated
5. The farm management practices employed
6. The management of the irrigation system

Irrigation waters transported in open, unlined canals can seep into adjacent soils, eventually carrying soluble pollutants into ground or surface waters.

Overirrigating results in a portion of applied waters running off the land into surface waters or seeping through the soil and eventually ending up in surface or groundwater. In either case, the excess water can carry these pollutants:

1. Sediment and particulate organic solids
2. Particulate-bound nutrients, chemicals, and metals
3. Soluble nutrients, a portion of the applied pesticides, soluble metals (i.e. selenium and iron) and salts, and many other major and minor nutrients
4. Bacteria, viruses, and other microorganisms

Any pollutants linked to irrigation water, i.e. salts, metals, or nutrients, can concentrate in the soil, infiltration, or runoff associated with an irrigation system. Reducing overall water use for irrigation leaves more water for natural stream flow and increases flow needed by marshes, wetlands, or other environmental uses. If the irrigation source is groundwater, reducing overall use maintains higher ground water levels, which could be important for sustaining base flow in nearby streams. Reduced diversion of surface water likewise lessens the salt or other pollutant load brought by irrigation water, thereby diminishing the mass of these pollutants that ultimately must be managed or discharged from the system. One way of managing these pollutants is through the implementation of water conservation and pesticide/nutrient best management practices (BMPs). Decreasing the diversion of water from streams and rivers also lowers amount of return flows from irrigated lands that might transport pollutants.

Guidelines for improving irrigation water use efficiency are available from many state Extension programs, and the U.S. Department of Agriculture. An example of such guidance is provided by Shock et al. (2013) from the Oregon State University Extension Service that identify a myriad of water saving techniques that include:

1. Avoiding over irrigation by eliminating deep watering of shallow-rooted crops and use frequent, light applications to keep water and mobile nutrients in the plant root zone.
2. Scheduling of water applications based on measured evapotranspiration and soil moisture.
3. Practicing deficit irrigation, irrigation that applies less water than the crop needs for full development. Some crops lose little yield and quality with modest irrigation deficits, saving water without compromising financial returns.
4. When water is limited, plant crops that require less water.
5. Conservation tillage practices such as minimum tillage, no till, and strip till to help conserve soil water. Reduction in tillage reduces surface soil

drying, and retention of crop residues cools the soil and reduces water loss due to evaporation.

6. Carefully managing surface irrigation. Surface or flood irrigation systems are inherently inefficient. Sprinkler irrigation or drip irrigation allow more efficient water application and often increase yields. Drip irrigation can, in many cases save 30% to > 50% of the amount used for furrow irrigation.

16.6 Water Conservation in the Domestic Sector

Domestic water use includes everyday uses that take place in residential homes, whereas commercial water uses are those which take place in office buildings, hotels, restaurants, civilian and military institutions, public and private golf courses and other nonindustrial commercial facilities. Typical categories of residential water use include normal household uses such as:

1. Drinking and cooking
2. Bathing
3. Toilet flushing
4. Washing clothes and dishes
5. Watering lawns and gardens
6. Maintaining swimming pools
7. Washing cars

When divided into indoor uses and outdoor uses, the amount of indoor water use remains fairly constant throughout the year. By far the largest percentage of indoor water use occurs in the bathroom, with 41% used for toilet flushing and 33% for bathing, with cleaning representing 21% and the kitchen area representing 5% of overall residential indoor water use (U.S. Environmental Protection Agency 1990). Outdoor residential water use varies greatly depending on geographic location and season. On an annual average basis, outdoor water use in the arid West and Southwest is much greater than that in the East or Midwest and can amount to more than 50% of total domestic water use in these arid regions.

Water conservation in the domestic sector is primarily focused on changes in behavioral practices by residential water users. Behavioral changes in water use habits are focused on using water more efficiently thus reducing the overall water consumption in a home. These practices require a change in behavior, not modifications in the existing plumbing or fixtures in a home. Behavioral practices for residential water users can be applied both indoors in the kitchen, bathroom, and laundry room, as well as outdoors (U.S. EPA 2011).

In the kitchen, for example, 10–20 gal of water a day can be saved by running the dishwasher only when it is full. If dishes are washed by hand, water can be saved by filling the sink or a dishpan with water rather than running the water continuously. An open conventional faucet lets about 5 gal of water flow every 2 min.

Water can be saved in the bathroom by turning off the faucet while brushing teeth or shaving. Water can be saved by taking short showers rather than long showers or baths and turning the water off while soaping. This water savings can be increased even further by installing low-flow showerheads. Toilets should be used only to carry away sanitary waste, and a wide range of low-flow or dual flow toilets are available through hardware chains to reduce even further toilet water use.

Water can be saved in the laundry room by adjusting water levels in the washing machine to match the size of the load. If the washing machine does not have a variable load control, water can be saved by running the machine only when it is full. If washing is done by hand, the water should not be left running. A laundry tub should be filled with water, and the wash and rinse water should be reused as much as possible.

Outdoor water use can be reduced by watering the lawn early in the morning or late in the evening and on cooler days, when possible, to reduce evaporation. Allowing the grass to grow slightly taller will reduce water loss by providing more ground shade for the roots and by promoting water retention in the soil. Growing plants that are suited to the area (native plants), particularly in lieu of green lawns can save more than 50% of the water normally used to care for outdoor plants (U.S. EPA 2011). Native grasses, shrubs, and trees may require irrigation to establish them and should be watered in times of severe drought, but generally do not require any irrigation at other times. Native landscapes can be exceedingly beautiful as they change in color and texture with the seasons, provide habitat for wildlife, and generally require only limited maintenance to flourish (Theodore and Theodore 2021). Outdoor water use restrictions for irrigation purposes are often instituted by municipalities during severe drought conditions. An example of such restrictions in the county of one of the author's residence follows (Incorporated Village of East Williston 2018):

> Outdoor Water Use Restrictions – Use of water for irrigation purposes for lawns, shrubs, trees, plants and vegetation of any type is regulated by hours set forth by the County of Nassau. Absolutely no watering between the hours of 10:00 a.m. and 4:00 p.m. Watering will be permitted all other hours under the following conditions:

1. Residents with even house numbers may water on even dates
2. Residents with odd house numbers may water on odd dates

3. Premises without numbered addresses may water on even dates
4. No watering is permitted on the 31st of any month

As much as 150 gal of water can be saved when washing a car by turning the hose off between rinses. The car should be washed on the lawn if possible to reduce runoff and reuse the wash water for lawn irrigation. Additional savings of water can result from sweeping or blowing off sidewalks and driveways instead of hosing them down. Washing a sidewalk or driveway with a hose uses about 50 gal of water every 5 min. If a home has an outdoor pool, water can be saved by covering the pool to reduce evaporation when it is not in use.

16.7 Water Conservation in the Industrial Sector

Water is a very important resource to many industries, and as a result, there are many opportunities to reduce water use in this sector. In the case of the high-tech industry, which uses considerable quantities of water to manufacture semiconductors and other components, water is vital to industry operations. Cleaning and rinsing silicon chips can require billions of gallons of water per year. Apparel also ranks high in the list of water-intensive industries, with cotton production and textile processing requiring substantial inputs of water. Among industrial manufacturers, the forest products sector is the third largest consumer of water in the U.S., requiring high volumes of water for pulp and paper manufacturing.

There is a great deal of variation among industrial sectors in relative water use, and the need for water varies greatly across the stages of the product or service value chain, from suppliers through production to product use. For example, the apparel industry uses high volumes of water in raw material production, whereas the biotech and pharmaceuticals industry uses most of its water resources in direct operations. Table 16.1 provides a summary of gal of water use per unit of product to demonstrate the quantity of water consumed and the variability across common product categories (USGS 2021).

In order to reduce water waste in industry, it is important to understand the many ways water is used within facilities. Understanding water end uses is critical to identifying water conservation opportunities. While end uses of water vary by industry and by facility, there are categories of water use that are present at most industrial facilities and can be classified into the following broad end use categories:

1. Production processing and in-product use
2. Auxiliary processes (e.g. pollution control, labs, and cleaning)
3. Cooling and heating (e.g. cooling towers and boilers)
4. Indoor domestic use (e.g. restrooms, kitchens, and laundry)
5. Landscape irrigation

Table 16.1 Typical Water Use Values Per Unit of Product (Modified from USGS 2021).

Product	Water Use, gal/unit
Paper	0.5 to 8/sheet
Coffee	35/cup
Egg	50/egg
Chicken	500/lb
Beef	1,840/lb
Steel	30/lb
Cotton shirt	650/shirt
Ethanol	10 to 340/gal
Gasoline	3 to 6/gal

Among the United States industrial customers, cooling operations (including cooling towers and open cooling systems) comprise the single largest category of industrial water end use, with more than 50% of industrial and commercial water demand combined going toward cooling. The amount of water required for the various other end uses differs by industry. Service and manufacturing facilities require the most water for washing and processing, while food and beverage facilities use most of their water intake in product preparation.

Major end uses often provide the greatest opportunities for water waste reduction and efficiency improvement. For example, in many food, beverage, and pharmaceutical companies, cleaning process equipment can account for as much as 50 to 70% of a facility's total water use, and represents a substantial opportunity to save water.

Industrial/commercial users can apply a number of conservation and water use efficiency practices. Some of these practices can also be applied by users in the other water use categories. Any reused water must be used for a beneficial purpose and in accordance with applicable rules (such as local ordinances governing water reuse).

Factors that should be considered in an industrial water reuse program include:

1. Identification of water reuse opportunities
2. Determination of the minimum water quality needed for a particular application
3. Identification of wastewater sources that satisfy these water quality requirements
4. Determination of how this reuse water can be transported to the new use

The reuse of wastewater or reclaimed water is beneficial because it reduces the demands on available surface and groundwater supplies. Perhaps the greatest benefit of establishing water reuse programs is their contribution in delaying or eliminating the need to expand potable water supply and treatment facilities. Is should be noted that recycled water may require treatment before it can be used again, so Step 2 is critical to identifying the level of pretreatment necessary to meet a specific reuse requirement.

There are also questions that can be asked to identify water savings opportunities in industrial operations and support processes. These include:

1. Cleaning – can process cleaning or facility cleaning be accomplished without using water (i.e. using pressurized air to clean products or containers, sweeping debris off the floor)?
2. Process Equipment – Can process equipment reuse water (closed loop) or use reclaimed water from other parts of the facility?
3. Cooling and Heating – Has the facility replaced once-through cooling systems with a multi-pass cooling tower or closed systems? Can the blowdown/bleed-off controls on boilers and cooling towers be optimized? Has switching to air-cooled equipment instead of water-cooled equipment been considered? Does the facility reuse condensate water?
4. Restrooms and Kitchens – Do restrooms have water-efficient fixtures (e.g. WaterSense labeled toilets and urinals faucet aerators, and showerheads? Do kitchens use new water- and energy-efficient dishwashers (e.g. ENERGY STAR qualified models)?
5. Leaks – Have leaks been identified and repaired throughout the facility?
6. Landscaping and Irrigation – Has the facility designed its landscape to consider the local climate and grouped plants by similar watering needs? Does the facility use drip irrigation, low-flow sprinklers, and optimized watering schedules to minimize water use?

Key specific industrial conservation measures that can be instituted across a wide range of industries are summarized in Table 16.2.

The term *utilities* generally designates ancillary services needed in the operation of any production process. These services are normally supplied from a central location on-site or from off-site utility providers and usually include:

1. Electricity
2. Steam for process heating
3. Water for cooling, potable use, and steam production
4. Refrigeration
5. Compressed air
6. Inert gas supplies

Table 16.2 Key Areas of Industrial Water Conservation.

Area	Water Conservation Opportunities
Process and equipment use	Cleaning, washing, rinsing
	Metal finishing
	Painting
	Dyeing and finishing
	Photo processing
	Process water reuse
	Product fluming (water transport)
	Pretreatment/filtration systems
	Pump and conveyor lubrication
	Water use in products
Cooling and heating	Single-pass cooling
	Cooling towers
	Boilers, hot water, steam systems
	Air washers
	Boiler scrubbers
Sanitary and domestic	Toilets
	Faucets
	Urinals
	Showers
	Wash-up basins
Kitchens	Food preparation and cleaning
	Dishwashers
	Ice machines
	Faucets
	Food disposals
Other facility support	Floor washing
	Air emission wet scrubbers
	Building washing
	QA/QC testing
	Laboratories
	Landscaping and irrigation
	Dust and particulate emission control
	Decorative fountains and ponds
	Vehicle washing
	Cooling water for air compressors and vacuum pumps
	Hazardous waste storage and effluent

The production and use of these utility services can have a significant impact on both the water and energy demand of a given process or facility. The petroleum industry provides some interesting information on water usage and conservation. Throughout the life cycle of crude oil, through extraction, transportation, and refining, roughly 3.6 to 7.0 gal of water is used for every gal of crude oil. Of this total, roughly 1 to 2.5 gal of water are used for every 1 gal of crude oil stock. In the United States, 1 to 2 billion gal of water are used per day in the refining industry. Water consumption varies for different regions and different qualities of crude oil. These water estimates do not include unconventional shale formations, where water usage is generally much higher. Water consumption in refining is a result of evaporation and drifting, blowdown, process discharge, and some incorporation into finished products. All water that is incorporated into the refining process is treated or filtered onsite before it is either reused onsite or discharged into the environment. Wastewater streams include separation process wastewater, tank bottoms, cooling tower water condensate blowdown, source water treatment, and storm water. American Petroleum Institute (API) separators are commonplace in most refineries and separate sludge and hydrocarbon components out of process wastewater. Refineries today have installed dike systems to keep process stream wastewater from draining into a storm water sewer.

Buildings are also significant users of water. Indoor water use reduction is typically achieved through efficient plumbing fixtures, fittings, appliances, and process equipment used to heat and cool a building. Outdoor water use reduction efforts should focus on water-wise landscaping and efficient irrigation. By using less potable water, a building will also reduce the amount of energy needed to transport, treat, and redistribute that water, along with cost savings associated with reduced potable water consumption. Innovative solutions for water efficiency should include not only a reduction in the amount of potable water used but also a reduction in the use of non-potable water where appropriate (for flushing toilets, watering the landscape, etc.).

16.8 Illustrative Examples

Six illustrative examples complete the presentation of water conservation materials highlighted in this chapter.

16.8.1 Illustrative Example 1

Preventing pollution is only one reason why using water efficiently makes sense. Provide additional benefits of improving water use efficiency.

Solution. Improving water use efficiency can provide both monetary and reliability benefits. Monetary benefits include the following:

1. Less water use results in lower pumping and treatment costs.
2. Money not spent on water and wastewater treatment can be used for improving water quality and public health.
3. Water saved also equates to energy and money savings that can be used elsewhere in the community.

Reliability benefits include the following:

1. Water conservation provides a hedge against drought and climate change impacts.
2. Improving water use efficiency may be implemented more quickly and less expensively than developing a new water supply.
3. Reduced water use will extend the life of water and wastewater treatment facilities.
4. Reduced water use will increase the efficiency of wastewater treatment and reduce combined sewer overflows during storm events.

16.8.2 Illustrative Example 2

Provide some strategies that can be applied for improving water use efficiency in government buildings.

Solution. Six water use reduction strategies appropriate for government buildings are provided here:

1. Use Federal Acquisition Regulation (FAR) environmental and energy guidelines for product and service purchases.
2. Use high-efficiency plumbing fixtures in lavatories, such as aerators on faucets (0.5 gpm), high efficiency showerheads (2 gpm), dual flush toilets (0.8/1.6 gal/flush) or high efficiency toilets (1.2 gal/flush) and provide training for occupants in their proper use.
3. Avoid automatic flushes on toilets because they often result in double flushing.
4. Consider waterless urinals to save significant amounts of water and ensure that cleaning and maintenance staff receive training in the specific maintenance regimes required.
5. Consider high efficiency water fountains, with a wider range of acceptable temperatures to encourage use of tap over bottled water. Incorporate taps into the design to make them user friendly.
6. Consider submetering of water systems to identify future opportunities for improvements.

16.8.3 Illustrative Example 3

Water conservation is claimed to be a highly cost-effective approach to extending existing water supplies and increasing the security and sustainability of water sources. Provide documentation that this is indeed the case (Theodore and Dupont 2019).

Solution. This is an open-ended question and can be answered by considering the reduction in water use by high efficiency water fixtures, quantifying the water savings on an annual basis, and determining the value of the water saved annually in a specific local setting. The payback of water conservation estimates will be dependent on the value of water at a specific location and the cost of implementing specific water conservation practices. Some practices, such as turning off a tap during the brushing of teeth, cost nothing and have an instant payback, others require the investment and installation of new fixtures that, depending on water costs, may have payback periods of 2 to 3 years.

16.8.4 Illustrative Example 4

Assume that the average water usage of a community is approximately 130 gal/person/day. After implementing water saving practices, the average residential water usage is expected to drop to 87 gal/person/day. A 10 million gal per day (MGD) wastewater treatment plant has 60% of the flow as residential wastewater. Calculate the % reduction in water usage. Also calculate the current wastewater flow in MGD and the future flow after water saving practices are implemented.

Solution. First, calculate the water savings, W_s, in gal/person/d:

$$W_s = (130 - 87 \text{ gal/person/d}) = 43 \text{ gal/person/d}$$

The % water use reduction:

$$\% \text{ water use reduction} = (43/130 \text{ gal/person/d})(100) = 33.1\%$$

The current residential wastewater flow, $Q_{residential}$, is:

$$Q_{residential} = (0.6)(10 \text{ MGD}) = 6 \text{ MGD}$$

After implementation of water saving practices in the residential sector, the total wastewater flow to the treatment plant, Q_{total}, is expected to be:

$$\begin{aligned} Q_{total} &= (1 - 0.331)Q_{residential} + Q_{other} = (0.669)(6 \text{ MGD}) + 4 \text{ MGD} \\ &= 4.04 + 4 = 8.04 \text{ MGD} \end{aligned}$$

16.8.5 Illustrative Example 5

Soap, detergents, and various toilet articles are produced in an industry having the sewer map given in Figure 16.3. Fly ash, tank bottoms, and spent caustic are discharged gradually from batch holding tanks into the sewer. Waste composition data given in Table 16.3 are mean values. Seven sampling stations within the sewer system are denoted in the sewer map. Using these data, carry out a material and flow balance for the plant as it currently exists as part of a water conservation study.

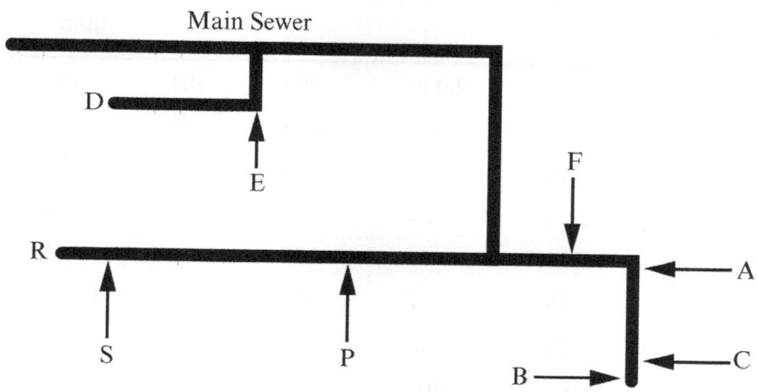

Figure 16.3 Sewer Map for Soap Manufacturing Facility for Illustrative Example 4.

Table 16.3 Waste Composition of Soap Manufacturing Facility Waste Streams for Illustrative Example 5.

Waste source	Sampling location	COD, mg/L	BOD, mg/L	TSS, mg/L	ABS, mg/L	Flow, gpm
Liquid soap	D	1,100	565	195	28	300
Toilet articles	E	2,680	1,540	810	69	50
Soap production	R	29	16	39	2	30
ABS production	S	1,440	380	309	600	110
Powerhouse	P	66	10	50	0	550
Condenser	C	59	21	24	0	1,100
Spent caustic	B	30,000	10,000	563	5	2
Tank bottoms	A	120,000	150,000	426	20	1.5
Fly ash	F	-	-	6,750	-	10
Main sewer		449	260	120	37	2,150

Solution: The first step in the solution to this problem is to determine the mass loadings at each location on the sewer map using the table of data given in the problem statement. This is done by multiplying the concentration by the flow rate to yield a mass flow rate, with proper unit conversions. The result is shown in Table 16.4. Also given in Table 16.5 are the percentages of the total mass of each contaminant associated with a given sampling location.

Table 16.4 Mass Loadings for Soap Manufacturing Facility Waste Streams for Illustrative Example 5.

Waste source	Sampling location	COD, lb/d	BOD, lb/d	TSS, lb/d	ABS, lb/d	Flow, gpm
Liquid soap	D	3,963	2,036	703	101	300
Toilet articles	E	1,609	925	486	41	50
Soap production	R	10	6	14	1	30
ABS production	S	1,902	502	408	793	110
Powerhouse	P	436	66	330	0	550
Condenser	C	779	277	317	0	1,100
Spent caustic	B	721	240	14	0	2
Tank bottoms	A	2,162	2,702	8	0	1.5
Fly ash	F	0	0	811	0	10
Main sewer		11,619	6,713	3,098	955	2,150

Table 16.5 Percent Mass Loadings for Each Waste Stream in the Soap Manufacturing Facility for Illustrative Example 5.

Waste source	Sampling location	COD, wt%	BOD, wt%	TSS, wt%	ABS, wt%	Flow, %
Liquid soap	D	34.2	30.1	22.7	10.8	13.9
Toilet articles	E	13.9	13.7	15.7	4.4	2.3
Soap production	R	0.1	0.1	0.5	0.1	1.4
ABS production	S	16.4	7.4	13.2	84.7	5.1
Powerhouse	P	3.8	1.0	10.7	0	25.5
Condenser	C	6.7	4.1	10.3	0	51.1
Spent caustic	B	6.2	3.6	0.4	0	0.1
Tank bottoms	A	18.7	40.0	0.2	0	0.1
Fly ash	F	0	0	26.2	0	0.5
Main sewer		100	100	100	100	100

16.8.6 Illustrative Example 6

Refer to Illustrative Example 5. Suggest how wastes could be segregated for optimal treatment. Prepare a new material and flow balance using your revised sewer layout.

Solution: A study of the sewer map and the loading data presented in Illustrative Example 5 should yield clues about which waste streams would be candidates for segregation. For example, the fly ash line, F, carries 26 wt% of the suspended solids but only 0.5% of the flow. This suggests that segregation of this source would reduce the solids loading significantly at a small cost, since the flow is so small. In fact, it may be possible to eliminate the water from this source directly and treat the fly ash in a dry state. Similarly, the tank bottoms contain a significant amount of organic matter but very little suspended solids and almost no flow. Segregation here would also be beneficial to eliminate this large source of organic material. In order to help decide which flows could be segregated, two indices can be used.

The first index is calculated based on the product of the mass loading (*MLi*) and the square of the concentration (*Ci*) for a given source, divided by the product of the mass loading (*MLs*) and the square of the concentration (*Cs*) in the main sewer (Equation 16.1). This index emphasizes the higher concentration sources as candidates because generally, higher concentration wastes are easier to treat. The values of the index for each contaminant from each source is given in Table 16.6.

$$Index1 = (ML_i)(C_i)^2 / [(ML_s)(C_s)^2] \tag{16.1}$$

The second index (Equation 16.2) is based not on the concentration in the main sewer, but on typical secondary wastewater treatment standards for each contaminant. The ratio of the actual concentration (*Ci*) to the concentration of the water quality standards (C_{ssi}) is used to determine which source has the most impact on the plant related to the plant's ability to meet these standards for pretreatment or direct discharge purposes. Index 2 is shown in Table 16.7.

$$Index2 = (C_i) / (C_{ssi}) \tag{16.2}$$

In each table, the very large numbers stand out for COD and BOD in the tank bottoms (A) and spent caustic (B), ABS in the ABS production stream (S), and suspended solids in the fly ash (F) line. This shows up consistently for both indices. These groupings suggest segregation as shown in Figure 16.4.

The new map takes the lines A, B, F, and S, and provides some pretreatment (say 90%) before discharge to the sewer. Lines R and C when combined are below secondary treatment standards and may be discharged directly to

Table 16.6 Index 1 Values Based on Main Sewer Composition for Each Waste Stream for Illustrative Example 6.

Waste source	Sampling location	COD	BOD	TSS	ABS
Liquid soap	D	2	1	1	0.1
Toilet articles	E	5	5	7	0.1
Soap production	R	0	0	0	0
ABS production	S	2	0.2	1	219
Powerhouse	P	0	0	0	0
Condenser	C	0	0	0	0
Spent caustic	B	277	53	0.1	0
Tank bottoms	A	13,276	134,422	0	0
Fly ash	F	-	-	828	-

Table 16.7 Index 2 Values Based on Secondary Wastewater Treatment Standards for Each Component for Illustrative Example 6.

Waste Source	Sampling Location	COD	BOD	TSS	ABS
Liquid soap	D	18	19	6.5	5.6
Toilet articles	E	45	51	27	14
Soap production	R	0.5	0.5	1.3	0.4
ABS production	S	24	13	10	120
Powerhouse	P	1.1	0.3	1.7	0
Condenser	C	1.0	0.7	0.8	0
Spent caustic	B	500	333	19	1.0
Tank bottoms	A	2,000	5,000	14	4
Fly ash	F	-	-	225	-
Standards		60	30	30	5

surface water or may be reused in the plant to reduce its overall water demand. The new waste stream composition and mass balance tables reflecting these changes are shown in Table 16.8 and Table 16.9.

A new set of segregation indices were calculated as earlier and reflect the changes indicated in Figure 16.4 and Tables 16.8 and Table 16.9. As seen in

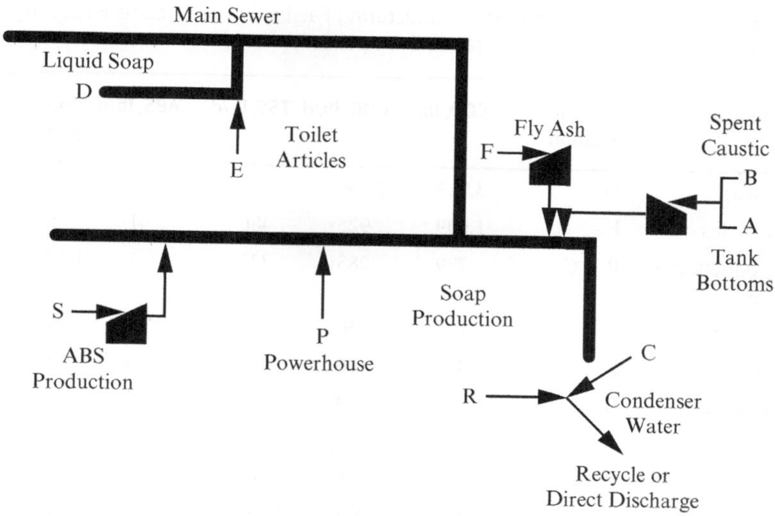

Figure 16.4 Revised Sewer Map for Soap Manufacturing Facility Showing Waste Streams Identified for Consideration of Waste Segregation for illustrative Example 6.

Table 16.8 Waste Composition of Soap Manufacturing Facility Waste Streams Following Waste Segregation Analysis for Illustrative Example 6.

Waste source	Sampling location	COD, mg/L	BOD, mg/L	TSS, mg/L	ABS, mg/L	Flow, gpm
Liquid soap	D	1,100	565	195	28	300
Toilet articles	E	2,680	1,540	810	69	50
Soap production + Condenser	R + C	58	21	24	0.1	1,130
ABS production	S	144	38	31	60	110
Powerhouse	P	66	10	50	0	550
Spent caustic	B	3,000	1,000	56	0.5	2
Tank bottoms	A	12,000	15,000	43	2.0	1.5
Fly ash	F	-	-	675	-	10
Main Sewer		528	274	134	18	1,024

Tables 16.10 and Table 16.11, further reductions in COD and BOD in Line A may be necessary, or it may be appropriate to isolate Line A for separate treatment by virtue of its low flow rate and very high BOD and COD concentrations, even after 90% reduction following treatment.

Table 16.9 Mass Loadings for Soap Manufacturing Facility Waste Streams Following Waste Segregation Analysis for Illustrative Example 6.

Waste source	Sampling location	COD, lb/d	BOD, lb/d	TSS, lb/d	ABS, lb/d	Flow, gpm
Liquid soap	D	3,963	2,036	703	101	300
Toilet articles	E	1,609	925	486	41	50
Soap production + Condenser	R + C	789	283	331	1	1,130
ABS production	S	190	50	41	79	110
Powerhouse	P	436	66	330	0	550
Spent caustic	B	72	24	1.4	0	2
Tank bottoms	A	216	270	0.8	0	1.5
Fly ash	F	0	0	81	0	10
Main sewer		6,486	3,371	1,643	221	1,024

Table 16.10 Index 1 Values Based on Main Sewer Composition for Each Waste Stream Following Waste Segregation for Illustrative Example 6.

Waste source	Sampling location	COD	BOD	TSS	ABS
Liquid soap	D	2.7	2.6	0.9	1.1
Toilet articles	E	6.4	8.7	11	2.7
Soap production + Condenser	R + C	0	0	0	0
ABS production	S	0	0	0	4.0
Powerhouse	P	0	0	0	0
Spent caustic	B	0.4	0.1	0	0
Tank bottoms	A	17	240	0	0
Fly ash	F	-	-	1.3	-

16.9 Problems

1. Provide some maintenance tips to reduce water use in the industrial sector.
2. Describe some engineering water conservation methods that can be easily implemented at the household scale.
3. There are numerous general water conservation practices that can be instituted at chemical plants. List some of these practices that would apply for both water and energy conservation.

Table 16.11 Index 2 Values Based on Main Sewer Composition for Each Waste Stream Following Waste Segregation for Illustrative Example 6.

Waste source	Sampling location	COD	BOD	TSS	ABS
Liquid soap	D	18	19	6.5	5.6
Toilet articles	E	45	51	27	14
Soap production + Condenser	R + C	1.0	0.7	0.8	0
ABS production	S	2.4	1.3	1.0	12
Powerhouse	P	1.1	0.3	1.7	0
Spent caustic	B	50	33	1.9	0.1
Tank bottoms	A	200	500	1.4	0.4
Fly ash	F	-	-	22	-
Standards		60	30	30	5

4. Describe some of the ways a municipal water supplier can reduce water demand and encourage water conservation.
5. Describe some of the ways a municipal government can reduce water demand and encourage water conservation within their jurisdictional boundaries.
6. Complete the water balance shown in Table 16.12.

Table 16.12 Example Water Balance Summary.

Water use	Input, gal/yr	Output, gal/yr	Loss, gal/yr (% of water use)
Cooling tower make-up and boiler make-up	8,484,000	8,248,000	?
Process use	6,298,000	?	112,000 (1.8%)
Cooling for air compressors and pumps	7,654,000	7,455,000	?
Landscaping	?	1,533,000	142,000 (8.5%)
General washing, sanitation, and maintenance	1,101,000	1,095,000	?
Food preparation, dishwasher	624,000	613,000	?
Subtotal	26,660,000	?	1,530,000 (5.7%)

References

Diehl, T.H., Harris, M., Murphy, J.C., Hutson, S.S., and Ladd, D.E. (2013). *Methods for Estimating Water Consumption for Thermoelectric Power Plants in the United States*. Science Investigation Report. 2013–5188. Reston, VA: U.S. Geological Survey. https://pubs.usgs.gov/sir/2013/5188/pdf/sir2013-5188.pdf. (accessed July 2, 2021).

Incorporated Village of East Williston (2018). *Drinking Water Quality Report*. East Williston, NY: Village of East Williston Water Department.

Perkins, A. (2018). Water usage chart. World and Health Organization. https://waterandhealth.org/safe-drinking-water/you-too-can-embrace-water-conservation-and-it-really-works/attachment/water-usage-chart. (accessed 2 July, 2021).

Shock, C.C., Shock, B.M., and Welch, T. (2013). Strategies for efficient irrigation water use. Sustainable Agriculture Techniques. EM 8783. Corvallis, OR: Oregon State University Extension Service. https://catalog.extension.oregonstate.edu/sites/catalog/files/project/pdf/em8783.pdf. (accessed July 2, 2021).

Solley, W.B., Pierce, R.R., and Perlman, H.A. (1998). *Estimated Use of Water in the United States in 1995: U.S. Geological Survey Circular 1200*. Reston, VA: U.S. Geological Survey. http://pubs.er.usgs.gov/publication/cir1200. (accessed July 2, 2021).

Stafford, E.R. (2021). Wind and solar electricity free up needed Utah water. Utah Policy. June 29. https://utahpolicy.com/politics/60066-wind-and-solar-electricity-free-up-needed-utah-water (accessed July 2, 2021).

Theodore, L. and Dupont, R.R. (2019). *Water Resource Management Issues. Basic Principles and Applications*. 2019. Boca Raton, FL: CRC Press/Taylor and Francis Group.

Theodore, M.K. and Theodore, L. (2010). *Introduction to Environmental Engineering*. Boca Raton, FL: CRC Press/Taylor and Frances Group.

U.S Department of Energy (2010). *Estimating Freshwater Needs to Meet Future Thermoelectric Generation Requirements, 2010 Update*. DOE/ NETL-400/2010/1339. Washington, D.C.: National Energy Technology Laboratory, U.S. Department of Energy. http://www.netl.doe.gov/energy-analyses/pubs/2010_Water_Needs_Analysis.pdf. (accessed July 2, 2021).

U.S. Environmental Protection Agency (1990). *Preventing Pollution through Efficient Water Use*. OW WH-5 56, OPPE PM-222. Washington, DC: Office of Water, Office of Pollution Prevention.

U.S. Environmental Protection Agency (2011). *Lean & Water Toolkit. Achieving Process Excellence Through Water Efficiency*. Washington, DC: Office of Policy. http://www.epa.gov/lean/environment/toolldts/water/resources/lean-water-toolkit.pdf. (accessed July 2, 2021).

U.S. Geological Survey (2021). The water content of things: How much water does it take to grow a hamburger? https://water.usgs.gov/edu/activity-watercontent.php (accessed July 4, 2021).

Water Footprint Network (2021). Water stress to affect 52% of the World's population by 2050. https://waterfootprint.org/en/about-us/news/news/water-stress-affect-52-worlds-population-2050 (accessed July 1, 2021).

17

Economic Considerations

17.1 Introduction

Regardless of the type of society or government within which a business is operated, it must either show a profit or perish in the normal course of events. What applies to a desalination process or an individual's business applies equally to a community or an entire economy. Without profit the structure is weak; with profit it *may* be strong.

Because of the necessity for combining economic considerations with environmental fundamentals and principles in any desalination operation or process, the environmental engineer or scientist must be aware of many different types of costs involved in a host of issues. Money must be paid out for many different plant expenses, such as those for raw materials, labor, and equipment. In addition, many other indirect expenses are incurred, and these must be included if a complete analysis of the total cost is to be obtained. Some examples of these other expenses are addressed later in this chapter.

Since this chapter was written primarily for the engineer it would be almost impossible at this level to cover all the various aspects of engineering economics. This chapter therefore introduces the philosophy of economics as a tool used in engineering. For the interested reader there are many texts available in the literature which give a much broader treatment of the subject.

Both the qualitative and quantitative viewpoints are emphasized in the material to follow. It should be realized that the broad subject of engineering economics cannot be fitted into any rigid set of formulas. However, the material presented in this chapter addresses applied engineering economics through a general review of economic theory, definitions, a discussion of the need for economic analysis, capital investment and risk, guidelines for purchasing equipment, and applications to industrial water usage. Six illustrative examples of concepts and implementation of engineering economics in desalination applications are provided at the end of the chapter.

Introduction to Desalination: Principles, Processes, and Calculations, First Edition. Louis Theodore and R. Ryan Dupont.

17.2 The Subject of Economics

Economics is a social science concerned with the production, distribution, exchange, and consumption of goods and services. Economists focus on the way in which individuals, groups, business enterprises, and governments seek to achieve efficiency and economic objectives they select.

Economic issues have occupied people's minds throughout the ages. Aristotle and Plato in ancient Greece wrote about problems of wealth, property, and trade. Both were prejudiced against commerce, feeling that to live by trade was undesirable. The Romans borrowed their economic ideas from the Greeks and showed the same contempt for trade. During the Middle Ages the economic ideas of the Roman Catholic church were expressed in the canon law, which condemned usury (the taking of interest for money loaned) and regarded commerce as inferior to agriculture.

Modern economics can be divided into two major fields. The first, price theory or microeconomics, explains how the interplay of supply and demand in competitive markets creates a multitude of individual prices, wage rates, profit margins, and rental charges. *Microeconomics* assumes that people behave rationally. Consumers try to spend their income in ways that give them as much pleasure as possible. As economists say, they maximize utility. For their part, entrepreneurs seek as much profit as they can extract from their consumer operations.

The second field, macroeconomics, deals with modern explanations of national income and employment. *Macroeconomics* dates from the British economist John Maynard Keynes's (1936) book *The General Theory of Employment, Interest, and Money*. His explanation of prosperity and depression centers on the total or aggregate demand for goods and services by consumers, business investors, and governments. Because, according to Keynes, inadequate aggregate demand increases unemployment the indicated cure is either more investment by businesses or more spending and consequently larger budget deficits by government.

17.3 Definitions

Before proceeding to the developmental material and applications later in the chapter, it would be wise to provide the reader with certain key definitions in the field. Fourteen concepts that often come into play in an economic analysis are given here. The definitions have been drawn from Theodore and Neuser (1996).

17.3.1 Simple Interest

This term can be defined as the money paid for the use of money. It is also referred to as the value or worth of money. Two terms of concern are simple interest and compound interest. Simple interest is always computed on the original principal. The basic formula to employ in simple interest calculation is:

$$F = P(1 + ni) \tag{17.1}$$

where F = sum of interest and principal after n years, also known as the future value; P = original principal; and i = annual interest rate. Normally, the interest period is 1 year, in which case i is referred to as the effective interest rate.

17.3.2 Compound Interest

Unlike simple interest, with compound interest, interest is added periodically to the original principal. The terms "conversion" or "compounding of interest" simply refer to the addition of interest to the principal. The interest period or conversion period in a compound interest calculation is the time interval between successive conversions of interest. It is the ratio of the stated annual rate to the number of interest periods in 1 year. Thus, if the given interest rate is 10% compounded semi-annually, the interest period is 6 months and the interest rate per interest period is 5%. Alternatively, if the given interest rate is 10% compounded quarterly, then the interest period is 3 months and the interest rate per interest period is 2.5%. One should always assume the interest is compounded annually unless otherwise stated. The basic formula to employ for compound interest is as follows:

$$F = P(1 + i)^n \tag{17.2}$$

If interest payments become due m times per year at compound interest, $(m)(n)$ payments are required in n years. The future value of this principle can be based on a nominal interest rate per period, $i' = i/m$:

$$F = P(1 + i')^{mn} \tag{17.3}$$

In this case, the effective annual interest, i_{eff}, is:

$$i_{eff} = (1 + i')^m - 1 \tag{17.4}$$

In the limit (as m approaches infinity), such payments may be considered required at infinitesimally short intervals, in which case, the interest is said to be compounded continuously. Numerically, the difference between compounding can be significant when applied to very large sums of money.

17.3.3 Present Worth

The present worth is the current value of a sum of money due at time *n* and at interest rate *i*. This equation is the compound interest equation solved for the present worth term, *P*.

$$P = F(1+i)^{1/n} \tag{17.5}$$

17.3.4 Time Value of Money

The value of a sum of money changes with time because of interest considerations: \$1,000,000 today, \$1,000,000 10 years from now and \$1,000,000 10 years ago all have different meanings when interest is taken into account. The \$1,000,000 today would be worth more 10 years from now because of the interest that could be accumulated in the interim. On the other hand, \$1,000,000 today would have been worth less 10 years ago because a smaller sum of money could have been invested then to yield \$1,000,000 today. Therefore, one must refer to the date as well as the sum of money when considering value and comparing costs.

Summarizing, evaluating single sums of money requires multiplying by $(1 + i)$ if the required date of evaluation is after the date associated with the obligation. If the required date of evaluation is before the date associated with the obligation, single sums of money must be multiplied by $(1 + i)^{1/n}$. The term n is always the time in periods between the date associated with the obligation and the date of evaluation.

The time value of money may be applied to the evaluation of a uniform series of payments. A uniform series of equal payments is made at equal intervals. Suppose R is invested at the end of every interest period for n periods. The total value of all these payments, F, as of the date of the last payment, may be calculated from the equation:

$$F = R\frac{\left[(1+i)^n - 1\right]}{i} \tag{17.6}$$

17.3.5 Depreciation

Depreciation refers to the decrease in the value of an asset. Two approaches that can be employed are the straight line and sinking-fund method. In the straight-line method of depreciation, the value of the asset is decreased each year by a constant amount. The annual depreciation amount, D, is given by:

$$D = \frac{(\text{Original cost} - \text{Salvage value})}{\text{Estimated life in years}} \tag{17.7}$$

In the sinking-fund method of depreciation, the value of the asset is determined by first assuming that a "sinking" fund consisting of uniform annual payments had been set up for the purpose of replacing the asset at the end of its estimated useful life. The uniform annual payment (UAP) may be calculated from the equation:

$$UAP = (\textit{Original cost} - \textit{Salvage value})(SFDF) \tag{17.8}$$

where $SFDF$ = the sinking-fund deposit given by:

$$SFDF = \frac{i}{(1+i)^n - 1} \tag{17.9}$$

The value of the asset at any time is estimated to be the difference between the original cost and the amount that would have accumulated in the sinking fund. The amount accumulated in the sinking fund is obtained by substituting the UAP into Equation 17.6 for R and calculating the future value of these regular payments at interest rate i over the corresponding number of payment periods, n.

17.3.6 Equipment Cost Indices

A simple process is available to estimate equipment and plant construction costs from historical data. The method consists of adjusting the earlier cost data to present values using factors that correct for inflation. A number of such indices are available; one of the most commonly used is the Chemical Engineering Plant Cost Index (CEPCI) available monthly in the *Chemical Engineering Magazine*. The CEPCI is indexed to 1957–1959 costs (Index = 100) and the January 2021 index was reported as 616.5, with a 2020 annual index of 596.2. The CEPCI consists of the weighted average of 41 industry and commodity indices and 12 labor cost indices, including construction labor;

engineering and supervision; heat exchangers and tanks; process machinery; pipes, valves, and fittings; etc., all reported by the Bureau of Labor Statistics (Maxwell 2021). Another relevant cost index for public works projects is the Corps of Engineering (2018) Civil Works Construction Cost Index (CWCCI). The CWCCI is updated regularly and compiles public works project cost information on a quarterly and annual basis and provides future cost index projections and regional cost adjustment factors for 19 categories including reservoirs; dams; power plants; pumping plants; buildings, grounds, and utilities; and permanent operating equipment. The CWCCI is indexed to 1967 (Index = 100), and reported a 2017 composite index of 835.57 and a projected 2020 index of 893.56.

All indices work in the same manner, that is, if a cost and index value for a given year are known, with the index value for another year the cost for that year can be estimated using Equation 17.10:

$$Cost_{Year\ B} = Cost_{Year\ A}\frac{Index\ Value_{Year\ B}}{Index\ Value_{Year\ A}} \tag{17.10}$$

17.3.7 Capital Recovery Factor

In comparing alternative processes or different options for a particular process from an economic point of view, one recommended procedure to follow is that the total capital cost can be converted to an annual basis by distributing it over the projected lifetime of the facility (or the equivalent). The sum of both the annualized capital cost (*ACC*), including installation, and the annual operating cost (*AOC*), is called the total annualized cost (*TAC*) for the project or facility. The economic merit of the proposed facility, process, or scheme can be examined once the total annual cost is available.

The conversion of the total capital cost (*TCC*) to an *ACC* requires the determination of an economic parameter known as the capital recovery factor (*CRF*). This parameter can be found in any standard economics textbook or calculated directly from the following equation:

$$CRF = \frac{i(1+i)^n}{(1+i)^n - 1} \tag{17.11}$$

The CRF is a positive decimal number. Once this factor has been determined, the ACC can be calculated from the following equation:

$$ACC = (TCC)(CRF) \tag{17.12}$$

The annualized capital cost reflects the cost associated with repaying the initial capital expenditure with interest over the depreciable life of the system.

17.3.8 Net Present Worth

There are various approaches that may be employed in the economic selection of the best of several alternatives. For each alternative in the net present worth (NPW) method of economic selection, a single sum is calculated that would provide for all expenditures over a common time period. The alternative having the least NPW of expenditures is selected as the most economical. The equation to employ this method of economic evaluation is:

$$NPW = CC + PN + PWD - PWS \tag{17.13}$$

where CC = total capital costs; PN = present worth of required new investments over the comparison period; PWD = present worth of operation and maintenance costs over the comparison period; and PWD = present worth of salvage value of investment at the end of the comparison period.

If the estimate lifetimes differ for the various alternatives, one should employ a period of time equal to the longest multiple of the different lifetimes for reinvestment calculations.

17.3.9 Perpetual Life

Capitalized cost can be viewed as present worth under the assumption of perpetual life. Computing capitalized cost involves, in a very real sense, finding the present worth of an infinite series of payments. To obtain the present worth of an infinite series of payments of R at the end of each interest period forever, one needs simply to divide R by i, where i is the interest rate per interest period. Thus, to determine what sum of money, P, would have to be invested at 8.0% to provide payments of $100,000 at the end of each year forever, P, would have to be such that the interest on it each period would be $100,000. Withdrawal of the interest at the end of each period would leave the original sum intact to again draw $100,000 interest at the end of the next period. For this example,

$$P = \frac{\$100,000}{0.08} = \$1,250,000$$

The $1,250,000 would be the present worth of an infinite series of payments of $100,000 at the end of each year forever, assuming money is worth 8%. To determine the present worth of an infinite series of payments of R at the end of each n periods forever, first multiply by the *SFDF* to convert to an equivalent single period payment and then divide by i to obtain its present worth.

17.3.10 Breakeven Point

From an economic point of view, the breakeven point of a process or operation is defined as that condition when the costs (C) exactly balance the income (I). The breakeven point is where the profit (Pr) is equal to zero or:

$$Pr = I - C = 0 \tag{17.14}$$

17.3.11 Approximate Rate of Return

Rate of return can be viewed as the interest that will make the present worth of net receipts equal to the investment. The approximate rate of return (ARR), may be estimated from:

$$ARR = \frac{\text{Average annual profit or earings}}{\text{Initial total investment}} \tag{17.15}$$

To determine the average annual profit, simply divide the difference between the total income and the total expenses by the number of years in the period of the investment.

17.3.12 Exact Rate of Return

Using the approximate rate of return as a guide, one can generate the exact rate of return (ERR). This is usually obtained by trial-and-error and interpolation calculations of the rate of interest that makes the present worth of the net income equal to the investments. The approximate rate of return will tend to overestimate the exact rate of return when all or a large part of the income occurs at the end of a period of investment. The approximate rate will also tend to underestimate the exact rate when the salvage value is zero and also when the salvage value is a high percentage of the investment.

17.3.13 Bonds

A bond is a written promise to pay both a certain sum of money (redemption price) at a future date (redemption date) plus equal interest payments at equal intervals in the interim. The holder of a \$1,000, 5% bond, redeemable at 105 (bond prices are listed without the last zero) in 10 years, with interest payable semiannually would be entitled to semiannual payments of $1,000 (0.025) or $25 for 10 years and 105% of $1,000, that is $1,050, at the end of 10 years when the bond is redeemed.

The interest payment on a bond is found by multiplying the face value of the bond by the bond interest rate per period. From the above, the face value is $1,000 and the bond interest rate per period is 0.025. Therefore, the periodic interest payment is the $25. Redeemable at 105 means that the redemption price is 105% of the face value of the bond.

The purchase price of a bond depends on the yield rate, i.e. the actual rate of return on the investment represented by the bond purchase. Therefore, the purchase price of a bond is the present worth of the redemption price plus the present worth of future interest payments, all computed at the yield rate. The bond purchase price, V, formula is:

$$V = \frac{C(1+i)^n + R}{i} \tag{17.16}$$

where C = the redemption price; and R = periodic interest payments.

17.3.14 Incremental Cost

The average unit incremental cost is the increase in cost divided by the increase in production. Only those cost factors which vary with production can affect the average unit incremental cost. In problems involving decisions as to whether to stay in production or (temporarily) shut down, the average unit incremental cost may be compared with the unit selling price.

17.4 The Need for an Economic Analysis

Before an engineer or scientist can successfully apply the principles of economics to operational expenses, they must first become familiar with the existing economics with which they are associated, as well as with the policies of the company itself. A company or individual hoping to increase its profitability must carefully assess a range of investment opportunities and select the most profitable options from those available. Increasing competitiveness also requires efforts to be expended to reduce costs of existing processes. In order to accomplish this, environmental engineers and scientists should be fully aware of not only technical but also economic factors, particularly those that have the largest effect on profitability.

In earlier years, engineers and scientists concentrated on the technical side of projects and left the financial studies to the economists. In effect, engineers involved in making estimates of the capital and operating costs have often left the overall economic analysis and investment decision-making to others. This approach is no longer acceptable.

Some engineers are not equipped to perform a financial and/or economic analysis. Furthermore, many engineers and scientists already working for companies have never taken courses in this area. This shortsighted attitude is surprising in a group of people who normally go to great lengths to obtain all the available technical data before making an assessment of a project or study. The attitude is even more surprising when one notes that data are readily available to enable an engineer to assess the prospects of both their own company and those of their particular industry (Holland et al. 1974).

Bridging the gap between theory and practice is often a matter of experience acquired over a number of years. Even then, methods developed from experience all too often must be reevaluated in the light of changing economic conditions for optimum results. The approach presented here represents an attempt to provide a consistent and reasonably concise method for the solution of problems involving evaluation of economic alternatives (Happel 1975).

The purpose of this section is to provide a working tool to assist the reader in not only understanding economics and finance but also in applying technical information to the economic design and operation of desalination processes and plants. The material to follow will primarily focus on environmental and/or plant applications. Hopefully this approach will provide the reader with a better understanding of some of the fundamentals and principles.

The term "economic analysis" in engineering problems generally refers to calculations made to determine the conditions for realizing maximum financial return for a design or operation. The same general principles apply, whether one is interested in the choice of alternatives for completing projects, in the design of plants or projects so that the various components are economically proportioned, or in the technique of economical operation of existing plants. General considerations that form the framework on which sound decisions must be made are often simple. Sometimes their application in the problems encountered in the development of a commercial enterprise involves too many intangibles to allow exact analysis, in which case judgment must be intuitive. Often, however, such calculations may be made with a considerable degree of exactness.

Finally, concern with the maximum financial returns implies that the criterion for judging desalination projects involves profit. While this is usually true, there are many important objectives which, though aimed at ultimate profit increase, cannot be immediately evaluated in quantitative terms. Perhaps the most significant of these has been some tendency in recent years to regard management of commercial processes as operations with environmental obligations and responsibilities, i.e. considerations other than the profit motive may govern some business decisions. However, these additional social objectives are for the most part often not inconsistent with the economic goal of satisfying human wants, i.e. safe, potable water, with the minimum effort required to achieve these desired goals. In fact, even in the operation of primarily nonprofit organizations, it is still important to determine the effect of various policies on profit (Theodore and Neuser 1996).

17.5 Capital Investment and Risk

A capital investment is required for any process and determination of the necessary investment is an important part of a desalination project. The total investment for any process consists of fixed capital expenditures for the physical equipment and facilities plus the working capital which must be available to pay salaries, keep raw materials and products on hand, and handle other special items requiring a direct cash outlay. Thus, in an analysis of costs of any industrial processes, investment costs, operations costs, and general expenses that include income taxes must be taken into consideration.

The following sequence of steps must be taken in general before a desalination plant can be put into operation:

1. A process must be planned and evaluated, or even invented.
2. The equipment is designed and the operational procedure worked out and examined.
3. Land is acquired for the site and services made available through roads, lines for gas, water, and electricity, sewage disposal systems, etc.
4. Requisite materials are purchased, and the buildings and all equipment erected.
5. Funds must be made constantly available to keep the venture in operation once the plant has been put "on line."

All five of these steps require money. The total money involved is called the *capital investment*. This consists of two parts; that required for steps (1), (2), (3), and (4), known as *fixed capital* and that needed for step (5) called *working capital*. In more graphic terms, the money required to bring the process from the moment of conception to the moment of when the "button is pushed" and the plant is first put into operation is fixed capital. All money required for the venture thereafter (with the exception of capital improvements or expansions of the plant) is working capital. Typical items of the fixed capital variety are salaries paid to researchers and engineers for conception and design; the cost of raw materials; salaries and wages paid to production engineers, foremen, and plant workers; packaging and shipping costs; the costs of water, gas, fuel, electricity, etc., and the salaries paid to clerical personnel involved with the process.

Many a process that looks good in an early stage of research and/or development has proven to be economically unattractive after closer and more intensive examination. The development process often does not go beyond step (2) listed earlier. Money invested up to this point is certainly partly lost and may be completely lost. In other cases, the process gets all the way to the operational stage before proving to be unprofitable, in which case the loss may be significant. Certain things should be obvious. All investment of capital in processing projects involve risk. Manufacturing profit cannot start unless and until the process is put into actual operation, and since for every process that proves profitable there are others that do not, the profitability of the successful processes should be sufficient to absorb the losses incurred by the unsuccessful ventures. What potential profit level should seem probable before capital is invested in a manufacturing process?

Money put into a savings account or invested in bonds may draw a 3% or 4% return without risk, effort, or worry. It follows then that the investment of capital in a manufacturing process that would yield as little as a 3–4% return would be unsound since this would represent the assumption of business risk with no profit at all. How far in excess and above the potential rate of return should it be to justify the investment of capital depends upon the nature of both the industry and the industrial process. When new facilities or novel processes are being considered, a potential profit of 30% or more may be essential to make a plant design attractive. On the other hand, starting a plant to produce potable water by a process with a long record of successful performance, e.g. reverse osmosis, for which detailed economic data are available is another matter and may be attractive at a much lower profitability potential.

17.6 Guidelines for Purchasing Equipment

Prior to the purchase of process equipment, experience has shown the following points should be emphasized:

1. Refrain from purchasing any equipment without reviewing certified independent test data on its performance under a similar application. Request the manufacturer to provide performance information and design specifications.
2. In the event that sufficient performance data are unavailable, request that the equipment supplier provide a small pilot model for evaluation under existing process conditions.

3. If applicable, request participation of the local control authorities in the decision-making process.
4. Prepare a good set of specifications. Include a strong *performance guarantee* from the manufacturer to ensure that the equipment will meet all applicable local, state, and federal codes under specific process conditions.
5. Closely review your process and economic fundamentals.
6. Conduct a careful material balance study before authorizing an equipment evaluation test or purchase equipment.
7. Refrain from purchasing any equipment until *firm* installation cost estimates have been added to the equipment cost. *Escalating installation costs are the rule rather than the exception.*
8. Give operation and maintenance costs high priority on the list of equipment selection factors.
9. Refrain from purchasing any equipment until a solid commitment from the utility supplier(s) is(are) obtained. Make every effort to ensure that the new system will utilize fuel, controllers, filters, motors, etc. that are compatible with those already available at the plant.
10. The specification should include written assurance of *prompt* technical assistance from the equipment supplier. This, together with a complete operating manual (with parts list and full schematics), is essential and is too often forgotten in the rush to get the equipment operating.
11. Schedules, particularly on projects being completed under a court order or consent judgment, can be critical. In such cases, delivery guarantees should be obtained from the manufacturers and penalties defined for noncompliance.
12. The equipment should be of fail-safe design with built-in indicators to show when performance is deteriorating.
13. Withhold 10–15% of the purchase price until compliance is clearly demonstrated.

17.7 Application to Industrial Users

Costs associated include industrial water production include more than the direct costs associated with water supply, but also the costs of water as it travels through the operations of the process. Some of these added water costs include energy, pollution control, regulatory compliance, and raw material costs that all combine to reflect the true cost of potable water production.

Estimating the many components of the total cost for an industrial facility include the cost of water purchased from utilities, and the cost of various steps required to process, use, and discharge the water. These costs can amount to much more than what appears on a utility bill. It is important to consider these and other indirect costs throughout all functions of a facility. Efforts should also include estimating the future water rates when assessing total water costs in order to project any level of savings that can arise when improvements through process modifications are made. Figure 17.1 provides a guide for estimating total water production and usage costs. Note that Items 2 and 3 represent the so-called traditional cost on a typical water bill.

Four common costs associated with industrial water production and usage include:

1. Raw Material Costs. These include costs of measures to ensure that produced or purchased water supply is of sufficient quality for use at a facility. For example, some facilities may bear costs for labor, equipment, and materials needed to improve water quality before use (e.g. when water is a key product ingredient), as well as for chemicals and other raw materials needed to treat and manage water at the facility and before discharging it. Water treatment costs associated with deionized water and reverse osmosis can be substantial.
2. Energy Costs. Costs for the energy to heat and cool water, pump or transfer water, operate water-consuming equipment, treat water and wastewater, and other functions.
3. Pollution Control Costs. These include wastewater service costs paid to a utility, and/or pretreatment costs for the energy, labor, materials, equipment, and chemicals needed to treat wastewater before discharging it from the facility.
4. Regulatory Compliance Costs. These include costs for regulatory compliance activities such as completing permit applications and tracking and reporting wastewater discharges to regulatory agencies. In the United States, industrial, municipal, and other facilities must obtain National Pollutant Discharge Elimination System (NPDES) permits if the facilities discharge water pollutants directly into U.S. waters. Pretreatment regulations govern industrial facilities that discharge into municipal sewer systems.

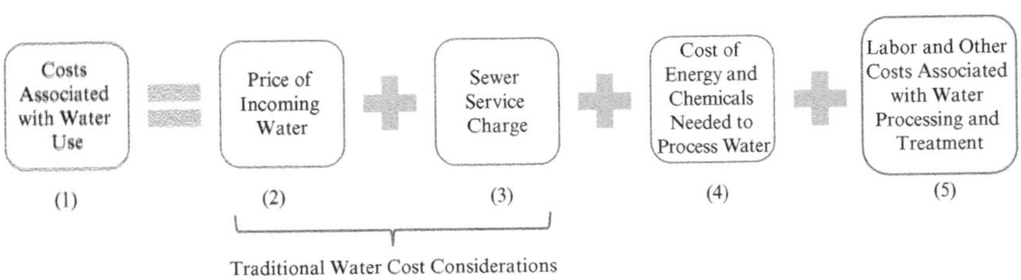

Figure 17.1 Costs Associated with Potable Water Production and Use.

17.8 Illustrative Examples

Six illustrative examples complete the presentation of materials on economic considerations highlighted in this chapter.

17.8.1 Illustrative Example 1

The concept of risk occasionally plays a role in some economic studies and analyses. Provide a qualitative definition of risk.

Solution. Risk can be defined as the product of two factors: (1) the probability of an undesirable event, and (2) the measured consequences of that event. Measured consequences may be stated in terms of financial loss, injuries, deaths, or other variables (Theodore and Dupont 2012).

17.8.2 Illustrative Example 2

The word "what" appears in numerous economic studies. List some questions/comments that are related to the use of this term in an economic studies context.

Solution. The word "what" is used to ascertain conditions or connections that may exist for a specific piece of equipment or process in a manufacturing facility. Some of the "is" and "is nots" related to conditions or connections of a system that can be related to this question of "what" are summarized in Table 17.1.

Table 17.1 What Questions for Illustrative Example 2.

What	Is	Is not
	What do we know?	What do we not know?
	What was observed?	What was not observed?
	What is a related problem?	What is unrelated?
	What are the constraints?	What is not a constraint?
	What is expected?	What is unexpected?
	What is the same?	What is different?
	What is important?	What is not important?
	What resources are needed?	
	What are the criteria?	
	What is the purpose?	

17.8.3 Illustrative Example 3

Describe the economics associated with the automated monitoring of water quality.

Solution. The economics of automatic monitoring are attractive in most cases. Usually, a fairly large capital expenditure must be made for the monitoring equipment, but operating expenses including maintenance usually are much less than the cost of analyzing repeated grab samples in controlled laboratory setting. The results from automated sampling are often more reliable because they are completely objective. Accuracy of the measurements can be determined easily with the proper calibration and checks against standards. Finally, because automated sampling provides a cost effective means of collecting continuous or semi-continuous data, the dynamics of water quality over time can be more accurately and more timely evaluated in the assessment of water quality changes due to intermittent discharges, upsets in process flows and operating conditions, variability in the quality of inflows being treated, etc.

17.8.4 Illustrative Example 4

Consider the following application. Three different desalination processes are under consideration. The service life is 10 years for each system. Capital and annual operating costs are shown in Table 17.2. All system's salvage value in Year 10 = $0. Which is the most economical system assuming straight-line depreciation?

Solution. To select the most economical system, a comparison can be performed among the three options based on the Total Annualized Cost (*TAC*). Table 17.3 can be used to simplify these calculations.

A comparison among Systems A, B, and C indicates that System C has the lowest *TAC* and should be selected as the most economical system of the three being evaluated. The same result would be obtained if a Return on Investment (*ROI*) method of analysis were employed to compare these system alternatives.

Table 17.2 Economic Cost Data for Desalination System Options for Illustrative Example 4.

System	Initial cost	Annual operating costs
A	$300,000	$50,000
B	$400,000	$35,000
C	$450,000	$25,000

Table 17.3 Total Annual Cost Calculations for Illustrative Example 4.

Cost item	System A	System B	System C
Capital investment	\$300,000	\$400,000	\$450,000
Depreciation	\$30,000	\$40,000	\$45,000
Operating costs	\$50,000	\$35,000	\$25,000
TAC	\$80,000	\$75,000	\$70,000

17.8.5 Illustrative Example 5

When considering the cost of desalination projects, it is often important to know how long it will take to recover the cost of an investment in process equipment. The time required to recover this cost is called the *payback period* (U.S. EPA 1993a). The commonly accepted formula for payback period is:

$$\begin{aligned} Payback\ Period &= (Capital\ Investment) \\ &/ (Annual\ Profit + Depreciation) \end{aligned} \tag{17.17}$$

A company spends \$10,000 for new desalination membranes for a small industrial facility. The manager decides to depreciate the equipment at \$1430/yr (7-year straight-line method depreciation) and estimates that the equipment will generate \$1500/yr in annual profit. What is the payback period for this investment?

Solution: Since the payback period is calculated as the fixed capital investment divided by the sum of the annual profit plus the annual depreciation, the following result is obtained:

$$\text{Payback Period} = (\$10{,}000)/(\$1500 + \$1430) = (\$10{,}000)/(\$2930) = 3.41\ \text{yr}$$

17.8.6 Illustrative Example 6

An engineer has compiled the data for two different project options (Table 17.4) – one that includes a new desalination technology for which there are government tax credits provided and one that does not. From an economic point of view, which project should the engineer select? The lifetime of the equipment is 10 years and the interest rate is 10%.

Solution: For the project with new membranes, calculate the total capital cost, *TCC*, as follows:

$$TCC = \$1{,}294{,}000 + \$786{,}000 = \$2{,}080{,}000$$

Table 17.4 Economic Data for Two Desalination Options for Illustrative Example 6.

Cost Item	Project with new membranes, $/yr	Project with conventional membranes, $/yr
Equipment cost	$1,294,000	$1,081,000
Installation cost	$786,000	$659,000
Operating labor	$39,900	$8,500
Maintenance	$43,000	$17,000
Utilities	$958,000	$821,000
Overhead	$51,300	$13,900
Taxes, insurance, & administration	$86,200	$72,600
credits	$380,000	$0

Calculate the Capital Recovery Factor, *CRF*, using Equation 17.11:

$$CRF = \frac{i(1+i)^n}{(1+i)^n - 1} = \frac{0.1(1.1)^{10}}{(1.1)^{10} - 1} = \frac{0.1(2.594)}{(2.594 - 1)} = 0.1627$$

Calculate the annualized capital cost, *ACC*, of the process with the new membrane option using Equation 17.12 as:

$$ACC = (TCC)(CRF) = (\$2{,}080{,}000)(0.1627) = \$338{,}416$$

Calculate the annualized capital cost, *ACC*, for the option with conventional membranes as follows, noting that the *CRF* for this option is identical to that calculated for the option with new membrane type:

$$ACC = (\$1{,}740{,}000)(0.1627) = \$283{,}098$$

The better choice for the project can then be chosen based on the summary of costs presented in Table 17.5. Based on these numbers, the project using the new membrane type is the better choice on a total annual cost basis by $79,300/yr.

Table 17.5 Final Economic Analysis Based on Annual Cost for Two Desalination Options for Illustrative Example 6.

Cost item	Project with new membranes, $/yr	Project with conventional membranes, $/yr
Annual capital cost	$338,416	$283,098
Operating labor	$39,900	$8,500
Maintenance	$43,000	$17,000
Utilities	$958,000	$821,000
Overhead	$51,300	$13,900
Taxes, insurance, & administration	$86,200	$72,600
Credits	–$380,000	$0
Total annual cost	$1,136,816	$1,216,098

17.9 Problems

1. Qualitatively define the following five risk terms:
   a. Substantial risk
   b. Unreasonable risk
   c. Insignificant risk
   d. *De minimis* risk
   e. Acceptable risk
2. Describe ten problems that can arise in the design/procurement/construction/startup of a desalination plant.
3. Comment on the results of Illustrative Example 6 in terms of the risk of the investment.
4. The total costs for a proposed desalination project were estimated to be $852,644 in 2014. Your company has now proposed a similar type of desalination project for your facility in 2020. Assume an annual rate of return of 10% and an operating period of 10 years. The composite CEPCI for 2014 is 576.1, and for 2020 was 596.2.
   a. What is the cost of the "new" project?
   b. If the total installation cost is 60% of the total equipment cost, what is the annualized capital cost, *ACC*, of this project?
5. As discussed in this chapter, the break-even point of a process operation is defined as that point at which the costs (C) exactly balance the income (I). The profit (P) is therefore $P = I - C = 0$ at the break-even point. The cost and income (in dollars) of a particular desalination operation are given as follows:

$$I = \$80{,}000 + 0.12N \tag{17.18}$$

$$C = \$1{,}008{,}000 + 0.09N \tag{17.19}$$

where N = the yearly production of potable water being manufactured, gal.

Calculate the break-even point, i.e. the value of N, for this operation.

6. What is the break-even point of the potable water production system described in Problem 5 if the unit production cost of a gallon of water increases to \$0.11? What is the break-even point if the value of the produced water increases to \$0.15/gal and the cost remains \$0.09/gal? What is the break-even point if the production cost is \$0.11/gal and the value of the produced water increases to \$0.25/gal? Discuss the implication of these results.

References

Happel, J. (1975). *Chemical Process Economics (Chemical Processing and Engineering), Enlarged*, 2nd Edition. New York: Marcell Decker.

Holland, F., Watson, F., and Wilkinson, J. (1974). Financing assets by equity and debt. *Chemical Engineering* 81 (9): 62–66.

Keynes, J.M. (1936). *The General Theory of Employment, Interest, and Money.* London, England: Palgrave Macmillan.

Maxwell, C. (2021). Cost Indices. https://www.toweringskills.com/financial-analysis/cost-indices (accessed June 7, 2021).

Theodore, L. and Dupont, R.R. (2012). *Environmental Health and Hazard Risk Assessment: Principles and Calculations.* Boca Raton, FL: CRC Press/Taylor and Frances Group.

Theodore, L. and Neuser, K. (1996). *Engineering Economics and Finance. A Theodore Tutorial.* East Williston, NY: Theodore Tutorials. Originally published by the U.S. EPA/APTI, RTP, NC.

U.S. Corps of Engineers (2018). *Civil Works Construction Cost Index System (CWCCIS).* EM 1110-2-1304. Washington, D.C.: Department of the Army. www.publications.usace.army.mil/Portals/76/Publications/EngineerManuals/EM_1110-2-1304_31_Mar_2018.pdf?ver=PIjVa1Vj-m6k-7hSzM9NDA%3d%3d.

U.S. EPA (1993). *A Primer for Financial Analysis of Pollution Prevention Projects.* EPA 600/R-93/059. Cincinnati, OH: Risk Reduction Engineering Laboratory, Office of Research and Development, U.S. Environmental Protection Agency.

18

Optimization Considerations

18.1 Introduction

Optimization? Here are some terms for *optimal* that can be found in a dictionary: most favorable or desirable; best, best possible option; most cheerful view of matters or expect the best outcome; bright side; to make the most of; develop or realize to the utmost extent; obtain the most efficient or optimum use of; the best or most favorable degree, condition, or amount; etc. Add to this are some of the classic sayings of yesteryear that have appeared in the literature:

- *The optimist proclaims that we live in the best of all possible worlds; and the pessimist fears this is true* (Cabell 1926, Chapter 26).
- *Optimism is a mania for maintaining that all is well when things are going badly* (Voltaire 1959 [1759], Chapter 19).
- *The optimist sees the doughnut but the pessimist sees the hole* (Wilson 2016 [c. 1915]).
- *An optimist is a guy that has never had much experience* (Marquis 1927).

As one might suspect, the term optimization has come to mean different things for different applications, i.e. it could involve a simple two-step calculation or one that requires the use of a detailed numerical methods. To take this a step further, the authors were undecided as to how to present optimization in this chapter. After much deliberation and meditation, it was decided to present both qualitative and quantitative material. This decision was primarily influenced by the desire that this be material not only on the introduction to optimization but also that addresses the topic quantitatively.

The optimization process has been described by Aris (1964) as "getting the best you can out of a given situation." Problems amenable to solutions by mathematical optimization techniques generally have one or more independent variables whose values must be chosen to yield one or more viable solutions and use a measure of "goodness" to distinguish among the many viable solutions generated by different choices of these variables. Mathematical optimization techniques are also used for

Introduction to Desalination: Principles, Processes, and Calculations, First Edition. Louis Theodore and R. Ryan Dupont.

guiding the problem solver to that choice of variables that maximizes a "goodness" measure (e.g. profit) or that minimizes a "badness" measure (e.g. cost).

As noted in its most elementary and basic form, one may say that optimization is concerned with the determination of the "best" solution to a given problem. This process is required in the solution of many general problems in engineering and applied science in the maximization (or minimization) of a given function(s), in the selection of a control variable to facilitate the realization of a desired condition, in the scheduling of a series of operations or events to control completion dates of a given project, or in the development of optimal layouts of organizational units within a given design space, etc.

One of the more important areas for the application of mathematical optimization techniques is in engineering design. The process of optimization in water treatment and environmental practice is required in the solution of many problems and often involves the maximization or minimization of a mathematical function. Once a particular subject or process scheme has been selected for study, it is common practice to optimize the process from a capital cost and operation and maintenance (O&M) standpoint.

Investing in all aspects of water resource management, services and infrastructure (development, operation, and maintenance) can generate significant social and economic benefits. The authors believe that appropriating capital toward drinking water supply, water management, and sanitation is highly cost-effective from a health perspective.

Traditional mathematical assessments of water supply and water resource management are often inadequate for addressing the optimization of water utilization and demands. Optimization studies are necessary to make informed investment and management decisions, facilitate cross-sector decision-making, and address compromises and trade-offs among the multitudes of stakeholder groups involved in water resource planning. Increased financial investment and support for water "development" is essential. When this support is efficiently managed and optimized, these investments can generate social, economic, financial, and other benefits that greatly outweigh the costs of this process. In addition to adequately financing the development, operation and maintenance of infrastructure, increased funding is also needed to develop institutional capacity and ensure well-functioning governance structures. Greater financial support is equally essential to enhance the knowledge base and provide reliable and objective information about the state of water resources. Their use can help water managers, engineers, and administrators inform decision-makers about policy/process options and their associated potential impacts. Finally, one's ability to understand optimization and any associated risks is the key to solving some of the more complex water related problems in the future.

This chapter attempts to provide a description of elementary optimization processes that can be applied to water supply and resource management.

Hopefully, it will allow water resource management administrators and technical personnel to treat and manage water more efficiently in the future. Chapter topics included are the scope of optimization, general analytical formulation of optimization problems, traditional optimization applications, linear programming applications, and optimization of existing systems. An application section provides six illustrative examples on traditional applications, and three illustrative examples of linear programming applied to water treatment and management related problems.

18.2 The Scope of Optimation

Optimization is mathematical in nature, and typically involves the maximization or minimization of a function that represents the "performance" of a "system." This is carried out by quantifying the values of the input variables that produce an optimal value for the function.

Perhaps the most important tool employed in optimization by engineers and scientists is *linear programming*. Linear programming consists of a linear, multivariable function which is to be optimized (maximized or minimized), subject to a particular number of constraints. The constraints are normally expressed in linear form. Integer linear programming refers to optimization problems in which at least some of the variables assume integer values. The reader should note that the terms *linear* programming and *nonlinear* programming are similar from an application perspective. There were problems around the middle of the last century because of difficulties that arose in attempting to solve nonlinear programming problems. However, the arrival of the modern-day computer and readily available software (e.g. Excel) have eliminated these problems. Scientific calculations previously estimated in lifetimes of computational time can currently be generated in nanoseconds. Unless a solution is presented graphically in this chapter, it was obtained directly from Excel so that the reader need not be concerned with whether a system's describing equations are linear or nonlinear. Thus, details regarding the solution methodology for both linear programming and nonlinear programming problems, and any corresponding complimentary examples, will not be presented. They are simply not necessary and beyond the scope of this chapter.

One of the major responsibilities in optimization is to construct an accurate objective function to be maximized or minimized. For example, in an environmental problem, the objective function might be the profit associated with the operation of a landfill that treats two categories of waste: "inert" and "non-inert" waste. Essentially, the optimization problem is to maximize profit (from handling the two waste streams) by employing an operating schedule subject to the landfill's capacity and environmental regulatory requirements. Rephrased, the objective function should represent a mathematical description of the combination of two wastes which will produce maximum profit subject to the constraints connected with the landfill's processing capabilities and the regulation requirements imposed on each type of waste.

The objective function might be simple to identify and quantify, or may be very illusive due to the presence of conflicting or dimensionally incompatible objectives. One might be asked to optimize the profit for the aforementioned landfill that not only minimizes air and water pollution, but is also *aesthetically pleasing*. Thus, it may not always be possible to quantify an objective function using strictly mathematical terms.

Optimization problems can sometimes be divided into parts, where it is possible to attain the optimum for the original problem by optimizing the constituent parts. This process is very powerful, as it allows one to solve a series of smaller, easier problems rather than one large one. One of the best-known techniques to attack these problems is *dynamic programming*. This approach is characterized by a process which is performed in stages. One such example is a manufacturing process. Rather than solving the problem as a whole, dynamic programming optimizes one stage at a time to produce an optimal set of decisions for the entire process. Although dynamic programming has application in some systems/processes, it is not reviewed in this chapter. If the assumption of linearity cannot be made, a collection of techniques called *nonlinear programming* is used. This is addressed later in this chapter.

Finally, most engineering optimization applications involve economics: maximizing profit, minimizing cost, or both. Environmental applications can involve the purification of water, minimizing toxic emissions, maximizing energy use efficiency and conservation, minimizing health and safety risks, etc. The reader should not lose sight, however, that most real-world industrial applications involving optimization often require simple solutions. Following is a comment from a retired process engineer after 30 years in industry (R. Altomare, personal communication), "Generally, the consumer goods industry where I worked does not rely on sophisticated models. Basically, optimization was conducted to determine how to use minimum resources to minimize costs, subject to constraints. In the determination of consumer preference, we typically designed experiments to examine the broad range of attributes that led to maximum 'product liking' or decision to purchase. Occasionally, we modeled processes and used regression models. In that case, we could search or compute the desired maximum. The last set of techniques used is what the business world calls *Operations Research.* For example, linear programming would be used in the production of America's favorite barbecue item – the hot dog. The objective function was to produce a combination of meats that maximized flavor, met or exceeded FDA requirements for protein, fat, and water content, while minimizing overall production costs."

18.3 General Analytical Formulation of Optimization Problems

When more than one independent variable is involved in determining the objective function, e.g. profitability, the objective function *y*, that depends on a system of $x_1, x_2, x_3, \ldots, x_n$ independent variables, is written as follows:

$$y(profit) = f(x_1, x_2, x_3, \ldots, x_n) \tag{18.1}$$

The problem is to specify $x_1, x_2, x_3, \ldots, x_n$ independent variables so that y will be a maximum. In general, this will involve some type of trial-and-error calculation.

It is possible in some cases to set up an explicit function which will relate the objective function *y* to the independent variables. Differentiation will then result in a series of partial derivatives,

$$\frac{\partial y}{\partial x_1}, \frac{\partial y}{\partial x_2}, etc. \tag{18.2}$$

Setting these partial derivatives equal to zero results in the same number of simultaneous equations as number of variables involved. These equations, often subject to "limitations" and "constraints," can be solved for the corresponding optimum values of these variables. The maximum occurs at a point where the first derivatives equal zero, accounting for restrictions on the second derivative, continuity of the function throughout the range of independent variables, and absence of optimum conditions at limiting values of one or more of the variables. These conditions can, on occasion, restrict the applicability of this approach (Happel 1955). For example, for the case of a two-variable function where $y = f(x_1, x_2)$, points $x_1 = a$ and $x_2 = b$ exist for a maximum value of $y = f(x_1, x_2)$ if the following conditions hold:

$$\frac{\partial y}{\partial x_1} = \frac{\partial y}{\partial x_2} = 0 \tag{18.3}$$

with,

$$\frac{\partial^2 y}{\partial x_1^2} < 0; \frac{\partial^2 y}{\partial x_2^2} < 0 \quad (18.4)$$

and,

$$\left(\frac{\partial^2 y}{\partial x_1^2}\right)\left(\frac{\partial^2 y}{\partial x_2^2}\right) > \left(\frac{\partial^2 y}{\partial x_1 \partial x_2}\right)^2 \quad (18.5)$$

In contrast, if $\partial^2 y / \partial x_1^2$ and $\partial^2 y / \partial x_2^2$ are positive, and Equation 18.5 holds, there will be a minimum. However, if,

$$\left(\frac{\partial^2 y}{\partial x_1^2}\right)\left(\frac{\partial^2 y}{\partial x_2^2}\right) < \left(\frac{\partial^2 y}{\partial x_1 \partial x_2}\right)^2 \quad (18.6)$$

the point concerned will be a "saddle" point. The case where,

$$\left(\frac{\partial^2 y}{\partial x_1^2}\right)\left(\frac{\partial^2 y}{\partial x_2^2}\right) = \left(\frac{\partial^2 y}{\partial x_1 \partial x_2}\right)^2 \quad (18.7)$$

is open and the value may be a maximum, a minimum, or neither.

Now consider an objective function specified as:

$$f(x_1, x_2) = \left(x_1^2 - 2x_1 + x_2^2 + x_1 x_2 + 1\right) \quad (18.8)$$

To determine if the function has a maximum or a minimum, proceed as follows

$$\frac{\partial f}{\partial x_1} = 2x_1 - 2 + x_2 = 0 \quad (18.9)$$

$$\frac{\partial f}{\partial x_2} = 2x_2 + x_1 = 0 \quad (18.10)$$

When Equations 18.9 and 18.10 are solved, the following equations result:

$$x_1 = \frac{4}{3} \quad (18.11)$$

$$x_2 = \frac{1}{3} \quad (18.12)$$

Since

$$\frac{\partial^2 f}{\partial x_1^2} = 2; \frac{\partial^2 f}{\partial x_2^2} = 2 \tag{18.13}$$

the solutions for x_1 and x_2 represent a *minimum* with

$$y = f(x_1, x_2) = 5 \tag{18.14}$$

Consider now a traditional mathematician's exercise to determine the maximum or minimum of a function dependent on three variables

$$P = 8x + 6y + 8z + xy + xz + yz - 2x^2 - 3y^2 - 4z^2 \tag{18.15}$$

Here

$$\frac{\partial P}{\partial x}|_{y,z} = 8 + y + z - 4x = 0 \tag{18.16}$$

$$\frac{\partial P}{}|_{x,z} = 6 + x + z - 6y = 0 \tag{18.17}$$

$$\frac{\partial P}{\partial z}|_{x,y} = 8 + y + x - 8z = 0 \tag{18.18}$$

The solution to these three linear simultaneous equations can be shown to be

$$x = -4.62;\ y = 0.94; z = 0.54 \tag{18.19}$$

for which

$$P = -84.5 \tag{18.20}$$

To determine whether the P is a maximum or minimum, calculate the three second order partial derivatives as:

$$\frac{\partial^2 P}{\partial x^2} = -4; \frac{\partial^2 P}{\partial y^2} = -6; \frac{\partial^2 P}{\partial z^2} = -8 \tag{18.21}$$

Since the three derivatives are negative, the value of $P = -84.5$ is a maximum, i.e. any other value of P for a different x, y, and z would produce a value of P less than (or more negative) than –84.5. The "proof" is left as an exercise for the reader. However, if one were to select values of x, y, and z straddling the supposed minimum set of values, P should be below –84.5. For example, the value of P for x = –5, y = 1, and z = –0.5 is – 95.0.

18.4 Mathematical Concepts in Linear Programming

Linear programming is a technique for optimization which has evolved into a useful procedure for the technical community. As noted earlier, the title results from the assumption inherent in the method that linear relationships describe the system under consideration. It deals with the determination of an optimum solution of a problem expressed in linear relationships where there are a large number of possible solutions. In general, the methods employed are trial-and-error, where the procedure systematically follows a mathematical treatment which minimizes the labor involved and ensures that a correct result has been obtained.

In order to apply linear programming to a problem, the relationships between variables must be expressed as a set of linear equations or inequalities. The variables must be otherwise independent and must exceed the number of equations. Mathematically expressed,

$$\sum_i a_{ij} = b_j; (i = 1, 2, \ldots m; j = 1, 2, n; n > m \tag{18.22}$$

where the sum of the coefficient a_{ij} times the value of each variable x_j equals the requirement b_j in the ith equation. The number of variables must exceed the number of equations, so that an infinite number of solutions is possible. The best one is selected. The relationship in Equation 18.22 is often expressed as inequalities. "Dummy" additional variables could be introduced in order to convert these inequalities into appropriate equations (Happel 1955). Another requirement is that no quantity can be negative. Thus, mathematically

$$x_j \geq 0; b_j \geq 0 \tag{18.23}$$

This requirement will ensure that only useful answers are obtained.

The solution of a problem by linear programming also requires, for example, a profit or objective function to be maximized (or minimized). This requirement is stated as:

$$P = \sum_{i} c_j x_i = maximum\left(or\ minimum\right) \tag{18.24}$$

where c_j is the profit (or cost) per unit x_i used. The solution is carried out by an iterative process, which at each stage of the calculation assigns either zero or positive values to all x_i variables. As the calculation proceeds, the values selected, while meeting the constraints imposed, will tend to increase the summation to be maximized at each stage. If a solution exists, a maximum (profit) case will be realized.

18.5 Applied Concepts in Linear Programming

Consider now an example that involves minimizing an objective function, i.e. E (expenditures):

$$E = a_1 x_1 + a_2 x_2 \tag{18.25}$$

subject to two constraints

$$B = b_1 x_1 + b_2 x_2 \tag{18.26}$$

$$C = c_1 x_1 + c_2 x_2 \tag{18.27}$$

In addition

$$x_1 \geq 0; x_2 \geq 0 \tag{18.28}$$

In terms of a solution, this system of equations can first be examined graphically. The problem can be viewed in Figures 18.1 and 18.2. First note that only the positive quadrant needs to be considered because Equation 18.28 must hold. Equations 18.26 and 18.27 can be rewritten in the following form:

$$x_2 = \frac{B}{b_2} - \left(\frac{b_1}{b_2}\right) x_1 \tag{18.29}$$

$$x_2 = \frac{C}{c_2} - \left(\frac{c_1}{c_2}\right) x_1 \tag{18.30}$$

Each appears in Figure 18.1a and 18.1b. It is apparent that the only values of x_1 and x_2 which will satisfy Equations 18.29 and 18.30 lie above the solid lines in Figure 18.1c, i.e. the shaded area. This shaded area is termed the

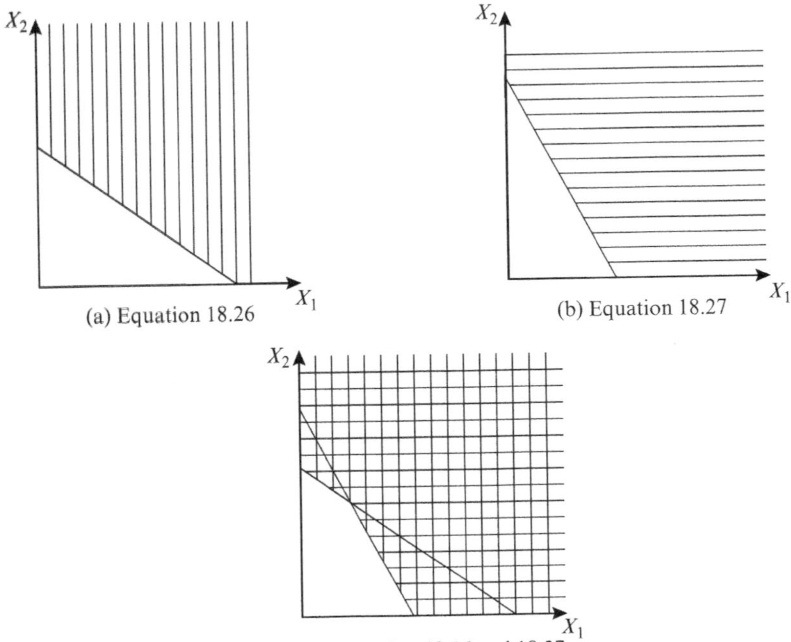

(a) Equation 18.26

(b) Equation 18.27

(c) Equation 18.26 and 18.27

Figure 18.1 Two-variable Minimization Problems.

constraint set. Now a series of straight lines or contours corresponding to the expenditure Equation 18.25 with a different value of E used for each line are placed in Figure 18.2a. Superimposing Figure 18.2a onto Figure 18.lc produces Figure 18.2b. Of all these lines, the smallest E for which x_1 and x_2 remain in the specified region in Figure 18.1c will be the minimum expenditure. This will yield that proportion of x_1 and x_2 which minimize the cost. Stated in other words, one desires the lowest-value contour having some point in common with the constraint set. For the particular values of a_1 and a_2 chosen for Equation 18.25, the minimum point is seen to correspond to E where $E_3 = E^*$. Further if the constants c_1 and c_2 in Equation 18.25 are changed, it is possible that the minimum point would occur elsewhere.

Which point is the minimum is determined by constraints a_1 and a_2 of the problem. It should be pointed out that if the expenditure equation was nonlinear the contour lines would be curved and if the constraint equations were nonlinear, the constraint lines would also not be straight. It should be intuitively obvious that locating the minimum point is more complicated in this case. The situation is essentially reversed for a maximization problem, i.e. for a profit P. This is demonstrated in Figure 18.3.

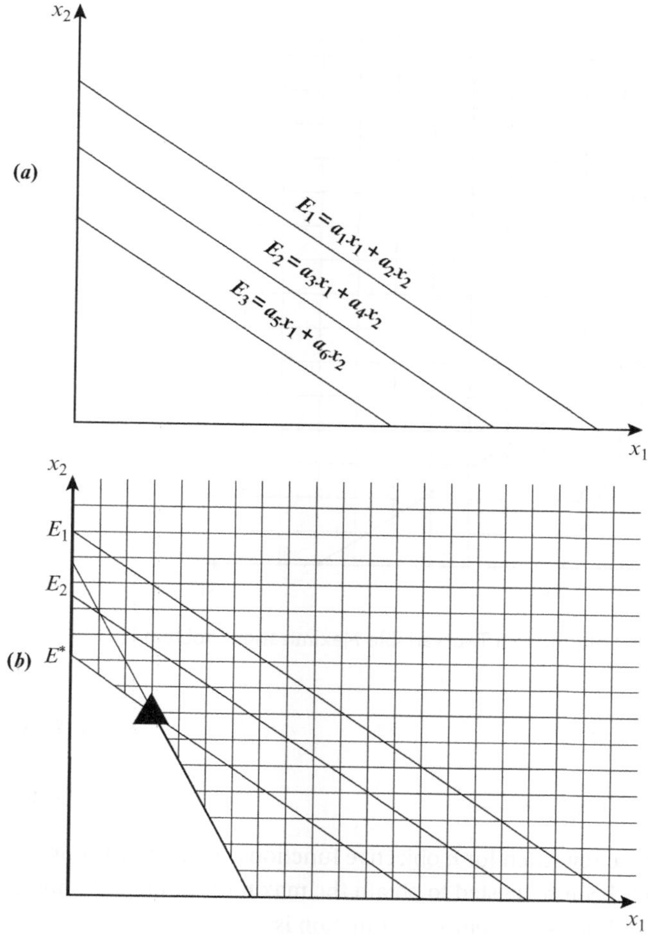

Figure 18.2 Solution to a Two-variable Minimization Problem.

The above methodology can be extended to more than two variables and more than two constraints. The following would apply for the three variables and four constraints:

$$P = a_1x_1 + a_2x_2 + a_3x_3 \tag{18.31}$$

$$B = b_1x_1 + b_2x_2 + b_3x_3 \tag{18.32}$$

$$C = c_1x_1 + c_2x_2 + c_3x_3 \tag{18.33}$$

$$D = d_1x_1 + d_2x_2 + d_3x_3 \tag{18.34}$$

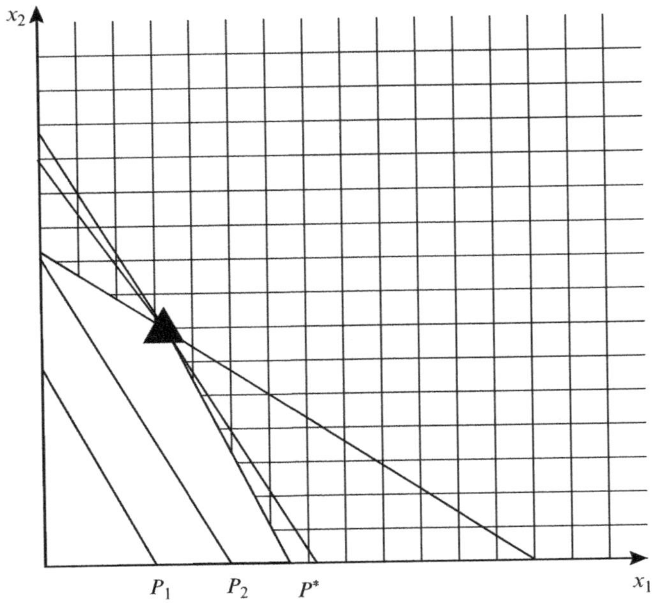

Figure 18.3 Possible Solution to a Two-variable Maximization Problem.

$$E = e_1x_1 + e_2x_2 + e_3x_3 \tag{18.35}$$

$$x_1 \geq 0; x_2 \geq 0;\ x_3 \geq 0 \tag{18.36}$$

Naturally, this can be rewritten for n objective functions and m constraints.

Suppose you have been requested to obtain the maximum value of P subject to the following conditions. The objective function is:

$$P = 3x_1 + 5x_2 \tag{18.37}$$

Subject to the following constraints:

$$x_2 = -0.5x_1 + 3 \tag{18.38}$$

$$x_2 \leq -3x_1 + 6 \tag{18.39}$$

$$x_1 \leq 3 \tag{18.40}$$

$$x_2 \geq 4 \tag{18.41}$$

Once again, the solution is first provided in graphical form. Refer to Figure 18.4. Conditions in Equations 18.40 and 18.41 require that the

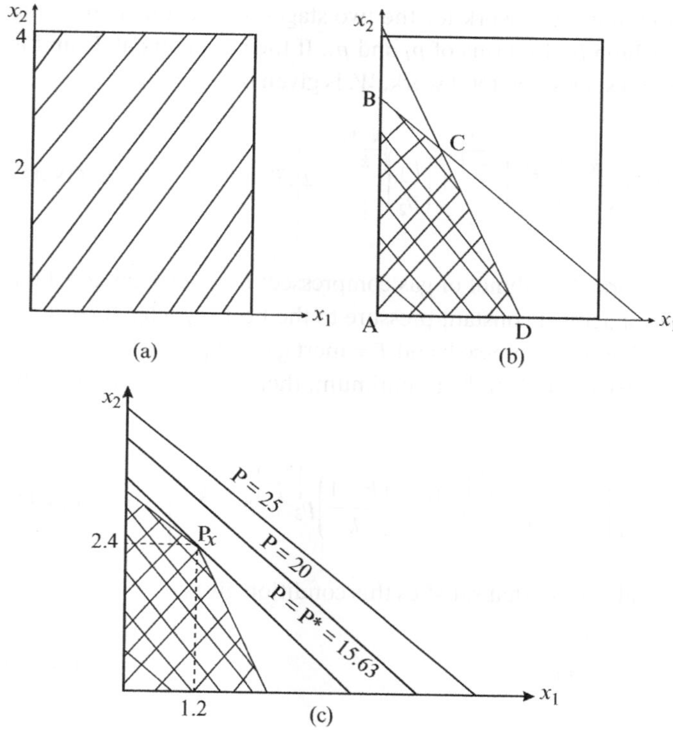

Figure 18.4 Graphical Solution for Maximization Problem.

solutions, i.e. the values x_1 and x_2 that will maximize P lie within the rectangle provided in Figure 18.4a.

Conditions in Equations 18.38 and 18.39 require that the solution be within the area ABCD in Figure 18.4b. When the objective function equation is superimposed on Figure 18.4b, Figure 18.4c results. The maximum P is located at $x_1 = 1.2$ and $x_2 = 2.4$, for which $P = 15.6$. A more detailed solution process is provided in Illustrative Example 3.

18.6 Engineering Applications in Optimization

In terms of real world application, consider the two-stage reversible adiabatic compression of a gas from an initial pressure, p_1, to a final pressure, p_3. If all fixed charges on the compressors are assumed to be essentially independent of the interstage pressure employed, i.e. they are constant, then the optimum operation involves the determination of the interstage pressure, p_2, for which the total power requirement is a minimum. For a fixed gas flow, this corresponds to the minimum work for the two stages and requires obtaining an equation describing p_2, in terms of p_1 and p_3. If the gas enters at T_1 and is cooled to T_i between stages, the total work, W, is given by:

$$W = NRT\left(\frac{k}{k-1}\right)\left[\left(\frac{p_2}{p_1}\right)^{\left(\frac{k-1}{k}\right)} + \left(\frac{p_3}{p_2}\right)^{\left(\frac{k-1}{k}\right)} - 2\right]; T = T_1 \tag{18.42}$$

where W = total work; N = lbmol of gas compressed; R = ideal gas constant; K = ratio of heat capacity at constant pressure to the heat capacity at constant volume for the gas being compressed; and T = inert gas temperature = $T_1 = T_2$. If the dependent variable W is to be a minimum, then the derivative must be zero, i.e.:

$$\frac{dW}{dp_2} = NRT_1\left(\frac{k}{k-1}\right)\left[\left(\frac{k-1}{k}\right)p_1^{\left(\frac{1-k}{k}\right)}p_2^{-\left(\frac{1}{k}\right)} - \left(\frac{k-1}{k}\right)p_3^{\left(\frac{k-1}{k}\right)}p_2^{-\left(\frac{1-2k}{k}\right)}\right] = 0 \tag{18.43}$$

Solving for the value of p_2 that satisfies this condition gives:

$$p_2^{\left(\frac{2k-2}{k}\right)} = p_3^{\left(\frac{k-1}{k}\right)}p_1^{\left(\frac{k-1}{k}\right)} \tag{18.44}$$

or

$$p_2 = \sqrt{p_1 p_3} \tag{18.45}$$

This solution is found in most handbooks and thermodynamics texts (Happel 1955; Theodore and Reynolds 1992; Theodore et al. 2009) for the optimum interstage pressure in two-stage compression. Thus, if p_1 = 1 atm, and p_3 = 10 atm, then the interstage pressure p_2 = 3.16 atm. The location of a maximum or minimum becomes obvious if the function in Equation 18.42 is plotted. This analytical method has the advantage of directness and simplicity in many cases. The *graphical* method involves calculation of the function at a number of points and requires more time to obtain the minimum work.

If one were to extend this and consider the three-stage reversible adiabatic compression of a gas from an initial pressure p_1 to a final pressure p_4 and if the fixed charges on the compressors are again assumed to be essentially independent of the interstage pressure employed, then the optimum operation involves the determination of the interstage pressures for which the total power requirement is again a minimum. (Note: Independent calculation indicates that, for optimum results, intercooling back to the inlet temperature is desirable). If the gas again enters as a temperature T and is cooled to T between stages, the total work is given by:

$$W = NRT\left(\frac{k}{k-1}\right)\left[\left(\frac{p_2}{p_1}\right)^{\left(\frac{k-1}{k}\right)} + \left(\frac{p_3}{p_2}\right)^{\left(\frac{k-1}{k}\right)} + \left(\frac{p_4}{p_3}\right)^{\left(\frac{k-1}{k}\right)} - 3\right] \quad (18.46)$$

If this quantity is to be a minimum

$$\frac{\partial W}{\partial p_2} = NRT\left[p_1^{\left(\frac{1-k}{k}\right)} p_2^{-\left(\frac{1}{k}\right)} - p_3^{\left(\frac{k-1}{k}\right)} p_2^{-\left(\frac{1-2k}{k}\right)}\right] = 0 \quad (18.47)$$

$$\frac{\partial W}{\partial p_3} = NRT\left[p_2^{\left(\frac{1-k}{k}\right)} p_3^{-\left(\frac{1}{k}\right)} - p_4^{\left(\frac{k-1}{k}\right)} p_3^{-\left(\frac{1-2k}{k}\right)}\right] = 0 \quad (18.48)$$

This leads to

$$p_2^2 = p_1 p_3 \quad (18.49)$$

$$p_3^2 = p_2 p4_3 \quad (18.50)$$

Whence

$$p_2 = \left(p_1^2 p_4\right)^{1/3} \quad (18.51)$$

$$p_3 = \left(p_4^2 p_1\right)^{1/3} \quad (18.52)$$

If one now differentiates $\partial W/\partial p_2$ with respect to p_2 and substitutes the values for p_2 and p_3 from Equations 18.51 and 18.52, the following is obtained:

$$\left(\frac{\partial^2 W}{\partial p_2^2}\right)_0 = 2NRT\left(\frac{k-1}{k}\right)\left(p_1^{\left(\frac{1-5k}{3k}\right)} p_4^{-\left(\frac{1+k}{3k}\right)}\right) \quad (18.53)$$

The subscript 0 indicates that the value of $\partial^2 W/\partial {p_2}^2$ is at the possible minimum which one is attempting to verify. Since the powers of p_1 and p_4 must be positive, and $k > 1$, it is clear that, from Equation 18.53, $\partial^2 W/\partial {p_2}^2$ is greater than zero, a positive quantity. Similarly, by differentiating Equation 18.48 with respect to p_3 and substituting, the following is obtained:

$$\left(\frac{\partial^2 W}{\partial p_3^2}\right)_0 = 2NRT\left(\frac{k-1}{k}\right)\left(p_1^{\left(\frac{1-3k}{3k}\right)} p_4^{-\left(\frac{1+3k}{3k}\right)}\right) \tag{18.54}$$

This, by similar reasoning also is positive. The product of the second derivatives is:

$$\left(\frac{\partial^2 W}{\partial p_2^2}\right)_0 \left(\frac{\partial^2 W}{\partial p_3^2}\right)_0 = 4\left[\frac{(NRT)(k-1)}{k}\right]^2 \left(p_1^{\left(\frac{2-8k}{3k}\right)} p_4^{-\left(\frac{2+4k}{3k}\right)}\right) \tag{18.55}$$

Similarly, the cross derivative is:

$$\left(\frac{\partial^2 W}{\partial p_2 \partial p_3}\right)_0 = \left[\frac{(NRT)(k-1)}{k}\right]^2 \left(p_1^{\left(\frac{2-8k}{3k}\right)} p_4^{-\left(\frac{2+4k}{3k}\right)}\right) \tag{18.56}$$

Thus, it has been demonstrated that:

$$\left(\frac{\partial^2 W}{\partial p_2^2}\right)\left(\frac{\partial^2 W}{\partial p_3^2}\right) = \left(\frac{\partial^2 W}{\partial p_2 \partial p_3}\right)^2 \tag{18.57}$$

If $p_1 = 1$ atm and $p_4 = 10$ atrn, the interstage pressures for minimum work are:

$$p_2 = 2.15 \text{ atm}; \; p_3 = 4.64 \text{ atm} \tag{18.58}$$

and

$$W = W(2.15, 4.64) = 2.57; \text{ consistent units} \tag{18.59}$$

18.7 Optimization of Existing Systems

The objective of this section is to provide a general discussion of the importance of proper operation and maintenance to overall plant performance. The plant, including all equipment, must be recognized as a highly specialized and complex facility producing an acceptable product, e.g. potable water from a desalination plant. It is the plant management's responsibility to produce this product at the lowest unit cost and at the highest quality possible. The key to fulfilling this responsibility is a sound operation and maintenance management program.

The initial cost objective when purchasing a desalination system is to achieve a technically reliable system capable of meeting potable water production goals and achieving all of this at the minimum annualized cost. Once the system has been purchased, this goal cannot be realized without a conscientious and determined operation and maintenance effort. It is obvious, therefore, that once the optimum or near optimum level of performance is achieved, it must be maintained day in and day out. Unfortunately, such achievements rarely happen on their own in actual practice. High levels of performance result from the application of not only sound engineering practice, but also operation and maintenance procedures, both which are addressed in this section.

Effective operation must be accompanied by a routine maintenance program. The program should provide for scheduled inspections of all equipment components as well as for necessary monitoring of operating parameters to ensure correct operation and optimum performance of all of the process equipment. Monitoring of the inlet and outlet streams, water quality, etc., will provide valuable information about the performance of the process.

Routine maintenance of equipment can also be important because equipment failure can be expensive in terms of lost production, degradation of resources, and potential health effects on employees. Each component in a process must operate properly to ensure steady, efficient output and desired results from the system.

Establishing an equipment maintenance program need not be elaborate or complicated. The actual work in routine inspection and servicing may, in large part, be performed by personnel operating the equipment. Of course, extensive repairs or rebuilding should be accomplished by skilled and trained maintenance personnel.

An adequate scheduled maintenance program will repay itself many times over by reduction in production time losses and costly major repairs. Effective maintenance is easily accomplished if the following basic rules are followed:

1. At initial installation, provide for easily accessible access for cleaning and removal of any part of the system. Provision for convenient maintenance begins on the drawing board. The designer should allow for the eventual removal of parts that may wear out. Special material should be designated for those parts normally exposed to the greatest wear or possible abuse.
2. Establish an information file for history of all equipment. The file should include actual operating data, such as flow volumes, operating temperatures, inlet and outlet velocities, pressure differentials, and any and all other data affecting operation.

3. Maintain conditions as designated on nameplate.
4. Maintain copies of arrangement or construction drawings, including parts lists and a spare-parts reference drawing. These contain commercial descriptions of such items as gaskets, bolts, nuts, and washers.

At least one copy of the information file should always be available to maintenance personnel, and a master copy should always be kept in the plant engineering department.

In a large, complex, and continuously operating process unit, these maintenance programs have evolved almost into a science, which cannot be discussed fully in this section. Two general areas of maintenance are recognized: emergency and preventive. Emergency maintenance follows an unexpected failure of a plant component which often disrupts plant operation, sometimes causes a major shutdown, and almost always results in an abnormally high repair cost. Preventive maintenance is scheduled so as not to disrupt operations, and it does not involve overtime use of maintenance personnel and the high costs of emergency procurement of parts or tools. The purpose of preventive maintenance is to minimize emergency maintenance, and a judicious program of preventive maintenance is vital to the continued satisfactory performance of a system.

An effective maintenance program is based on an inventory of every component of the plant, categorized with respect to its potential for failure and its availability for service. Some components wear more rapidly than major process vessels or utility piping. Preventive maintenance programs must assign a frequency of attention to all these items; although the components least likely to fail will be examined less frequently, they must not be ignored entirely. Forced outages to repair defective equipment are costly. For this reason, specifications demand high component reliability and design redundancy, with the exception that sufficient performance reliability will be attained so that major maintenance can be performed during regularly scheduled outages. Any action on a piece of equipment that is planned and executed to prolong availability by detecting or preventing impending failures is preventive maintenance. In addition to potential reduction in downtime cost and improved reliability, a major advantage of preventive maintenance is that since it is planned, it can be scheduled. As noted earlier, key elements in a successful preventive maintenance program are promoting the concept of recognizing trouble before it happens, following the preventive maintenance schedule without exception, and recording data at appropriate times.

Operation and maintenance problems have plagued users since the installation of the first device in a process, and over time there are now several reasons for an operation and maintenance program (Theodore and Buonicore 1992). The most important are:

1. Continuously meeting present quality requirements
2. Maintaining productivity (reducing production interruptions) of the process
3. Prolonging equipment life
4. Prolonging process life
5. Reducing operating costs
6. Promoting better public relations and avoiding community alienation
7. Promoting better relations with regulatory officials

The following is a list of the basic features that should be included in any operation and maintenance program, and should be applicable to all plants, including desalination plants, regardless of type or size (Theodore and Buonicore 1992):

1. Equipment record system
2. Planning and scheduling
3. Storeroom and inventory system
4. Maintenance personnel
5. Cost and budgets for maintenance operations
6. Special tools and equipment
7. Lubrication
8. Major equipment information
9. Warranty and contract provisions

To troubleshoot equipment performance problems adequately, a multiphase program should be implemented. As a guideline, the following procedures are recommended (Theodore and Buonicore 1992):

- Phase 1 – Problem identification
- Phase 2 – Implementation
- Phase 3 – Testing and sampling
- Phase 4 – Preventive maintenance

A well-conceived maintenance program replete with appropriate spare equipment and trained personnel, coupled with a systematic approach to monitoring and control of process operations, will pay significant dividends to overall process performance and cost management.

Additional information on the role of operation, maintenance and inspection (O&M&I) in optimization is available in the work of Theodore and Buonicore (1992). The American Water Works Association (AWWA) provides extensive details on operation and maintenance standards and practices specifically for drinking water facilities (AWWA 2017).

18.8 Illustrative Examples

Six illustrative examples complete the presentation of optimization principles highlighted in this chapter.

18.8.1 Illustrative Example 1

In many industrial applications, the "factor" being optimized is a function of a single variable, which greatly simplifies the problem. The RDLT Corporation is considering three approaches to optimizing an existing, evaporative water purification unit. The three modifications are focused on energy conservation (Dupont et al. 2017; Theodore 2010).

- Option A – The installation of a double pipe heat exchanger
- Option B – The installation of a shell-and-tube heat exchanger (Dupont et al. 2017)
- Option C – Process modification of the existing water purification unit

Based on the data provided below, you have been asked to use the "system of criteria" method to determine which option is most attractive. RDLT Corporation has determined that annualized cost is the most important criterion, employing a weighting factor of 100 (out of a maximum of 100). Other significant criteria include regulations (weight of 80), potential recovery/reuse of water (weight of 70), and reliability (weight of 50). Options A, B, and C have also been assigned effectiveness factors. Options A is given a rating of 80 for the annualized cost, 60 for regulations, 40 for recovery/reuse, and 20 for reliability. The corresponding effectiveness factors for B and C are 60, 30, 40, 20, and 30, 80, 50, 80, respectively.

The overall rating (OR) for each option is defined as follows:

$$OR_j = \sum_{1}^{n} (\text{Criteria})_i (\text{Weight})_i \tag{18.60}$$

Solution. In solving this example, first generate the overall rating (OR) for each option:

$$OR_A = (100)(80) + (80)(60) + (70)(40) + (50)(20) = 16{,}000$$

$$OR_B = (100)(60) + (80)(30) + (70)(40) + (50)(20) = 12{,}200$$

$$OR_C = (100)(30) + (80)(80) + (70)(50) + (50)(80) = 16{,}900$$

The results show that Option C, process modifications of the existing water purification unit, produces the highest OR with a score of 16,900, and may be viewed as the most optimal of the three plans with respect to energy conservation for the water purification unit.

18.8.2 Illustrative Example 2

A stream of 100,000 gal/d of seawater is to be desalinated. You have been asked to find the optimal process to accomplish this task. A reputable vendor has provided information on the cost of three process units, as well as installation, operation, and maintenance costs that are summarized in Table 22.1. Determine which process you would select in order to minimize cost on an annualized basis.

Solution (Theodore 2010). From the cost information in Table 18.1, the equipment, installation, and operating costs can be converted to total costs by multiplying each by the seawater flow, 100,000 gal/d. For example, the total costs for the evaporation process, *EV*, are calculated in the following manner:

$$\text{Equipment Cost} = 100{,}000\,\text{gal/d}\left(\$3.10/\text{gal/d}\right) = \$310{,}000$$

$$\text{Installation Cost} = 100{,}000\,\text{gal/d}\left(\$0.80/\text{gal/d}\right) = \$80{,}000$$

$$\text{Operating Cost} = 100{,}000\,\text{gal/d}\left(\$0.06/\text{gal/d}\right) = \$6{,}000$$

A summary of results is provided in Table 18.2. Note that the operating cost is on an annualized basis. The equipment cost and the installation cost are

Table 18.1 Desalination System Cost Data.*

	Evaporation (EV)	Reverse osmosis (RO)	Crystallization (CR)
Equipment cost	\$3.10/(gal/d)	\$1.90/(gal/d)	\$2.50/(gal/d)
Installation cost	\$0.80/(gal/d)	\$1.40/(gal/d)	\$1.00/(gal/d)
Operating cost	\$0.06/(gal/d-yr)	\$0.06/(gal/d-yr)	\$0.095/(gal/d-yr)
Maintenance cost	\$14,000/yr	\$28,000/yr	\$9,000/yr
Lifetime of equipment	20 years	15 years	20 years

*Costs are based on comparable performance. Interest is 10% and there is zero salvage value.

Table 18.2 Desalination System Cost Calculations.

	Evaporation (EV)	Reverse osmosis (RO)	Crystallization (CR)
Equipment cost	\$310,000	\$190,000	\$250,000
Installation cost	\$80,000	\$140,000	\$100,000
CRF	0.11746	0.13147	0.11746
Annual equipment cost	\$36,425	\$22,325	\$29,375
Annual installation cost	\$9,400	\$16,450	\$11,750
Annual operating cost	\$6,000	\$6000	\$9500
Annual maintenance cost	\$14,000	\$28,000	\$9000
Total annual cost	\$65,825	\$72,775	\$59,625

capital costs and must then be converted to an annual basis using the CRF as shown in Equation 18.67.

$$CRF = \frac{(0.1)(1+0.1)^{20}}{\left[(1+0.1)^{20}-1\right]} = 0.1175 \tag{18.61}$$

The annual cost for the equipment and the installation is given by the product of the CRF and the total costs of each as shown for the evaporation unit:

$$\text{Annual Equipment Cost} = (\$310{,}000)\,(0.1175) = \$36{,}425$$

$$\text{Annual Installation Cost} = (\$80{,}000)\,(0.1175) = \$9{,}400$$

The calculation for the RO and CR desalinators are performed in the same manner. The three desalinators can then be compared after the annual costs are added. The tabulated results are provided in Table 18.2. According to the analysis, the RO process is the least economically attractive process since the annual cost is the highest. Crystallization is determined to be the optimal process for this application since the total annual cost is lowest.

18.8.3 Illustrative Example 3

In the environmental field, optimization and linear programming are pushing engineers and industry to address the unknowns and uncertainties associated with water use and waste resource management. There is increased recognition and appreciation for developing long-term water management plans that include solutions to problems such as water scarcity, water quality, climate change, stakeholder expectations, and government regulations.

Linear programming has the capability to impact the way in which both society and industry use and impact water resources. For instance, many industries, such as oil refining, need large volumes of low-quality water primarily for cooling purposes, while others such as the electronic industry, need smaller amounts of ultra-pure water. Differences in the nature of water use can therefore influence an industry's reliance on a high quality or large quantity water supply.

Based on a water resource optimization problem, you are required to obtain the maximum value of P, subject to the following conditions,

$$\text{Objective function}: P = 3x_1 + 5x_2 \tag{18.62}$$

subject to the following constraints:

$$x_2 \leq 4;\ x_1 \leq 3 \tag{18.63}$$

$$x_2 \leq -3x_1 + 6 \tag{18.64}$$

$$x_2 \leq -0.5x_1 + 3 \tag{18.65}$$

Solution. The graphical solution for this problem is provided in Figure 18.5. Constraints (1) and (2) indicate the range of values for x_1 and x_2 that will maximize P, and lie within the rectangle provided in Figure 18.5a. Constraints (3) and (4) require that the solution be within the area ABCD shown in Figures 18.5c. When the objective function is superimposed on Figure 18.5d, Figure 18.5e results. The maximum, P, is located at $x_1 = 1.2$ and $x_2 = 2.4$, for which $P = 15.6$. Microsoft Excel can provide the same solution mathematically.

18.8.4 Illustrative Example 4

The AMF Water Purification Company has two RO desalination plants on its grounds that treat brackish water of different compositions from Lake 1 and Lake 2, which are located 20 and 50 miles from the AMF plant, respectively. The processed water from each of the two lakes provide different quantities and various levels of water purity. This includes *Ultrapure* (UP) water used in semiconductor and pharmaceutical industries; *Pure* (PU) water used in hospitals; and *Semipure* (SP) water used primarily in areas such as household work. However, production of UP, PU, and SP is limited. Production information is

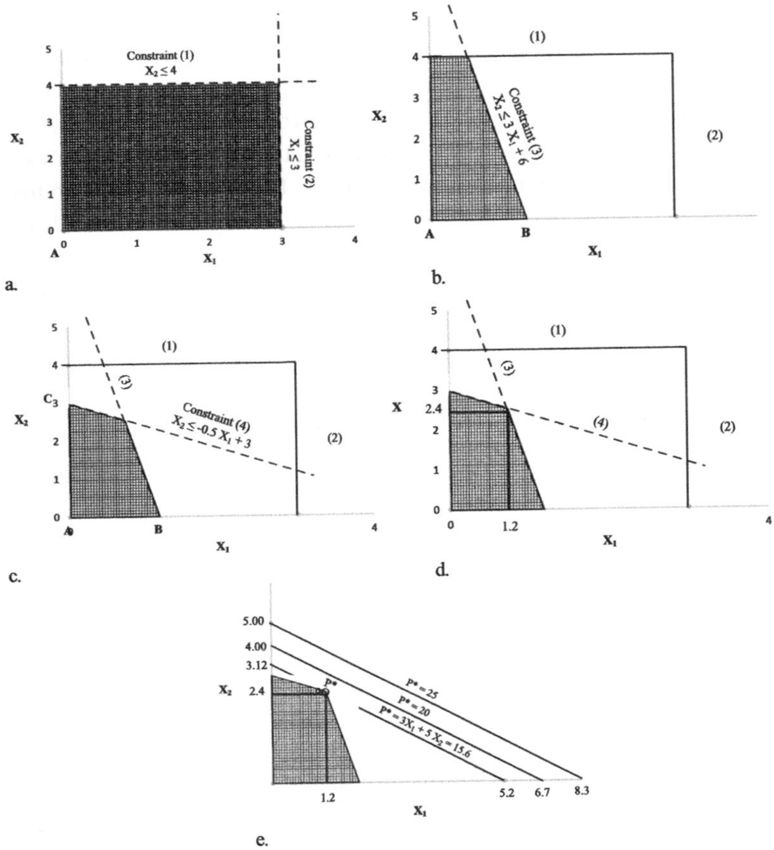

Figure 18.5 Graphical Solution for Illustrative Example 3.

provided in Table 18.3. The profit, *P*, of processing water from Lake 1 and Lake 2 are \$2.00/ft^3 and \$1.60/ft^3, respectively. Determine the water processing ratio of the two lakes that maximizes profit, *P*.

Solution. Set up the following expressions,

$$q_1 = \text{ft}^3 \text{of Lake 1 water/d}$$

$$q_2 = \text{ft}^3 \text{of Lake 2 water/d}$$

The objective function for the daily profit, *P*, is,

$$P = 2.0q_1 + 1.6q_2 \tag{18.66}$$

Table 18.3 Desalination Plant Information for Illustrative Example 4.

Water quality	Water Source (% Distribution by Volume of Processed Water)		Maximum production (ft^3/d)
	Lake 1	Lake 2	
Ultrapure	8	11	1,500
Pure	29	54	5,500
Semipure	63	35	11,000

under the following constraints:

$$0.08q_1 + 0.11q_2 \leq 1{,}500 \tag{18.67}$$

$$0.29q_1 + 5.4q_2 \leq 5{,}500 \tag{18.68}$$

$$0.63q_1 + 3.5q_2 \leq 11{,}000 \tag{18.69}$$

$$q_1 \geq 0; \quad q_2 \geq 0 \tag{18.70}$$

The problem can be set up in Microsoft Excel and solved using its Solver function. Figure 18.6 shows the original spreadsheet with the profit function calculated as the combination of objective coefficients and values for q_1 and q_2 according to Equation 18.66, and the five constraint equations given earlier. Initial values of 1 for both variables is shown in Figure 18.6. Figure 18.7 shows

	A	B	C	D	E	F
1						
2	**LP Illustrative Example 18.4**					
3						
4		Profit				
5	Max P=	$ 4				
6						
7	Variables	q1	q2			
8	Objective Coefficients	2	1.6			
9	Values	1	1	<--Optimal values appear here after solving		
10						
11	Coefficients	q1	q2	Total		Limit
12	Constraint Equation 1	0.08	0.11	0.19	<=	1500
13	Constraint Equation 2	0.29	0.54	0.83	<=	5500
14	Constraint Equation 3	0.63	0.35	0.98	<=	11000
15	Constraint Equation 4	1	0	1	>=	0
16	Constraint Equation 5	0	1	1	>=	0

Figure 18.6 Excel Worksheet to Solve the Linear Programming Problem of Illustrative Example 4.

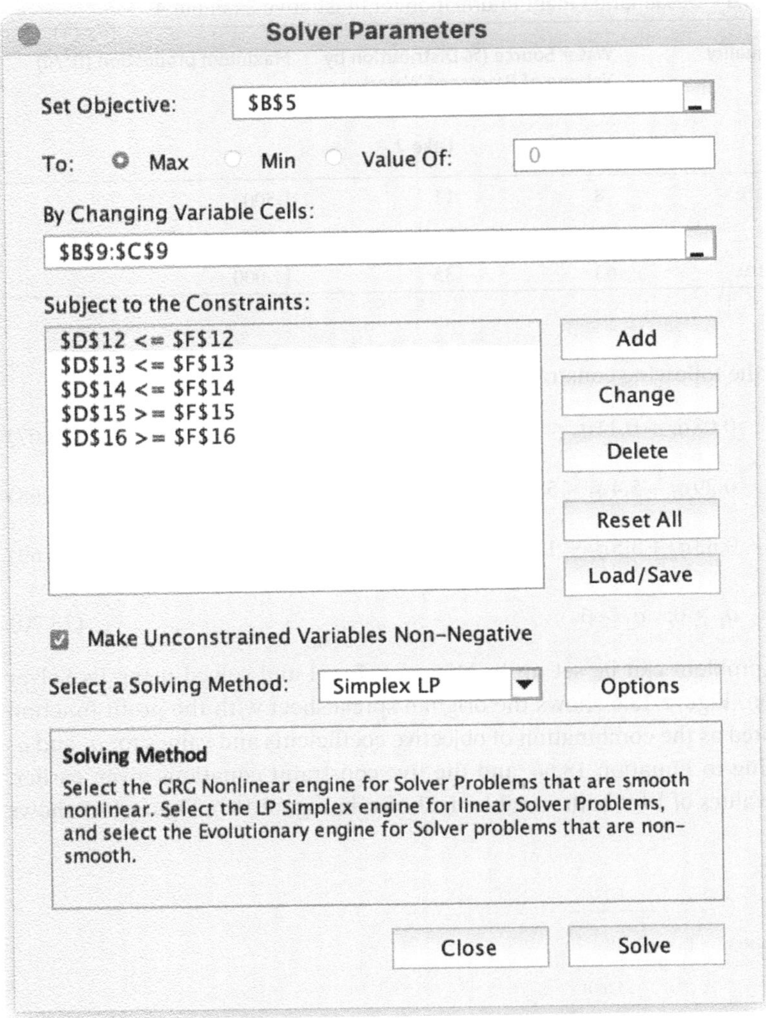

Figure 18.7 Excel Solver Parameters Window Used to Solve the Linear Programming Problem of Illustrative Example 4.

the Solver window used to generate an optimal solution for this problem, and Figure 18.8 shows the final optimal solution to this problem under the specified constraints. As indicated in Figure 18.8, 16,820 ft^3/day of water from Lake 1 and 1,152 ft^3/day from Lake 2 should be processed to result in a maximum total daily profit of $35,484.

	A	B	C	D	E	F	G	H
1								
2	**LP Illustrative Example 18.4**							
3								
4		Profit						
5	Max P=	$ 35,484						
6								
7	Variables	q1	q2					
8	Objective Coefficients	2	1.6					
9	Values	16,820	1,152	<--Optimal values appear here after solving				
10								
11	Coefficients	q1	q2	Total		Limit		
12	Constraint Equation 1	0.08	0.11	1472.4	<=	1500		
13	Constraint Equation 2	0.29	0.54	5500	<=	5500		
14	Constraint Equation 3	0.63	0.35	11000	<=	11000		
15	Constraint Equation 4	1	0	16820	>=	0		
16	Constraint Equation 5	0	1	1152.1	>=	0		

Figure 18.8 Excel Worksheet Showing the Optimal Solution for the Linear Programming Problem of Illustrative Example 4.

18.8.5 Illustrative Example 5

Consider now the following environmental application. A statewide utility has proposed to convert some of its 50 coal-fired boilers to oil-fired to meet new environmental regulations. DuPont Associates was hired to generate a model (objective function) that would minimize the cost of the conversion to oil-fired boilers.

The project also requested that an attempt be made to salvage as many of the old boilers already firing coal as possible while still meeting new environmental regulations. The cost associated with each oil varies with its composition, heating value, etc. The engineers at DuPont Associates proposed the following cost model in units of $10,000 for making the conversion:

$$DA = 3x_1 + 4x_2 - x_3 \tag{18.71}$$

where x_1= number of boilers burning Oil 1, x_2= number of boilers burning Oil 2, and x_3= number of remaining coal-fueled boilers. Equation 18.71 indicates that salvaging existing coal-fired boilers results in a $10,000 credit each.
The constraints, primarily due to emission rules for NOx (nitrous oxide), SOx (sulfur dioxide), and ROx (particulates) are:

$$1.1x_1 + x_2 + 1.5x_3 \leq 100;\ \textit{due to NOx} \tag{18.72}$$

$$1.2x_1 + x_2 + 3x_3 \leq 70;\ \textit{due to SOx} \tag{18.73}$$

$$x_1 + x_2x_3 + x_3^2 \leq 1000; \; due\ to\ ROx \tag{18.74}$$

$$x_1 + x_2 + x_3 = 50 \tag{18.75}$$

Solution. Excel provides the following solution

$$x_1 = 20; x_2 = 22; x_3 = 8$$

With these values, the objective function is a minimum at:

$$DA = \$1,400,000$$

and the solution requires the purchase of 20 new boilers that will burn Oil 1, 22 new boilers that will burn Oil 2, and the continued use of eight of the old coal-fueled boilers, all at a cost of $1,400,000 that represents the minimum expenditure meeting all of the upgrade and pollution control constraints.

18.8.6 Illustrative Example 6

The Canadian Rocky Mountain Water Plant is located in the foothills of the Canadian Rockies near Calgary. The plant "produces" bottled water and is the hub of their operation. The bottled water is delivered to two local municipalities – Edmonton and Winnipeg. Locality Edmonton and Winnipeg are approximately 200 and 650 miles from the plant, respectively.

Henry Hydroxic was recently assigned the task of maximizing profits for the plant's operation. The following information is available. The plant can only produce a maximum of 64,000 bottles/day. Income derived from the sale of water to Edmonton and Winnipeg is 27 cents and 38 cents per bottle, respectively. However, Harry is confronted with two constraints – one associated with energy costs and one associated with environmental regulations. The energy constraint is based on company policy that energy operating costs – 0.015 and 0.035 $/bottle for Edmonton and Winnipeg, respectively, be limited to $1,700/day. The environmental constraints arise because the water contains 3 ppm dissolved sulfite, SO_3. Local policy at Edmonton and Winnipeg require a "tax" of 1.1 cents and 1.9 cents/bottle, respectively, because of this chemical. Company policy limits this charge to $1,000/day. Unable to arrive at a solution to the company's request, Harry has requested that you provide a weekly delivery schedule that will maximize profits, P, for the company.

Solution. Set x_1 and x_2 equal to the daily delivery schedule of bottled water for Edmonton and Winnipeg, respectively. The profit objective function P "in units of $/d" is then:

$$P = 0.27x_1 + 0.38x_2 \tag{18.76}$$

under the following constraints:

$$\text{Energy}: 0.015x_1 + 0.035x_2 \leq 1{,}700 \tag{18.77}$$

$$\text{Environmental}: 0.011x_1 + 0.019x_2 \leq 1{,}000 \tag{18.78}$$

$$x_1 + x_2 \leq 64{,}000 \tag{18.79}$$

$$x_1 \geq 0; \quad x_2 0 \tag{18.80}$$

Solving this in Excel using the objective function and constraints listed previously, the following solution results:

$$x_1 = 27{,}000\,\text{bottles/d};\ x_2 = 37{,}000\,\text{bottles/d};\ P = \$21{,}350\,/\,\text{d}$$

To optimize profit at $21,350/d, The Canadian Rocky Mountain Water Plant should distribute 27,000 bottles of their product to Edmonton, and 37,000 bottles to Winnipeg, meeting all energy and environmental constraints under maximum daily plant capacity.

18.9 Problems

1. Provide some normal, everyday, domestic applications involving the general topic of optimization.
2. A company manufacturers two potable water products, A and B, with accompanying unit profit of P_A and P_B, respectively. A maximum quantity of water products, N, can be produced on a daily basis. The daily production quantity of A and B is x_A and x_B, respectively. In addition, the fractional defective rate of producing A and B are a and b, respectively. Each unit of defective product reduces the profit associated with product A and B by c and d, respectively. Determine the objective function describing profit from the manufacture of products A and B, and define the constraints allowing a determination of a production schedule to maximize profits.
3. A process emits q = 50,000 ft^3/d of water containing dissolved solids at a loading of c = 2.0 gr/ft^3. Reverse osmosis (RO) is employed for dissolved solids recovery, as the dissolved solids recovered from the unit are worth DV = $0.03/lb. Experimental data have shown that the dissolved solids

separation efficiency, E, is related to the RO process system pressure drop, ΔP, by the formula,

$$E = \frac{\Delta P}{\Delta P + 15.0} \tag{18.81}$$

where, ΔP = pressure drop across the RO membrane, lb/ft^2.

a. If the pump efficiency, E_p = 55% (overall) and electric power costs CE = \$0.18/kW·hr, at what separation efficiency is the cost of power $CP = [(q)(\Delta P)(CE)/E_p]$ equal to the recovered value of the material $\{RV = (q)(c\,(DV)(E)\}$?
b. What is the pressure drop in inches of water (in H_2O) at this condition?
c. What pressure drop will maximize profits ($P = RV - CP$)?

4. Graphically solve Illustrative Example 4 and verify your answer by comparing it to the linear programming solution provided through Excel.
5. Graphically solve Illustrative Example 6 and verify your answer by comparing it to the linear programming solution provided through Excel.
6. The drinking water supply for the town of Floral Park in Nassau County, New York is provided by two groundwater wells, W_1 and W_2. The water from W_1 contains 25 mg of 1, 4-Dioxane, 60 mg of Hexavalent Chromium, and 68 mg of Dibromochloromethane in each MG of water. The second well, W_2, contains 75 mg of 1, 4-Dioxane, 60 mg of Hexavalent Chromium, and 34 mg Dibromochloromethane in each MG. The owner of the wells generates a profit of \$30/MG and \$40/MG from W_1 and W_2, respectively, for each MG purchased by the town. Due to stringent local Nassau County water regulations, no more than 450 mg/day of 1, 4-Dioxane, 480 mg/day of Hexavalent Chromium, and 476 mg/day of Dibromochloromethane are allowed in the water sold by the supplier. How many gallons of W_1 and W_2 should be sold daily by the provider to maximize profits? In addition, calculate the corresponding maximum daily profit.

References

American Water Works Association (2017). *Water Treatment Plant Operation*, 1, 7th Edition. Sacramento, CA: Sacramento State Office of Water Programs.

Aris, R. (1964). *Discrete Dynamic Programming: An Introduction to the Optimization of Staged Processes*. Waltham, MA: Blaisdell Publishing Co.

Cabell, J.B. (1926). *The Silver Stallion*. New York: Robert McBride & Company.

Dupont, R.R., Ganesan, K., and Theodore, L. (2017). *Pollution Prevention: Sustainability, Industrial Ecology, and Green Engineering*, Second edition. Boca Raton, FL: CRC Press/Taylor & Francis Group.

Happel, J. (1955). *Chemical Process Economics*. Hoboken, NJ: John Wiley & Sons.

Marquis, D. (1927). *Archy and Mehitabel*. New York: Doubleday & Company, Inc.

Theodore, L. (2010). *Heat Transfer Applications for the Practicing Engineer*. Hoboken, NJ: John Wiley & Sons.

Theodore, L. and Buonicore, T. (1992). *Air Pollution Control Equipment: Selection, Design, Operation, and Maintenance. A Theodore Tutorial*. East Williston, NY: Theodore Tutorials.

Theodore, L. and Reynolds, J. (1992). *Thermodynamics. A Theodore Tutorial*. East Williston, NY: Theodore Tutorials.

Theodore, L., Ricci, F., and VanVliet, T. (2009). *Thermodynamics for the Practicing Engineer*. Hoboken, NJ: Wiley.

Voltaire (1959 [1759]). *Candide* (ed. L. Bair) (trans. L. Bair; with an appreciation by A. Maurois; illustrations by S. Beckett). New York: Bantam Dell.

Wilson, M. (2016 [c. 1915]). Optimist and Pessimist. In: *Oxford Essential Quotations* (ed. S. Ratcliff). Oxford England: Oxford University Press.

19

Ethical Considerations

19.1 Introduction

The primary responsibility of an engineering professional is to protect public health and safety. However, engineering professionals also have a responsibility to their employers or clients, to their families, to themselves, to the environment, and for Civil and Environmental Engineers, to "...strive to comply with the principles of sustainable development in the performance of their professional duties" (ASCE 2008). Meeting all of these responsibilities will challenge the practicing engineer to draw upon a system of ethical values and ethical professional practice.

Well, what about ethics? Ethics means "doing the right thing" as opposed to "what you have the right to do." But doing the right thing is not always obvious or easy. In fact, ethical decisions are often difficult and may involve a certain amount of self-sacrifice. Doing the right thing for a practicing engineer can be especially challenging. Furthermore, the corporate and government world has confused this concept by developing ethics programs that emphasize only what you have the right to do. An organization, for example, may have a list – often called a Code of Conduct – of what an employee can and cannot get away with. Employees are required to sign an acknowledgment that they have read and understood the list. The company unfortunately calls this "ethics training" (Tabak 2002).

One difficulty in some situations is recognizing when a question of ethics is involved. Frequently, in the area of water resource management, a breach of ethics involves a practice that endangers public health and safety or covers up a violation of a rule or regulation. Occasionally, however, a breach may involve a case of the exact opposite. This might seem an unlikely scenario. How can someone be too honest, too caring, or too professional?

One example of this might be lying to save a life. Suppose you are standing on a street and a woman runs past you chased by two men. She screams, "They are trying to attack me!" as she dashes into the entry of a building around a corner. The men ask you, "Which way did she go?" What do you tell them?

Introduction to Desalination: Principles, Processes, and Calculations, First Edition. Louis Theodore and R. Ryan Dupont.

Clearly, the right thing is to lie. In this case, the value of caring overrides the value of honesty. This situation is exaggerated to illustrate that sometimes it is appropriate to violate certain values to protect public health and safety. In doing the right thing ideally one should not have to make snap decisions and should take the time to investigate all of the facts, for instance, in this case whether or not the woman was a thief and the men were police.

Sometimes one must decide how much to sacrifice to ensure public health and safety. In establishing environmental regulations, the regulating agency must decide how safe and how stringent to make the regulations. For example, in the case of drinking water regulations, one standard may result in 10 cancer cases per one million people. But why isn't the standard established for one or 0 cases per one million people? Who should decide this acceptable level of risk?

The role of ethics in water resource management can be complex and confusing in the context of the engineer's role in providing safe drinking water to the public. This chapter provides a discussion of ethics with illustrative examples specifically focused on drinking water production and safe drinking water supply. Topics covered include the present state of ethics, ethics of water access, the do's and don'ts, integrity, moral issues, guardianship, engineering and environmental ethics, environmental justice, and future trends in professional ethics. An applications section provides four illustrative examples related to applied ethics that have been primarily drawn from the work of Wilcox and Theodore (1998).

19.2 The Present State of Ethics

Since this chapter is focused on ethics, there is a need for definitions. Here are four that might help the reader:

1. *Ethics* is defined as a that branch of philosophy dealing with the rules of right conduct.
2. *Environmental ethics* deal with the moral issues of conduct with respect to the environment.
3. *Ethical theory* is an attempt to answer certain questions about standards of conduct or ethics. It also attempts to provide a framework for decisions regarding what moral principles are correct and how one should treat one another, the environment, other species, etc.
4. *Morality* is concerned with reasons for the desirability of certain kinds of actions and the undesirability of others. To say that an act is right is not to express a mere feeling or bias, but instead to assess the best moral reasons of support. Moral reason is a reason that requires individuals to respect other people. In addition, moral reasons are such that they set limits to the

legitimate pursuit of self-interest. They can, for example, be used to evaluate, praise, and criticize laws.

With respect to these points listed, MBAs and CPAs of industry have spawned financial abuses and excesses. These individuals have turned what used to be moral and ethical questions into legal technicalities. Industry executives are more likely to ask what one can get away with legally rather than to be concerned with what is right, honest, and/or fair. The abuses and excesses have spread, particularly with lawyers and individuals in government. Lawyers have become adept at creating controversy, smokescreens, and/or obstruction of the truth, and using the complexity of the law for their own aggrandizement. Elected officials at all levels of government on the other hand, have used their entrusted position of power to further and maintain their careers rather than to represent their constituency. And, as most know, nearly all elected officials are lawyers (Theodore 2003).

Consider if an environmental engineer, acting as an expert witness in a court of law, knowingly conjures up an idea that creates reasonable doubt even though it is highly unlikely. This type of conduct is clearly defined as fraud in any engineering Code of Ethics. Given these, the reader should consider a lawyer's present-day conduct in and out of court (Theodore 2003).

Well, what's the answer? Is there an answer? The reality may be that the referenced corruption is just too widespread and amorphous – like humidity in South Florida in the summer. It seems that the only thing that can ultimately turn things around on this issue is for each individual to take a stand for basic values and virtues, and return to the likes of integrity, responsibility, and selflessness. If a grass-roots movement is to succeed, it will ultimately rest on each individual's ability and willingness to discard the practices of so many lawyers and elected officials. As Albert Schweitzer put it, "Man must cease attributing his problems to his environment and learn again to exercise his will and his personal responsibility in the realm of faith and morals" (Theodore 2004).

19.3 The Ethics of Water Access

Matters of ethics cover every aspect of life, from a personal view of right and wrong to a country's treatment of and responsibility to its people. Since water is a basic necessity for life, then, access to clean water for everyone becomes an important and timely ethical issue. In fact, in 2010, the United Nations formally declared that access to safe, drinkable water and sanitation was a basic human right (United Nations 2010).

Ethical issues often arise whenever a basic human right such as access to water is threatened. Governmental oversight and public policy are among the tools that can ensure public health and safety when it comes to access to clean water and sanitation. However, limited funding and competing priorities can affect access and water supply even for governments with the best resources, infrastructure, and intentions. Climate change, and the increasingly frequent extreme weather events that come with it, threatens to exacerbate already-tenuous accessibility and public health issues in many poverty-stricken and developing areas around the world. These threats create a range of new challenges even where water access is reliable and infrastructure is currently well developed.

In the absence of regulation and oversight, even where water scarcity is not a current problem, water pollution in its many forms threatens the health and livelihood of just about every human and living creature on the planet. The detrimental effects of water pollution are well documented and the sources are numerous, i.e. toxic runoff seeping into water supplies, illegal discharge of manufacturing waste into rivers and streams, wastewater from hydraulic fracturing contaminating groundwater, agricultural runoff causing algal blooms and dead zones in marine environments, etc., etc. There are many causes of water pollution, and the solutions for each will be quite different, whether necessitating new technological implementation, the mobilization of individuals or communities to speak out, or political involvement and policy interventions for change to be achieved.

Water scarcity is a growing concern throughout the world due to ever increasing demands and dwindling supplies of usable water resources. Water scarcity can be due to the natural state of the climate, such as in permanent desert and arid regions, but can be further exacerbated by droughts or water re-allocation. Sometimes droughts are temporary but can last for years. This has caused great suffering in regions of Africa, Asia, and the Middle East, as well as areas in the western United States, especially recently in Southern California. These dry seasons ultimately can lead to the suffering and death of millions, not just through lack of water, but also due to famine, and fires as was recently seen in Northern and Central California in 2018. Enhancing water resources through development of new supplies through desalination, and protection of water quality through effective waste management and treatment, are two ways environmental engineers can address water access issues and meet their professional ethical obligations to protect public health and the environment.

19.4 Do's and Don'ts

The need to make ethical decisions does not start the day one enters his or her first job. Encounters with ethical concepts begin early in life. Early on, before most have any active memory, parents almost certainly encourage their children to tell the truth. The conflict between choosing the ethical act of telling the truth and accepting the punishment that may follow, or choosing the unethical path of lying and going unpunished introduces virtually everyone to the struggle for ethical behavior that lasts a lifetime. In school, it is not long before students learn that it is unethical to copy homework and to cheat on tests. The unethical nature of stealing from a classmate is another area that surely received some attention early in life. As one grows up, ethical situations often become more complex. Conflicts develop. Situations arise in which one may have to decide if it is more ethical to hide an unpleasant truth that will hurt someone or stay loyal to the truth one was taught to value so highly.

As one moves into professional life, an individual learns a long list of ethical do's and don'ts. Some examples include:

- Never falsify data.
- Never offer or accept a bribe.
- Judge colleagues, employees, employers, clients, and others, on their professional merits and not on any other basis such as race, gender identification, age, religion, national origin, or disability.
- Never draw misleading conclusions from the data you have collected.
- Avoid conflicts of interest.
- Accept work only in those areas in which you are properly trained.
- Obey all laws, rules, and regulations.
- Never break confidentiality.
- Always do the best possible work of which you are capable.
- Statements to the public must always be the truthful and objective.
- Health and safety of the public must always be the highest priority.
- Always act to protect the environment.

Ethics becomes a dilemma for the professional when confronted with difficult choices or apparent conflicts between ethical imperatives. A modern term for the outcome of such dilemmas is "situational ethics" which refers to the temptation to bend the rules when a situation seems to have no satisfactory or satisfying solution. Examples of difficult choices are:

- A choice between keeping one's job and income that is needed to care for the children who depend upon you and refusing to certify substandard work as ordered by your employer (job versus family).
- A choice between whether or not to get involved when you become aware of unethical behavior on the part of one's colleagues or employer even though not personally touched by it (job versus conscience).
- Conflicts between two ethical imperative situations could include:
  - The need to observe the confidentiality of a client after learning of the client's unethical behavior (protection of the client versus unethical the behavior).

- The ethical imperative to faithfully serve an employer or client versus the ethical imperative to protect the public (protection of the employer versus the public). At what point does a professional stop working within the system and report a violation of regulations to authorities?

The second conflict of ethical imperatives stated here introduces issues that a water resources management professional is likely to encounter. Most professionals belong to one or more organizations which commonly adopt a code of ethics for their members. Each code is designed provide members with an ethical foundation from which to work. The ASCE (2008) Code of Professional Conduct is one such guide for practicing civil and environmental engineers.

19.5 Integrity

Scenarios are, for the most part, designed to reflect ambiguity in work situations. The ethicist hopes to get his or her hands dirty, dealing with the bottom-line motives of survival, competitiveness, and profitability as well as the mixed motives of self-interest, respect for the rights of others, and altruism. Obtaining an ethical solution to a difficult moral problem or dilemma is based on much more than choosing the correct ethical framework with its normative frame of reference. One must also be ready to examine fundamental assumptions and the values to which the assumptions give rise. Carter (1996) made this point in a discussion of "integrity."

- *Honesty in relation to integrity*. Carter explores integrity in relation to the value society places on honesty. On this subject, one of the best-known and most popular ethics books of the last four decades is Bok's (1978) *Lying: Moral Choice in Public and Private Life*. Without taking away from the merits of Lying, Carter notes: "Plainly, one cannot have integrity without being honest (although, as we shall see, the matter gets complicated), but one can certainly be honest and yet have little integrity." Honesty is far easier to practice than the tough work of figuring out what it takes to have integrity in a situation. Integrity requires a high degree of moral reflectiveness. Honesty may result in harm to another person. Furthermore, "...if forthrightness is not preceded by discernment, it may result in the expression of an incorrect moral judgment." The racist may be transparently honest, Carter declares, but he certainly lacks integrity because his beliefs, deeply held as they might be, are wrong. He has not engaged in the hard work of examining his fundamental assumptions, values, beliefs.
- *Personal integrity without public responsibility?* It would appear that one cannot have integrity without responsibility since any consideration of integrity addresses the effects of our conduct on other people. In our work life and our community life, we have public responsibilities for our clients and fellow citizens.

19.6 Moral Issues

The conflict of interest between Chief Seattle (and Native Americans in general) and President Pierce (and the European–American expansion) provides a perfect example of how ethics and the resulting codes of behavior they engender can differ drastically from culture to culture, from religion to religion, and even from person to person. This enigma, too, is noted again and again by Seattle (Martin and Schinzinger 1989):

> I do not know. Our ways are different from your ways ... But perhaps it is because the red man is a savage and does not understand ... The air is precious to the red man, for all things share the same breath ... the white man does not seem to notice the air he breathes ... I am a savage and do not understand any other way. I have seen a thousand rotting buffaloes on the prairie, left by the white man who shot them from a passing train. I am a savage and I do not understand how the smoking iron horse can be more important than the buffalo we kill only to stay alive.

Chief Seattle sarcastically uses the European word "savage" and all its connotations throughout his address. When one finishes reading the work, it becomes obvious which viewpoint (President Pierce's or his own) Chief Seattle feels is the savage one. What his culture holds dearest (the wilderness), the whites see as untamed, dangerous, and savage. What the whites hold in highest regard (utilization of the Earth and technological advancement), the Native Americans see as irreverent of all other living things. Each culture maintains a distinct and conflicting standard for the welfare of the world. Opposing viewpoints and moralities such as these are prevalent throughout the world and have never ceased to present a challenge to international, national, state, community, and interpersonal peace.

It is generally accepted, however, that any historical ethic can be found to focus on one of four different underlying moral concepts:

1. *Utilitarianism* focuses on good consequences for all.
2. *Duties Ethics* focus on one's duties.
3. *Rights Ethics* focus on human rights.
4. *Virtue Ethics* focus on virtuous behavior.

Utilitarians hold that the most basic reason why actions are morally right is that they lead to the greatest good for the greatest number. "Good and bad consequences are the only relevant considerations, and, hence all moral principles reduce to one: 'We ought to maximize utility'" (Martin and Schinzinger 1989).

Duties Ethicists concentrate on an action itself rather than the consequences of that action. To these ethicists there are certain principles of duty such as "Do not deceive" and "Protect innocent life" that should be fulfilled even if the most good does not result. The list and hierarchy of duties differs from culture to culture, religion to religion. For Judeo-Christians, the Ten Commandments provide an ordered list of duties imposed by their religion (Martin and Schinzinger 1989).

Often considered to be linked with Duties Ethics, *Rights Ethics* also assesses the act itself rather than its consequences. Rights Ethicists emphasize the rights of the people affected by an act rather than the duty of the person(s) performing the act. For example, because a person has a right to life, murder is morally wrong. Rights Ethicists propose that duties actually stem from a corresponding right. Since each person has a right to life, it is everyone's duty to not kill. It is because of this link and their common emphasis on the actions themselves that Rights Ethics and Duty Ethics are often grouped under the common heading *Deontological Ethics* (Barbour 1993).

The display of virtuous behavior is the central principle governing *Virtue Ethics*. An action would be wrong if it expressed or developed vices, for example, bad character traits. Virtue Ethicists, therefore, focus upon becoming a morally good person.

To display the different ways that these moral theories view the same situation, one can explore their approach to the following scenario that Martin and Schinzinger (1989) present:

On a midnight shift, a botched solution of sodium cyanide, a reactant in organic synthesis, is temporarily stored in drums for reprocessing. Two weeks later, the day shift foreperson cannot find the drums. Roy, the plant manager, finds out that the batch has been illegally dumped into the sanitary sewer. He severely disciplines the night shift foreperson. Upon making discreet inquiries, he finds out that no apparent harm has resulted from the dumping. Should Roy inform government authorities, as is required by law in this kind of situation?

If a representative of each of the four different theories on ethics just mentioned were presented with this dilemma, their decision-making process would focus on different principles.

The Utilitarian Roy would assess the consequences of his options. If he told the government, his company might suffer immediately under any fines administered and later (perhaps more seriously) due to exposure of the incident by the media. If he chose not to inform authorities, he risks heavier fines (and perhaps even worse press) in the event that someone discovers the cover-up. Consequences are the utilitarian Roy's only consideration in his decision-making process.

The Duties Ethicist Roy would weigh his duties and his decision would probably be more clear-cut than his utilitarian counterpart. He is obliged foremost by his duty to obey the law and must inform the government.

The Rights Ethicist mind-frame would lead Roy to the same course of action as the Duties Ethicist, not necessarily because he has a duty to obey the law, but because the people in the community have the right to informed consent. Even though Roy's inquiries informed him that no harm resulted from the spill, he knows that the public around the plant has the right to be informed of how the plant is operating.

Vices and virtues would be weighed by the Virtue Ethicist Roy. The course of his thought process would be determined by his own subjective definition of what things are virtuous, what things would make him a morally good person. Most likely, he would consider both honesty and obeying the law virtuous and withholding information from the government and the public as virtueless and he would, therefore, tell the authorities.

Numerous case studies of this nature are provided by Wilcox and Theodore (1998).

19.7 Guardianship

There are two important questions that must be kept in mind when an ethical problem is being assessed:

1. Who are the guardians of the system? This question addresses the issue of who, among engineering or science professionals, is responsible for the ethical standards in the organization. If professionals point the finger at senior management, the legal department, the Environmental Protection Agency, or the Department of Justice, they have indeed misunderstood the nature of a professional calling. The first line of defense is the willingness of professionals themselves to maintain and enhance the integrity of the engineering or scientific profession through their own personal adherence to the highest standards of conduct and to assume responsibility for commitment to these standards within the companies where they work. Moreover, ethics is a positive task, not a list of do's and don'ts. To achieve excellence in one's work presumes a commitment to the client's contract, public safety, and environmental integrity, among several factors that are all too often thought of as "management" issues. They are, in reality, the ethical standards of the work itself. Thus, the ethical engineer or scientist is the one who identifies with the profession and all that is involved in the work assigned or contracted.

2. Who gives support to the guardians of the system? This second issue goes to the heart of the assessment problem, but also has an impact on the first issue. Unless the organization backs those who assume positive responsibility for the ethical tenor of the group, very little will change. Why would someone risk ostracism or retaliation by confronting a person engaging in unethical behavior or illegal behavior if there is no institutional support for the one assuming responsibility?

Effective guardianship is facilitated if:

- There are clear-cut standards of behavior and high expectations of the membership.
- The standards are brought to the attention of the members through a well-developed training program.
- The standards are taken seriously by the senior leadership team of the firm. They must demonstrate that seriousness by taking an active role in the training, without, at the same time, creating a chilly climate stifling discussion and participation in the training. The ethics program must be seen not as frosting on the cake or as a value added to forestall legal problems through better compliance. The CEO needs to demonstrate a commitment to the values and principles that drive the business. Ethics training is no add-on. Ethics is what drives the organization: trust, integrity, fidelity to the client, to the public, and to the environment.
- It is evident that the leadership "walks the talk" in all aspects of its decision making and actions.
- There are mechanisms in place to address the concerns of the members, mechanisms such as an ombudsperson, a hotline, etc.
- Those who adversely affect the integrity of the business are effectively and fairly disciplined.

Another way of addressing the question of who supports the guardians is to emphasize the importance of organizational or corporate culture. A positive response to the six points just raised has a great impact on the culture of the organization. Unless there is what is sometimes called a "thick" culture, wherein respect for and adherence to guardianship and the tenets of integrity and trust are palpable, individually ethical persons can do very little to raise the moral climate of the organization. An organization is more than the sum total of the individuals who constitute the membership. The attitudes conveyed, values expressed, and ways of doing business in an organization profoundly affect the perceptions of the members therein and set the tone of the company. Having a positive impact on culture is a great challenge which is not easily achieved. Culture is so subtle that one often does not even realize or understand its dimensions until a significantly different culture is experienced.

An organization will not have effective guardianship of the system unless there is a concerted attempt to create, enhance, or reinforce a culture where values and ethics are clear and fully supported. There is little doubt, however, that the twin issues of guardianship and culture are much more difficult to address than the institutionalization of the ethics program itself.

19.8 Engineering and Environmental Ethics (Theodore and Theodore 2021)

In the ethical theories presented here, established hierarchies of duties, rights, virtues, and desired consequences exist so that situations where no single course of action satisfies all of the maxims can still be resolved. The entry of environmentalism into the realm of ethics raises questions concerning where it falls in this hierarchy. Much debate continues over these questions of how much weight the natural environment should be given in ethical dilemmas, particularly in those where ecological protection seems to oppose economic profitability and technological advances.

Those wrapped up in this technology/economy/ecology debate can generally be divided into three groups:

1. Environmental extremists.
2. Technologists to whom ecology is acceptable provided it does not inhibit technological or economic growth.
3. Those who feel technology should be checked with ecological responsibility.

Each is briefly discussed here.

After his year-and-a-half of simple living on the shores of Walden Pond, Henry David Thoreau rejected the pursuit of technology and industrialization. While most would agree with his vision of nature as being inspirational, few would choose his way of life. Even so, the movement rejecting technological advances in favor of simple, sustainable, and self-sufficient living is being embraced by more and more people who see technology as nothing but a threat to the purity and balance of nature. Often called environmental extremists by other groups, they even disregard "environmental" technologies that attempt to correct pollution and irresponsibilities, past and present. They see all technology as manipulative and uncontrollable and choose to separate themselves from it. To them, the preservation of the environment is the top priority.

On the other extreme are the pure technologists. They view the natural world as a thing to be subdued and manipulated in the interest of technological and economic progress. This is not to say one won't find technologists wandering in a national park admiring the scenery. They do not necessarily deny the beauty of the natural environment but they see themselves as separate from it. They believe that technology is the key to freedom, liberation, and a higher standard of living. It is viewed, therefore, as inherently good. They see the environmental extremists as unreasonable and hold that even the undeniably negative side effects of certain technologies are best handled by technological solutions. The technologists place environmental responsibility at the bottom of their ethical hierarchy.

Somewhere in the middle of the road is the third group. While they reap the benefits of technology, they are concerned much more deeply than the technologists with the environmental costs associated with industrialization. It is in this group that most environmental engineers find themselves. They are unlike the environmental extremists since, as engineers, they inherently study and design technological systems and have faith in the ability of such systems to have a positive effect on the state of the environment. They also differ from the technologists as they scrutinize the effects of technologies on human health and the environment much more closely and critically than the technologists do. While they may see a brief, dilute leak of a barely toxic chemical as an unacceptable side effect of the production of a consumer product, the technologists may have to observe destruction the magnitude of that caused by Chernobyl before they consider rethinking a technology they view as economically and socially beneficial. In general, this group sees the good in technology but stresses that it cannot be reaped if technological growth goes on unchecked at the expense of human health or environmental quality.

The ethical behavior of engineers is more important today than at any time in the history of the profession. The engineers' ability to direct and control the technologies they master has never been greater. In the wrong hands, the scientific advances and technologies of today's engineers could become the worst form of corruption, manipulation, and exploitation. Engineers, however, are bound by a code of ethics that carry certain obligations associated with the profession. Some of these obligations are to:

- Support one's professional society.
- Guard privileged information.
- Accept responsibility for one's actions.
- Employ proper use of authority.
- Maintain one's expertise in a state-of-the-art world.
- Build and maintain public confidence.
- Avoid improper gift exchange.
- Practice conservation of resources and pollution prevention.
- Avoid conflict of interest.
- Apply equal opportunity employment.
- Practice health, safety, and accident prevention.
- Maintain honesty in dealing with employers and clients.

There are many codes of ethics that have appeared in the literature. The preamble for one of these codes is provided here (Martin and Schinzinger 1989):

> Engineers, in general, in the pursuit of their profession, affect the quality of life for all people in our society. Therefore, an Engineer, in humility and with the need for Divine guidance, shall participate in none but honest enterprises. When needed, skill and knowledge shall be given without reservation for the public good. In the performance of duty and in fidelity to the profession, Engineers shall give utmost.

Engineering ethics is tricky enough when it concerns only how engineers relate to each other and the public; it becomes trickier still when one also considers how engineers ought to react to the nonhuman environment. That is, when one asks "what is the engineers' environmental ethic?"

Environmental ethics, to an even greater degree than ordinary ethics, is a subject without definition and without consensus. And yet, every person on this planet makes everyday decisions that relate to environmental ethics. Questions as simple as "What should I eat?" or "How should I move from place to place?" all raise environmental and ethical issues.

Environmental ethics is especially important for engineers because so much of their work affects the environment. How should the engineer balance the human gains of development against environmental damage? When should the engineer maintain client confidentiality in the face of potential environmental problems?

Now that a regulatory framework and general guidelines are in place, it is up to the individual to follow these procedures and strive to provide a safer, cleaner environment for the future of the planet. In any ethical dilemma, it will always be up to individuals, such as the engineers and scientists working on the ground, whom best understand the shifting dynamic between the changing climate, the local environment, and the specific work they do to speak up if and when they see something that goes against their knowledge and belief.

As indicated earlier, ethics is the branch of philosophy dealing with the rules of right conduct. When defining environmental ethics, right conduct with respect to the environment must be evaluated. The American Society of Civil Engineers (ASCE 2008) states as its first Fundamental Principle that engineers should uphold and advance the integrity, honor and dignity of the engineering profession by "using their knowledge and skill for the enhancement of human welfare and the environment." The society's first Fundamental Cannon goes on to says "Engineers shall hold paramount the safety, health and welfare of the public and shall strive to comply with the principles of sustainable development in the performance of their professional duties."

This statement of ethics can be extended to all individuals and to all aspects of engineering and technology. Whether one's particular area of responsibility is with the government, industry, or a public utility, the actions of an engineer, attorney, politician, or scientist should promote the general health and well-being of the community and the surrounding environment. If every individual makes even a small contribution to the reduction of pollution and the protection of the environment, the sum of the contributions will be great and the goal of long-term sustainable development one step closer.

19.9 Environmental Justice

Closely tied to the concept of environmental ethics is the concept of *environmental justice*. EPA defines environmental justice as the "fair treatment of people of all races, cultures, and incomes with respect to development, implementation, and enforcement of environmental laws and policies, and their meaningful involvement in the decision-making process of government" (Cole 1992) to address the disproportionately adverse human health and environmental impacts in overburdened communities. The environmental justice movement was started by individuals, primarily people of color, who sought to address the inequity of environmental protection in their communities. Professor Robert Bullard (1993, p. 17) wrote, "whether by conscious design or institutional neglect, communities of color in urban ghettos, in rural 'poverty pockets', or on economically impoverished Native-American reservations face some of the worst environmental devastation in the nation." The Civil Rights Movement of the 1960s sounded the alarm about the public health dangers for their families, their communities and themselves, and this movement has slowly made gains to protect underrepresented communities from unequitable environmental impacts for the last 60 years (U.S. EPA 2020b). From the Memphis Sanitation Strike of 1968, to the sit-in against a PCB Landfill sited in a minority community in Warren County, North Carolina; to the First National People of Color Environmental Leadership Summit in October 1991 establishing 17 Principles of Environmental Justice (Environmental Justice Network 2020); to the establishment of EPA's Office of Environmental Equity in 1992; to the creation of an EPA/HUD/DOT federal partnership for Sustainable Communities in 2009; to the Tribal and Indigenous Peoples Environmental Justice Policy signed in 2014; to the development of Technical Guidance for Assessing Environmental Justice in Regulatory Analysis (U.S. EPA 2016a); and EPA's EJ 2020 Action Agenda (U.S. EPA 2016b) the struggle for environmental justice has finally moved into the regulatory process and Agency consciousness.

The environmental policies of the EPA have historically had two main points of focus: defining an acceptable level of pollution and creating the legal rules to reduce pollution to a specified level. Understandingly, it seems that the programs have also been concerned with economic costs and efficiency. Thus, it appears to some that these programs lack consideration of equity, both in the distribution of pollution and pollution control systems, and the economic burden placed on some unequally (the recent Flint Michigan lead contamination case in point (NRDC 2018)). While EPA's two main points of focus are important considerations, relying on such criteria in the formation of environmental protection policy can potentially neglect potential inequalities of capitalism and its effects throughout the policy process.

The history of environmental policymaking illustrates to some the incompatibility of equity and efficiency. Economic pressures of environmental regulations have motivated corporations to seek new ways to reduce costs. Industries have attempted to maximize profits by "externalizing" their environmental costs (Lazarus 1987). It has been suggested that this redistribution of costs is more regressive in its effects than the general sales tax (Cole 1992). To date, big corporate polluters sometimes have more to gain financially by continuing pollution practices than in obeying regulations, particularly in minority communities that have few resources, less political clout, and that may have many community members that rely heavily on the polluting industry for their livelihoods. In some instances, the result of increased environmental costs has paradoxically caused negative impacts on environmental degradation. In effect, they have little incentive other than altruism to end these debilitating practices.

Environmental protection policy has attempted to reduce environmental risks overall; however, in the process of protecting the environment, risks have been redistributed and concentrated in particular segments of society. Although federal regulations to protect the environment are not explicitly discriminatory, some argue that environmental protection policies have not been sensitive to distributional inequalities. Others insist that they have not adequately addressed specific minority environmental concerns and have impacted minority and underrepresented communities inequitably and unfairly.

As indicated earlier, to address the disproportionately adverse human health and environmental impacts in these disadvantaged communities EPA created the Office of Environmental Justice (OEJ) in 1992. The OEJ (www.epa.gov/environmentaljustice) has the mission to decrease the environmental burdens, increase the environmental benefits, and work collaboratively to build healthy, sustainable communities of all people regardless of race, color, national origin, or income (U.S. EPA 2017). Some of the tangible products and policies that EPA has developed in the recent past to address these inequitable environmental burdens include the following:

- *Guidance on Considering Environmental Justice During the Development of Regulatory Actions* (U.S. EPA 2015) – the purpose of this guidance is to help rule-writers and decision-makers understand and identify potential EJ concerns, advises on how to integrate the consideration of EJ into the rule development process and to meaningfully engage minority populations, low-income populations, tribes, and indigenous peoples to minimize potentially disproportionate environmental and public health impacts experienced by minority populations, low-income populations, and/or indigenous peoples as a result of environmental rulemaking.

- *Technical Guidance for Assessing Environmental Justice in Regulatory Analysis* (U.S. EPA 2016a) – the purpose of this guidance document is to outline particular technical approaches and methods to help Agency analysts (including economists, risk assessors, and others) analyze potential EJ concerns for regulatory actions, and ensuring that potential EJ concerns are appropriately considered and addressed in the development of regulatory actions.
- *EJ 2020 Action Agenda* (U.S. EPA 2016b) – This action plan developed by the then EPA Administrator Gina McCarthy, established three major goals for advancing environmental justice in the United States through EPA actions. These goals included: (1) Deepen environmental justice practice within EPA programs (i.e. Rulemaking, Permitting, Compliance and Enforcement, and Science) to improve the health and environment of overburdened communities; (2) Work with partners (States and Local Governments, Federal Agencies, Community-based Work, and Tribes and Indigenous Peoples) to expand EPA's positive impact within overburdened communities; and (3) Demonstrate progress on significant national environmental justice challenges with focus on four areas: Lead Disparities, Drinking Water, Air Quality, and Hazardous Waste Sites.
- *EJSCREEN* (U.S. EPA 2019) – this environmental justice mapping and screening tool provides users with a nationally consistent dataset and approach for combining environmental and demographic indicators. EPA made this tool publicly available online to be more transparent about how they consider environmental justice in their decision making, to assist stakeholders in making informed decisions, and to create a common starting point for dialogue with partners and the public. This software and user support information can be accessed at https://www.epa.gov/ejscreen.

Extensive lead contamination in Flint Michigan starting in 2014 because of cost-cutting measures imposed on the financially struggling majority African American city suggests, however, that there remains a great deal more to do to move toward environmental equity and environmental justice for all Americans.

19.10 Future Trends in Professional and Environmental Ethics

Although the environmental movement has grown and matured, its development is far from stagnant. To the contrary, change in individual behavior, corporate policy, and governmental regulations are occurring at a dizzying pace.

Because of the Federal Sentencing Guidelines, the Defense Industry Initiative, as well as a move from compliance to a values-based approach in the marketplace, corporations have inaugurated company-wide ethics programs, hotlines, and senior line positions responsible for ethics training and development. The Sentencing Guidelines allow for mitigation of penalties if a company has taken the initiative in developing ethics training programs and codes of conduct.

In the near future, these same guidelines will likely apply to infractions of environmental law. As a result, the corporate community will undoubtedly welcome ethics integration in engineering and science programs generally, but more so in those that emphasize environmental issues. Newly hired employees, particularly those in the environmental arena who have a strong background in ethics education, will allay fears concerning integrity and responsibility. Particular attention will be given to the role of public policy in the environmental arena as well as in the formation of an environmental ethic.

Regulations instituted by federal, state, and local agencies have over the years become more and more stringent. The deadlines and fines associated with these regulations encourage corporate and industrial compliance of companies (the letter of the law) but it is the personal conviction of the corporate individuals that lies in the spirit of the law, and the heart of a true ecological ethic.

To bolster this conviction of the heart, there must be the emergence of a new *dominant social paradigm* (Barbour 1993). This is defined as "...the collection of norms, beliefs, values, habits, and survival rules that provide a framework of reference for members of a society. It is a mental image of social reality that guides behavior and expectations" (Barbour 1993). The general trend in personal ethics is steadily "greener" and is being achieved at a sustainable pace with realistic goals.

It has been suggested (Gleick 1987) that "The flap of one butterfly's wings can drastically affect the weather." While this statement sounds much like one conceptualized by a romantic ecologist, it is actually part of a mathematical theory explored by the contemporary mathematician Gleick (1987) in his book *Chaos, Making a New Science*. The "butterfly" theory illustrates that the concept of interdependence, as Chief Seattle professed it, is emerging as more than just a purely environmental one. This embracing of the connectedness of all things joins the new respect for simplified living and the emphasis on global justice, renewable resources, and sustainable development (as opposed to unchecked technological advancement) as the new, emerging social paradigm. The concept of environmentalism is now *widely* held; its future is it becoming *deeply* held.

Finally, one should note that an ethical analysis generally produces no absolute answers. This can be quite disconcerting because engineers expect valid, correct, and useful answers to problems. When one studies statics, for example, one knows what the rules are, and all agree on the right answer. In ethics, however, the best an individual can do is argue that some answers are better than others, and, of course, these answers are always open to disagreement.

Engineers, as professionals, have a special responsibility to the public, and this responsibility is often expressed in terms of professional ethics. Engineers invariably face situations where values become variables in the decision-making process. Indeed, the ethical aspects of a decision often prove more difficult than the technical ones. Recognizing this, the engineering profession has strongly encouraged engineering schools to introduce more professional ethics into engineering curricula. Professional ethics has rightly become an integral part of engineering education.

The progress EPA has made in the environmental justice area is significantly threatened, however, by the Trump Administrations priorities for EPA as reflected in the FY 2021 EPA budget (U.S. EPA 2020a). Despite statements by EPA that "In FY 2021, the Agency will enhance coordination across its programs and with federal partners to ensure alignment of mutual efforts, including consideration of vulnerable groups and communities when making decisions. EPA will leverage resources with new and existing partners to deliver services efficiently and effectively," the FY 2021 budget document clearly indicates the Trump Administration has largely abandoned environmental justice actions with the proposed reductions in environmental justice and environmental justice related programs across the board at EPA that includes the following:

- *Environmental Programs and Management – Environmental Justice Enforcement* – Cut by 71.4% to $2.7 million.
- *Hazardous Substance Superfund – Environmental Justice Enforcement* – Cut by 100% to $0.
- *Research: Sustainable Communities – Research: Sustainable and Healthy Communities* – Cut by 50% to $70.9 million.
- *Research: Chemical Safety and Sustainability* – Cut by 20.9% to $67 million.
- *Research: Human Health and Environmental Risk Assessment* – Cut by 38% to $30.9 million.

- *Research: Safe and Sustainable Water Resources* – Cut by 24.4% to $78.9 million.
- *Infrastructure Assistance: Alaska Native Villages* – Cut by 89.7% to $3.0 million.
- *Eliminated Programs* – Cut by 100% the following programs:
  - *Nonpoint Source Section 319 Program.*
  - *Pollution Prevention Program.*
  - *Indoor Air Radon Program.*
  - *Reduce Risks from Indoor Air Program.*
  - *Environmental Education Program.*
  - *Regional Science and Technology Program* – laboratory analysis, field monitoring and sampling, to build Tribal capacity for environmental monitoring and assessment.
  - *Safe Water for Small and Disadvantaged Communities.*
  - *Toxic Substances: Lead Risk Reduction Program.*

With such actions at the federal level it becomes even more imperative that professional ethics drive engineering firms and engineers to meet their obligations for equity and the social science component of sustainable development in all of the projects they are associated with.

19.11 Illustrative Examples

Four illustrative examples complete the presentation of ethics principles highlighted in this chapter. The last two are case studies that have been adapted from Wilcox and Theodore (1998).

19.11.1 Illustrative Example 1

Provide some normal, everyday, domestic applications involving the general topic of ethics.

Solution. Ethical decisions crop up in our daily lives on a frequent basis and take the form of deciding between lying and telling the truth, following the rules or cheating, being honest and fair to others or manipulating them for your benefit, etc. Some issues are clear, i.e. it is not ethical to cheat on an exam, plagiarize or take credit for another's work. Other decisions are a bit less clear, i.e. telling my friend less than the truth about how his haircut looks so his feelings aren't hurt. Perhaps the Hippocratic oath is relevant when considering ethical behavior in normal, everyday life, that is, "primum non nocere," or "first, do no harm."

19.11.2 Illustrative Example 2

Provide an historical overview of the trends in corporate policies related to their environmental responsibility and stewardship.

Solution. In the early 1990s, the defense industry led the initiation of the Federal Sentencing Guidelines for Organizations (FSGO), which arose from an increase in white-collar crime during the previous decade. The FSGO allowed for increased oversight into corporations' fraudulent and other unethical behaviors. The fact that senior management could now be prosecuted for the behavior of its employees led to widespread implementation of company ethics training and ethics programs. Compliance programs covered a host of topics including antitrust, equal opportunity, and environmental compliance. In today's corporate culture, 30 years after these initial efforts began, ethics and compliance programs and employee training have become commonplace in most companies due to the legacy of FSGO and other legislation. Companies no longer treat ethics and compliance as an extra drain on resource allocation because proper handling of environmental issues and serving as positive influences in their communities have helped many companies raise their visibility among consumers, which in turn can improve brand loyalty and sales. Walmart, for example, annually publishes a corporate Global Compliance and Ethics Report (now titled the Environmental, Social, and Governance Report; https://corporate.walmart.com/esgreport), it has published a Global Statement of Ethics (https:// www.walmartethics.com/content/dam/walmartethics/documents/statement_of_ethics/Walmart_Statement_of_Ethics_English.pdf) and it has a stated commitment to enhance the sustainability of its operations and global supply chains through goals of zero waste, operations with 100% renewable energy, and selling products that sustain resources and the environment to achieve significant and lasting improvement in environmental and social outcomes, in a way that also improves their business (https://corporate.walmart.com/global-responsibility/sustainability).

19.11.3 Illustrative Example 3

Vincenzo has a PhD in Environmental Engineering, a Professional Engineer (P.E.) license, and a job as an Assistant Professor at a local university. To learn more about ISO 14000 and extra money, Vincenzo became a certified ISO 14000 auditor. During an audit of a desalination facility, Vincenzo observed a significant violation of an EPA regulation related to brine management at the facility. (Note that ISO 14000 is voluntary standard of environmental management systems, i.e. it is not a regulation and ISO 14000 audits are confidential).

What does Vincenzo do? Does he forget about the significant violation since he is not an EPA inspector and it does not appear to violate the ISO 14000 standard he is there to audit? Does he insist it be part of the audit report? Does he call it to the attention of the facility management? Does he report it to the EPA?

Solution. The question of professional ethics presented here can be generalized as follows: does a professional have an ethical duty to report any violation of which he/she becomes aware? Most licensing organization have codes of ethics to which their professional must subscribe. These codes generally call upon the professional to serve their client/employer with loyalty and to protect the welfare of the public. The codes, however, are sufficiently vague so as to not provide a clear-cut answer to Vincenzo's situation. In the absence of such a clear-cut code, there is no one right answer and one's answer may vary from the one presented here.

Vincenzo must first investigate further to make certain of his facts. If the violation is confirmed and has never been reported, he must insist that it be part of the findings presented to management at the exit meeting and in the report to the ISO Registrar who makes the final decisions on awarding certification. Although ISO 14000 does not require compliance, knowingly failing to comply does affect the evaluation of the facility's commitment to good environmental management system practices and continuous improvement that the standard does require.

Having done all that, does Vincenzo go the last step and report it to the U.S. EPA? One aspect of the problem that Vincenzo must evaluate is whether the violation poses an imminent threat to public health or the environment. Such a report would violate the confidentiality of the audit and might even precipitate a lawsuit. Vincenzo is also concerned that he may never be hired again as an auditor. He senses a conflict between two sets of ethics: the ethics of a professional bound to protect public health and the environment, and the ethics of a professional bound to confidentiality. Keeping silent would be easier but ethics is not easy.

It would be easy to answer "report it!" but the real question is what would YOU (the reader) do?

19.11.4 Illustrative Example 4

Fact Pattern. Joe Murray, an environmental engineer at the Pristine Water Company, is responsible for the production of potable drinking water the company distributes to its municipal and industrial customers. Joe has been reviewing results of routine water quality monitoring conducted within the plant for process monitoring purposes, but that are not required to be reported to the local regulatory agency. Joe notices that there is a disturbing trend in pH that while not currently violating permit requirements, has potentially significant implications for water quality within the distribution system for the mobilization of lead and for the production of undesirable disinfection by-products during treated water chlorination.

Joe decides to go to his manager and proposes that expensive source water and distribution system monitoring be carried out, first to attempt to determine if changes in source water quality could be the cause of these process changes, and secondly to ensure that undesirable conditions within the distribution system are not occurring. His manager, concerned about quarterly profits and control of operating costs, feels that if no permit violations are occurring then no actions are warranted so no more money or personnel time are wasted.

Ethically, Joe feels strongly that this is wrong and feels this situation needs special attention before conditions within the distribution system become unstable and their customers are subjected to unnecessary risk if the plant's response to water quality changes are delayed. Joe's only alternative is to go above his manager and try to convince upper management to expend the necessary funds to conduct the required monitoring and perhaps process modifications to ensure the quality of the produced water and the health and safety of their customers. However, doing this could anger his manager and could also lead to Joe's dismissal. But it might also make him appear to be confident and conscientious, could prevent a major water crisis such as happened in Flint, Michigan in 2014, and could also serve to publicize his abilities as an engineer.

Questions for Discussion

1. What are the facts in this case?
2. What issues are involved?
3. What are the choices Joe must make?
4. What could be the consequences of each decision he could make?
5. What were the outcomes of the Flint, Michigan, water crisis, and what can be learned for this real-life event that pertains to Joe's problem?

Solution. This is an open ended question and there may be a range of responses to these questions. Put your answer in the context of professional ethics and refer to ethics guidelines from ASCE (2008) for appropriate responses and actions you suggest as a Professional Engineer. Also refer to National Resources Defense Council (2018) for a detailed summary of events and consequences of the Flint, Michigan, water system crisis.

19.12 Problems

1. Provide some normal, everyday, domestic applications involving the general topic of ethics.
2. As this text was being written, the United States and the rest of the world was struggling to deal with the COVID-19 pandemic. As in many other areas of health and environmental justice, there were indications that there were significant racial disparities in the infection, hospitalization, and mortality rates of the disease in minority populations. Explore the realities of the disproportionate impact of COVID-19 on communities of color, discuss the underlying causes of such disparities, and suggest ways in which these disparities can be addressed as the Nation attempts to ethically prepare for future pandemics.
3. As the COVID-19 pandemic raged throughout the United States in 2020, the use of masks and social distancing, best practices prescribed by Public Health and Medical Professionals, became a political talking point, and a debate about "personal freedoms." Discuss the ethics of public health measure mandates versus infringements on individual freedoms and individual choice during a declared Public Health Emergency.
4. On February 11, 1994, Executive Order 12898 was issued to direct Federal agencies to incorporate achieving environmental justice into their mission. This Executive Order made reference to Title VI of the 1964 Civil Rights Act in its statement of purpose. Describe Title VI of the Civil Rights Act, discuss why it was referred to in an Executive Order referring to environmental justice, and discuss how Title VI and environmental justice are different and complement one another.
5. Discuss the difference between moral obligation, personal responsibility, and professional ethics.
6. ***Fact Pattern.*** Laura is an engineer working in a chemical plant. She has recently received a job offer from another company, which she accepts because she knows that the new job could be a big step in her career.

   Laura is responsible for one of the production lines in the plant she will soon be leaving. She has always been a reliable worker and an effective manager. However, having handed in her letter of resignation, she has been less attentive to her work over the past couple of weeks. She figures that there is no need to worry about this job anymore; she has to concentrate on her future.

   On Laura's next-to-last day of work at the plant, Harry a coworker on the same production line, finds out that there is a problem with the purity of a wastewater discharge stream. The level of contaminants is a little higher than acceptable. Harry decides to consult with Laura.

He says, "The effluent stream coming out of the process exceeds the required water quality limit. I think you should investigate it so we can solve this problem."

Laura replies, "I would love to help you, Harry, but tomorrow is my last day here. I don't want to start dealing with this problem because it could take a while to solve. Let my replacement worry about it."

Harry answers, "Laura, if we let this problem go, we'll continue to have a discharge that doesn't meet regulations. The problem could also get worse. You are the expert here, so you could easily fix this mess."

"Harry, you're a friend of mine. Please don't ask me to get involved in this problem. It's not my concern anymore. I just want to relax during my last 2 days at work," pleads Laura. "It's not like the plant will blow up. Wait for 2 days. You can pretend that you didn't notice anything until then."

Reluctantly, Harry agrees. "I know you're really looking forward to your new job. It's just that I'll feel guilty knowing that something is wrong, and I'm not doing anything about it. But I guess I can wait for 2 days."

"Harry, don't worry. Take it easy for a couple of days. Just think of it as a minor delay," replies Laura.

Questions for Discussion

a. What are the facts in this case?
b. Do you think Laura should stay focused on her current job?
c. Should Laura handle the problem?
d. Do you think it's okay for Harry to ignore the problem for the next 2 days?
e. Should Harry consult someone else now that Laura has refused to deal with the problem?
f. What are the regulatory requirements to report a permit violation and who will be held responsible if the violation is responsible for harm to human health and/or the environment?

Put your answer in the context of professional ethics and refer to ethics guidelines from ASCE (2008) for appropriate responses and actions you suggest as a Professional Engineer. Also consider ethical issues that may arise as a citizen in a minority community that lies downstream from this plant's discharge.

References

American Society of Civil Engineers (2008). *Ethics Guidelines for Professional Conduct for Civil Engineers*. Reston, VA.: ASCE.

Barbour, I. (1993). *Ethics in the Age of Technology*. San Francisco, CA: Harper.

Bok, S. (1978). *Lying: Moral Choice in Public and Private Life*. New York NY: Pantheon.

Bullard, R.D. (1993). *Confronting Environmental Racism: Voices from the Grassroots*. Boston, MA: South End Press.

Carter, S.L. (1996). The insufficiency of honesty. *Atlantic Monthly* (February), 74–76.

Cole, L. (1992). Empowerment as the key to environmental protection: The need for environmental poverty law. *Ecology Law Quarterly* 19 (4): 619–683.

Environmental Justice Network (2020). [1991]. *The Principles of Environmental Justice (EJ)*. Washington, D.C: Policy Statement from Delegates to the First National People of Color Environmental Leadership Summit. October 24–27, http://www.ejnet.org/ej/principles.pdf (accessed June 17, 2020).

Gleick, J. (1987). *Chaos: Making a New Science*. New York, NY: Viking Penguin Inc.

Lazarus, R. (1987). *Pursuing Environmental Justice: The Distributional Effects of Environmental Protection*. Chicago, IL: Northwestern University Law Review. Spring.

Martin, M. and Schinzinger, R. (1989). *Ethics in Engineering*. New York NY: McGraw-Hill.

National Resources Defense Council (2018). *Flint water crisis: Everything you need to know*. https://www.nrdc.org/stories/flint-water-crisis-everything-you-need-know (accessed June 19, 2020).

Tabak, H. (2002). Ethics corner: Doing the right thing. *Environmental Management* 29 (1):15-16.

Theodore, L. (2003). *On Ethics I. Williston Times* (November 28).

Theodore, L. (2004). *On Ethics II. Williston Times* (March 26).

Theodore, M.K. and Theodore, L. (2021). *Introduction to Environmental Management. 2nd Edition*. Boca Raton, FL: CRC Press/Taylor & Francis Group.

United Nations (2010). *Resolution A/EWS/64/292: The Human Right to Water and Sanitation*. New York, NY: United Nations General Assembly. https://undocs.org/E/A/RES/64/292.

United States Environmental Protection Agency (2015). *Guidance on Considering Environmental Justice During the Development of Regulatory Action*. Washington, D.C.: Office of Environmental Justice. U.S. EPA. https://www.epa.gov/sites/production/files/2015-06/documents/considering-ej-in-rulemaking-guide-final.pdf.

United States Environmental Protection Agency (2016a). *Technical Guidance for Assessing Environmental Justice in Regulatory Analysis*. Washington, D.C.: Office of Environmental Justice. U.S. EPA. https://www.epa.gov/sites/production/files/2016-06/documents/ejtg_5_6_16_v5.1.pdf.

United States Environmental Protection Agency (2016b). *EJ 2020 Action Agenda*. EPA-300-B-1-6004. Washington, D.C.: Office of Enforcement and Compliance Assurance, Office of Environmental Justice. U.S. EPA. https://www.epa.gov/sites/production/files/2016-05/documents/052216_ej_2020_strategic_plan_final_0.pdf.

United States Environmental Protection Agency. (2017). *Office of Environmental Justice in Action*. Washington, D.C.: Environmental Justice Factsheet. U.S. EPA. https://www.epa.gov/sites/production/files/2017-09/documents/epa_office_of_environmental_justice_factsheet.pdf.

United States Environmental Protection Agency (2019). *EJSCREEN, Environmental Justice Mapping and Screening Tool, EJSCREEN Technical Documentation*. Washington, D.C: Office of Policy. U.S. EPA. https://www.epa.gov/sites/production/files/2017-09/documents/2017_ejscreen_technical_document.pdf.

United States Environmental Protection Agency (2020a). *FY 2021 EPA Budget in Brief*. EPA-190-5-20-002. Washington, D.C.: Office of Chief Financial Officer. U.S. EPA. https://www.epa.gov/sites/production/files/2020-02/documents/fy-2021-epa-bib.pdf.

United States Environmental Protection Agency (2020b). *Environmental justice*. https://www.epa.gov/environmentaljustice (accessed June 17, 2020).

Wilcox, J. and Theodore, L. (1998). *Engineering and Environmental Ethics: A Case Study Approach*. Hoboken, NJ: John Wiley & Sons.

20

The Future of Desalination

20.1 Introduction

This final chapter attempts to revisit several of the earlier chapters from the perspective of the future of desalination. Seven additional sections complement the presentation addressing the role of a wide range of topics pertaining to the future of water that include: the role of conservation; the role of energy use; the role of operation and maintenance; the role of government; the role of risk assessment, terrorism, and water security; and predicting future water demands.

20.2 The Role of Conservation

The role of water conservation was addressed in some detail in Chapter 16. In a general sense, future consideration would simply involve expanding many of the suggestions presented earlier. Some examples of domestic water conservation could include:

1. Turning off the faucet when brushing one's teeth or washing one's hands.
2. Taking showers instead of baths.
3. Reducing the time of a shower.
4. Refraining from taking unnecessary showers.
5. Running the dishwasher only when the dishwasher is full.
6. Using dual flush or low flush toilets.
7. Reducing outside water use via drip irrigation, xeriscaping, and best practices for lawn watering (avoiding over watering, watering during peak evaporation, limiting water application rates, etc.).

There are also numerous water conservation measures that can be implemented at the industrial level. These too are detailed in Chapter 16. New conservation measures will have to be employed with the development of new

Introduction to Desalination: Principles, Processes, and Calculations, First Edition. Louis Theodore and R. Ryan Dupont.

products, new and modified processes, and new social activities. For example, nanotechnology (Theodore 2006; Theodore and Kunz 2005) will have to be fully understood to evaluate its true impact on water consumption, water quality demands, and water conservation efforts.

A not-so-well-understood impact of all present and new water conservation measures will be the impact on future water planning activities that could include:

1. Projected water demand requirements
2. Projected water demand variations
3. Required diversion flows
4. Expansion of water supplies without changing local reuse options

In any event, the effect of traditional and future water conservation will definitely have a significant effect on water usage. A companion benefit of water conservation will no doubt be the improvement of the quality of life for many in society.

20.3 The Role of Energy Use

Energy systems must be evaluated in light of their impact on the total environment and economic aspects, including all hazards and risks involved during all facets of their respective production methods and end uses. For example, as more and more water pollution control devices are employed, water discharges are reduced considerably; however, this may be at the expense of increasing the amount of solid waste generated. More land will then be needed for the disposal of this waste and energy use could also increase; this could very well offset the reduction in waste discharges. Although the damages from water pollution are much less severe with controls, the need to avoid unintentionally shifting environmental problems from one medium or location to another must be recognized, i.e. a multimedia approach is warranted (Theodore and Theodore 2021).

Finally, much of the environmental damage from the use of energy lies in those systems that provide the energy to the consumer that require significant quantities of water. If these systems were to function more efficiently, then their adverse environmental effects and water demand associated with energy production would decrease. Similarly, if the consumer were to use energy more efficiently (i.e. if less energy were to be expended to achieve domestic needs), then both energy-associated environmental impacts and water usage would be reduced. The environmental impacts of various energy policies and of new energy technologies must also be taken into consideration. An awareness of the water-energy-environment relationship and the importance of each in society is critical for a sustainable future to be realized.

For example, one can reduce the use of cooling systems. Many cooling systems, either in total or in part, provide cooling in spaces that do not need cooling. These could be eliminated with little or no sacrifice – for example, in entrances, hallways, storerooms, and other such spaces that people use either infrequently or for very short periods. Many spaces are unnecessarily cooled continuously throughout periods of non-occupancy, for example, homes and apartments unoccupied all day, meeting rooms, auditoriums, etc., which are used only occasionally. In these areas, the cooling system can be either turned off or the thermostats set at the highest possible settings. Energy savings and the associated water used for energy production can be realized by turning off unneeded cooling systems, and while site specific and variable depending on the type of building, this savings can amount to 5–10 percent of the original energy demand. The same analysis may be applied to heating systems.

Turning off or reducing to the lowest acceptable levels all lights and heat-producing appliances can also reduce cooling loads significantly, such as with the cooling load of the lighting in some commercial buildings which can amount to as much as 50 percent of the total cooling load. In homes, lights, and appliances (TV sets, fans, etc.) also increase the cooling load and should be turned off when not actually needed. Some careful considerations can lead to other energy "conservation" methods that not only can help ease energy shortages but can also save money.

Since it is unlikely that the United States will completely solve its present energy problems by dramatically increasing domestic supply, increased efforts should be directed to decreasing the demand for energy and associated water use in the energy sector by:

1. Developing conservation strategies to reduce residential and commercial energy demand through insulation, better maintenance, new building standards, and other measures to reduce consumption of fossil fuels.
2. Design of more efficient machines and processes. This will require a comprehensive effort of the consumer sector and government incentives for industry to redesign processes and prime movers that would consume energy and use water and waste heat more efficiently.
3. Penalizing wasteful consumption of all forms of energy (e.g. for inefficiently using energy in industrial and commercial settings or in other consumer products such as automobiles, etc.).

All of these measures require public support and public will. Some government incentives may be necessary, where the cost of conservation is too burdensome. Decisions on saving energy will come if a widespread conservation ethic is accepted by the public. There are those who feel that price increases and free market forces can by themselves spur the public to curb excessive and wasteful uses of energy and its associated water demand and environmental impacts. It remains to be seen whether price pressures alone will be as effective in the energy sector as they have been in reducing demand in certain commodities and for certain consumer goods. One should keep in mind that high prices may slow the growth of energy demand and promote conservation of energy but they must be vetted and scrutinized publicly

National goals in the energy field should be compatible with desirable and sustainable growth and full employment, with maximum concern for energy and water conservation and the public interest. A protracted energy shortage cannot be tolerated for long without risk of severe unemployment. Energy conservation is not a panacea, but only one of the things that can be done; it is not a substitute for increasing production of fossil fuels and green energy alternatives.

As noted earlier, sustainability measures could be implemented for homes by the use of better insulation, installation of double paned windows and doors, new utility rate systems, use of solar energy for electricity and hot water production, reduction of air flights on lightly served routes, reform in transportation (carpooling and mass transit), etc. Penalties for wasteful consumption of energy (e.g. the inefficient use of energy in automobiles, unfavorable state and local building codes) and incentives (e.g. to upgrade insulating properties of existing structures) would also help. Defining standards for thermal efficiency in new structures is also a must.

Green engineering approaches can include incentives for industry to redesign processes and prime movers with excessive energy consumption and to use waste heat more efficiently, e.g. by granting tax write-offs for investment in energy-saving machinery, or by taxation of excessively energy intensive uses. Government interference with private enterprise and the marketplace is likely to be challenged in any conservation strategy (e.g. how the government should deal with industries that manufacture "throw-away," e.g. paper and plastic products, or how to improve or redesign overland trucks or convert totally to steel-belted radial tires). Still, the urgency and the extent of change or adjustment, the magnitude of capital investment needed and the institutional problems, that arise may not be resolved without financial and expediting assistance from the government or outside sources (Dupont et al. 2017).

And finally, another topic that needs to be addressed is the future of global carbon markets and the role they will play in the energy business, climate change, environmental issues, and economics. Carbon markets have become a key component of the efforts to reduce global warming associated with the steady rise in carbon dioxide levels in the atmosphere worldwide. Private forces are attempting to constrain the rise by proactively reducing individual carbon footprints and public forces are passing 1egislation like cap and trade rules to minimize the release from industrial processes and mobile sources. How effective these efforts will be remains to be seen but the topic of global carbon markets is impacting the energy sector now and larger impacts are likely in the future as climate warming, sea level rise, extreme weather events, etc., continue. The reader is referred to the continuous stream of information available on the internet, trade publications, governmental publications, etc. for additional details on current and future carbon markets and the role of carbon in managing global warming or climate change processes.

20.4 The Role of Operation and Maintenance

This topic was addressed in Chapter 18, the optimization chapter. The role of operation and maintenance on improving and optimizing the performance of existing processes was detailed there. This section extends that discussion related to the future of water and desalination.

The best way to ensure the operation and performance of any desalination process is to initially design the process to meet all required criteria under all operating conditions relevant to the process. It is often difficult – and in some cases impossible – to modify a desalination process once it has been improperly designed or if the operating conditions have changed appreciably outside of those used during the design process. Within the constraints of an existing system, improving operation and performance refers to maintaining operation and original or consistent performance. There are several factors previously mentioned which are critical to the design and performance of a desalination process. Most equipment manufacturers will provide designs for alternative conditions as a guide for estimating the cost of improving efficiency via process variables as well as alternative process configurations. Keep in mind that the initial cost objective when purchasing a system is to achieve a technically reliable system capable of meeting production goals at the minimum annualize cost. Once a system has been purchased, efforts should be directed toward improving its operation and performance. It is assumed for the sake of this discussion the user is at a point where a system has no major problems and is operating smoothly; however, there is often a desire to improve operational performance and reduce operating costs. As one orients to optimizing costs it often becomes clear that general cost rules can be very misleading because the performance of the system can vary greatly from one unit to another, and therefore operating, maintenance, and capital cost relationships can change dramatically. If the approach to optimization is uniform, one may simply include trading off additional capital investment versus increased maintenance costs to reduce overall operating costs.

Knowledge that a desalination process is controlled by many and often interrelated variables may discourage the operator into thinking that there is little that can be done to improve the performance of a desalination process already in commercial operation. To the contrary, the sensitivity of units within the system can provide the operator with controls over performance that can even overcome some design deficiencies. At the very least, examination of operating parameters in the future can permit rapid and accurate performance diagnosis of a system that is not performing in an optimal way.

Efforts at improving operation and performance are as dependent upon approach as they are upon technique. The first steps should always be the complete correction of mechanical/component deficiencies. To adequately troubleshoot future system and equipment performance problems, implementation of a multiphase program should take place that includes:

1. Phase 1 – Problem identification. This phase should incorporate a detailed inspection of the system and culminate with a report listing all observations (positive and negative), provide interpretation of these observations (why things were the way they were), and finally include recommendations of methods and items to be implemented to improve performance.
2. Phase 2 – Implementation. After thorough analysis and discussion, the recommendations provided in Phase 1 should be implemented. These recommendations may be in the form of design modifications, component and accessory part replacement, and fabrication of any new equipment. Following this aspect of the program, repair and replacement of the procured and fabricated components must be instituted. Finally, the entire system must be started up and debugged.
3. Phase 3: Testing and sampling. A performance test must be executed to evaluate the work done. This may be a water sampling program and/or other measurement of system performance in continuous operation.
4. Phase 4: Preventive maintenance. When the modified and overhauled system has been shown to demonstrate satisfactory performance, a preventive maintenance program should be implemented to help maintain continuous and reliable operation.

In general, normal maintenance procedures should be followed in routine cleaning of electrical contacts, lubrication of all bearings, compressed-air components, air cylinder shafts, replacement of obviously broken or worn parts, and housekeeping practices around the facility. In the final analysis, common sense is the best maintenance tool available in view of the fact that a large percentage of equipment failures can be traced to neglect, improper operation, or just plain abuse. Finally, an adequate supply of spare parts should also be part of a preventative maintenance program.

20.5 The Role of Government

Political institutions in the United States have been strongly criticized over the past few years (2005–2019) for "failing to develop a water policy." The role of these institutions is to protect constituents from dramatic and possibly adverse consequences of the present water dilemma. Many believe the proper role is one of aggressive leadership in imposing polices, some of which may be quite painful in the development of a popular "consensus" on a water policy. There will obviously be winners and losers. The resulting "consensus" is not on who should win and who should lose in a transition from the current set of policies for dealing with water and any future water policy developments. There is no evidence that the electoral process will provide a clear mandate in this regard. Opposition leaders have suggested a change in administration will signal a new direction in water policy. Many believe that a continuation of present policies is likely.

The extent of government intervention through a national water policy into the operation of the market-based, capitalist approach to water supply and demand may come under fire in the near future. The federal government, through tax policy, regulation, trade policy, fiscal and monetary methods, antitrust activity, research and development funding, and public advocacy, strongly influences the domestic water situation. Two questions arise:

1. Is a water policy too important to let it remain out of the purview of the federal government?
2. In what sense is it too important to let the government get into it at all?

In particular, if both accelerated water development and conservation require truly substantial infusion of capital as nearly all studies seem to indicate, the federal government must be the instrument for securing this money. Alternately, partnerships between federal, state, and private institutions may be acceptable as part of the response to the water supply and demand problems. As noted earlier, the present water debate has a substantial state rights component. Indeed, the policy desires of certain states have substantial impact on the choices available to other states. In addition, the policy choices made by some may be entirely inconsistent with the thrust of a national water policy. At present, only Congress is in a position to provide a resolution of these issues. However, a variety of new approaches may be needed to deal with these issues.

In the final analysis it seems almost certain that any water plan or policy will have Congress' mark on it. The standard approach is to have a policy or plan first developed by an interested party and sent to Congress. Congress then passes a law that (hopefully) reflects the objectives of the policy. In the process, Congress must identify an agency or regulatory body to oversee and implement the law. The agency or regulatory body then may pass regulations to ensure that the intent of the law is achieved. However (and here is where it gets tricky), the regulations may set a policy without going through the normal process associated with passing a regulation. The difference between laws, regulations, and policy can be summarized in the following manner:

1. Laws provide the vision, scope, and authority for environmental protection and restoration. They are enacted by a legislative body and serve as a basis for administrative implementation.
2. Regulations establish general requirements that must be met by a regulated community. These requirements generally apply at a national, state, or local level and are usually adopted by an administrative entity that has received an authorization from a law. These regulations generally interpret the enabling legislation and explain how the law is to be implemented.
3. Policy is developed by an administrative entity to explain and further interpret how laws and regulations are to be implemented and to resolve issues and conflicts of interpretation.

Matystick (2011) provides the following:

> Policy represents a general framework or plan for a course of action embodying desired principles. Policy can form the basis for law (e.g. a nondiscrimination policy can form the basis for civil rights legislation). Laws can directly follow from policy or be the rationale for legislation. Additionally, in some cases the law is the policy. Policy can also stand alone without a supporting law, e.g. foreign policy while setting a general direction for diplomacy.

Many in the technical community have divergent views on where U.S. water efforts should be placed. As a result, the legislative road forward to water security is likely to be one with many turns and roadblocks and, as a result, may end up costing both the taxpayer and consumer dearly. Furthermore, the process of deciding policy direction becomes more complex when alternatives to traditional water resources are introduced.

Unfortunately, planning processes at almost all levels of government are still based upon traditional methods. These methods treat water (and other environmental factors) only as secondary effects or incidental results of a given policy or strategy. The result is that water problems occur and then a great deal of money is spent on corrective actions. In reality a small percentage of the U.S. gross national product is expended on water measures, most of which are concerned with water treatment.

In summary, the authors are convinced that it is both necessary and practical to examine the interrelated aspects of water resource and usage problems. The methodology that is ultimately employed will become more complicated in situations in which many variables prove to be significant. Also, for many states or planning regions, the solution may need to be more sophisticated than those used with some success in relatively small and isolated urban areas. However, the solution to most water problems appear to be far less costly than the corrective measures that are likely to be necessary if the United States continues to ignore the possibilities of and the need for an objective water resources and usage analysis.

20.6 The Role of Risk Assessment

Future trends in environmental risk assessment will likely be focused on hazard accident prevention, not hazard analysis. Employee-training programs are key to promoting hazard accident prevention. These programs should be designed to alert staff and employees about the hazards they are exposed to on the job and should cover company safety policies and proper procedures to follow in case an accident does occur. A major approach to reducing risk will involve source reduction of hazardous materials through material substitution and redesign of process chemistry, which will also reduce the potential for process and equipment related hazards in desalination systems. Communicating health risk and hazard risk to employees has emerged as a major area of concern. Understanding risk communications dynamics is essential to successful risk communication efforts (Theodore and Dupont 2012).

Different assumptions can change any quantitative risk characterization by several orders of magnitude. The uncertainty that arises is related to how well (and often, consistently) input data are obtained, generated, or measured, and the degree to which judgment is involved in developing risk scenarios and selecting input data. Simply put by some, uncertainty arises from how data/evidence was both measured and interpreted. Despite these limitations and uncertainties, risk characterizations provide the practitioner with assessment capabilities that are critical to improving process safety, sustainability, and risk reduction.

20.7 Terrorism and Water Security

Perhaps the key action taken by the U.S. government related to water security was the Bioterrorism Act of 2002. This act requires all Public Water Supplies (PWS) serving populations of more than 3,300 persons to conduct assessments of their vulnerabilities to terrorist attack or other intentional harmful acts, and to defend against adversarial actions that might substantially disrupt the ability of a system to provide a safe and reliable supply of drinking water. In addition to the vulnerabilities assessment, the act requires a PWS to certify and submit a copy of the vulnerability assessment to the EPA Administrator, prepare or revise an emergency response plan based on the results of the vulnerability assessment, and certify that an emergency response plan has been completed or updated within 6 months of completing the assessment (Theodore and Dupont 2019).

Improving the security of the nation's drinking water and wastewater infrastructure has become a top priority. Significant actions are underway to assess and reduce vulnerabilities to terrorist attacks; to plan for and practice response to emergencies and incidents; and to develop new security technologies to detect and monitor contaminants and prevent security breaches.

Finally, here are alternative water sources that could provide necessary water supplies during short-term events following a terrorist attack:

1. Bottled water provided by outside sources.
2. Bottled water provided by local retailers.
3. Bulk water provided by certified water haulers.
4. Bulk water transported or provided by military assets, i.e. National Guard or U.S. Army.
5. Bulk water provided by neighboring water utilities by truck or via pipeline.
6. Bulk water from hospitals, universities, and local industry that maintain backup water supplies.
7. Interconnections with nearby public water systems.
8. Water treated by the plant and hauled to distribution centers in the case of water distribution system contamination.
9. Water pumped from surface water sources treated at a nearby plant and hauled to distribution centers.
10. Water for firefighting from Federal agencies such as the US Army Corps of Engineers and FEMA.
11. Water from unaffected wells owned by local citizens and businesses.
12. Water previously stored for emergency purposes.

20.8 Predicting Future Water Demands

It is obviously difficult to predict future demands of water. Other than intuitive approaches, one simple approach is to generate mathematical equations (models) for predictive purposes based on earlier/past data. There are, of course, other approaches that can be employed, particularly when little-to-no information is available. A viable and popular method is to employ a Delphi panel approach; one of the authors modestly refers to it as the Theodore panel approach (Theodore and Dupont 2012). In order to estimate water demands in the future, several knowledgeable individuals are asked to provide independent estimates, with explanatory details, on the future demand. Each individual in the panel is then allowed to review all responses independently. The cycle is then repeated until the group's responses approach convergence. This section, however, presents details of a mathematical equation/approach, specifics of which follow (Shaefer and Theodore 2007).

The practicing engineer often encounters applications (e.g. describing future water demands) that require the development of a mathematical relationship between data for two or more variables. For example, if Y (a dependent variable, e.g. water demand) is a function of or depends on t (an independent variable, e.g. time), that is:

$$Y = f(t) \tag{20.1}$$

one may be required to express these (Y, t) data in equation form. This process is referred to as *regression analysis*, a statistical method used to evaluate data and provide forward predictions based on historical data and records. The regression method most often employed is the method of *least squares*. An important step in this procedure, which is often omitted, is to prepare a plot of Y versus t. The result, referred to as a scatter diagram, could take on any form. Three such plots arc provided in Figure 20.1(a-c). The first plot Figure 20.1a suggests a linear relationship between Y and t, that is:

$$Y = a_o + a_1 t \tag{20.2}$$

The second graph Figure 20.1b appears to be best represented by a second-order (or parabolic) relationship, that is:

$$Y = a_o + a_1 t + a_2 t^2 \tag{20.3}$$

The third plot suggests a linear model that applies over two different ranges; that is, it should represent the data where:

$$Y = a_o + a_1 t; t_o < t < t_M \tag{20.4}$$

and

$$Y = a_o' + a_1' t; t_M < t < t_L \tag{20.5}$$

This multi-equation model finds application in representing adsorption equilibria, multiparticle size distributions (Theodore 2009), quantum energy relationships, etc. In any event, a scatter diagram and individual judgment can suggest an appropriate model at an early stage in the analysis.

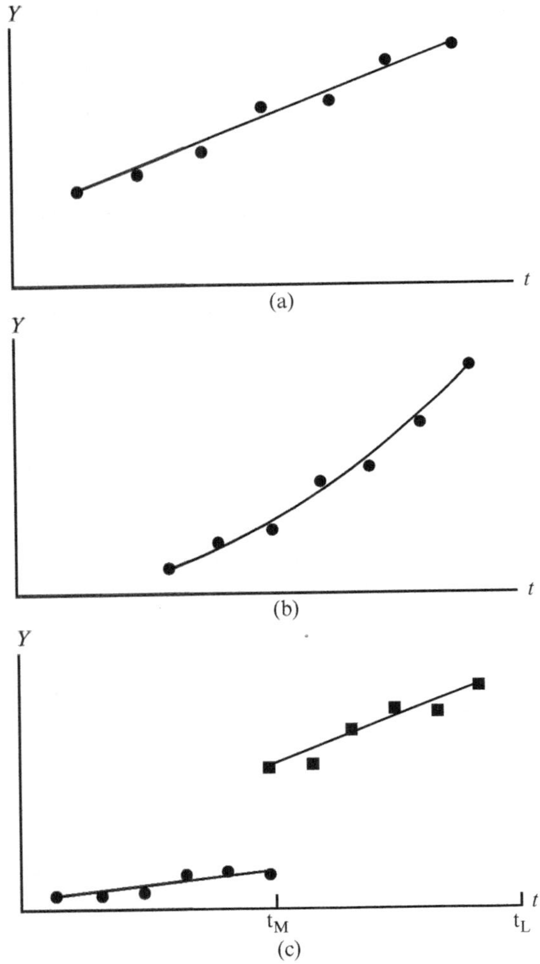

Figure 20.1 Scatter Diagrams: (a) Linear Relationship, (b) Parabolic Relationship, and (c) Dual-Linear Relationship.

Some of the models often employed by technical individuals are as follows:

$$Y = a_o + a_1 t \quad \text{Linear} \tag{20.2}$$

$$Y = a_o + a_1 t + a_2 t^2 \ \text{Parabolic (second order)} \tag{20.3}$$

$$Y = a_o + a_1 t + a_2 t^2 + a_3 t^3 \quad \text{Cubic (third order)} \tag{20.6}$$

$$Y = a_o + a_1 t + a_2 t^2 + a_3 t^3 + a_4 t^4 \ \text{Quadratic (fourth order)} \tag{20.7}$$

Procedures to evaluate the regression coefficients a_o, a_1, a_2, etc., are provided of as follows. The reader should note that the analysis is based on the method of least squares. This technique provides numerical values for the regression coefficients such that the sum of the square of the difference (error) between the actual Y value and that predicted by the model is minimized. This is shown in Figure 20.2.

In Figure 20.2, the dots (experimental value of Y) and triangles (equation or model value of Y, i.e. Y_e) represent the data and model values, respectively. On examining the two figures, one can conclude that the error $(Y - Y_e)$ squared and summed for the four points is less for the lower figure than the upper one. Also note that a dashed line represents the error. The line that ultimately produces a minimum of the sum of individual errors squared, that is the smallest possible value, is the appropriate regression model (based on the method of least squares).

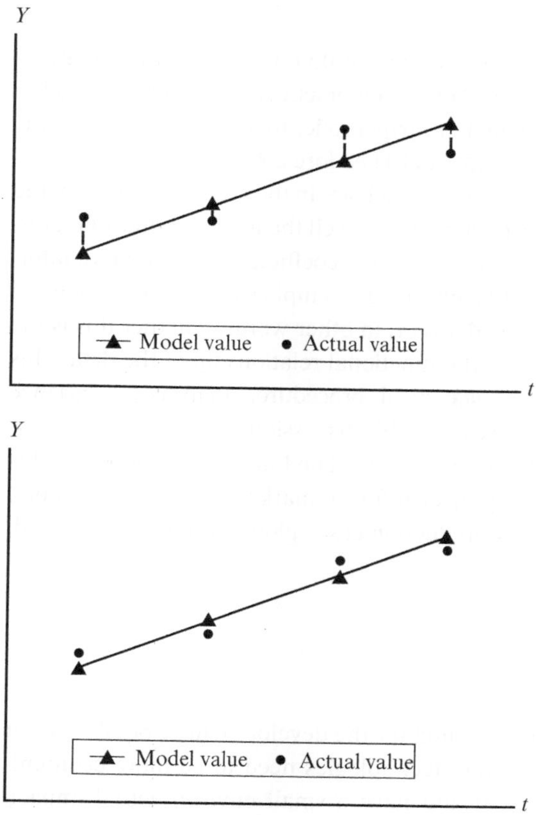

Figure 20.2 Error Difference: Actual and Predicted Values.

A cubic model takes the form of:

$$Y = a_o + a_1X + a_2X^2 + a_3X^3 \tag{20.6}$$

For n pairs (more than four) of (Y, X) values, the constants a_o, a_1, a_2, and a_3 can be obtained by the method of least squares the so that $\sum(Y - Ye)^2$ again has the smallest possible value (i.e. is minimized). The coefficients a_o, a_1, a_2, and a_3 are the solution to the following system of four linear equations:

$$a_o n + a_1\sum X + a_2\sum X^2 + a_3\sum X^3 = \sum Y \tag{20.7}$$

$$a_o\sum X + a_1\sum X^2 + a_2\sum X^3 + a_3\sum X^4 = \sum X\,Y \tag{20.8}$$

$$a_o\sum X^2 + a_1\sum X^3 + a_2\sum X^4 + a_3\sum X^5 = \sum X^2Y \tag{20.9}$$

$$a_o\sum X^3 + a_1\sum X^4 + a_2\sum X^5 + a_3\sum X^6 = \sum X^3Y \tag{20.10}$$

Because there are four equations and four unknowns, this set of equations can be solved for a_o, a_1, a_2, and a_3. This development can be extended to other regression equations, e.g. exponential, hyperbolic, higher order models, etc. (Theodore and Dupont 2012, Shaefer and Theodore 2007).

The correlation coefficient, details of which are in the literature (Shaefer and Theodore 2007) provides information on how well the model, or line of regression fits the data. Denoted by r, the correlation coefficient only provides information on how well the model fits the data. It is emphasized that r provides no information on how good the model is or, in other words, whether this is the correct or best model to describe the functional relationship of the data. This topic is addressed in another statistical procedure, analysis of variance (ANOVA), a subject beyond the scope of this discussion.

Finally, it should be noted that any water-related models developed in this manner should be adjusted for projected future market effects (if applicable) such as cost of water, increased production costs, global supply and demand, etc.

20.9 Epilogue

For desalination to be a viable method for the development of potable water sources in the future, desalination technologies need to be more economically and operationally efficient and have a small environmental impact. Concerted efforts on these fronts will result in improvement in the performance and design of traditional technologies, the development of innovative non-desalination potable water production systems, and the development of new desalination technologies.

Finally, it is reasonable to state that many, if not all, of our nation's economic problems would be solved with a cheap and plentiful supply of water that is sustainable and environmentally neutral. No coordinated action has been taken in the past 40 years to achieve this state of water sustainability. At this point in time, the only way to achieve this position is with the development and implementation of a sound water policy. This text partially attempts to address this issue by proposing a policy that will hopefully reshape the nation's water future, one that will positively impact the rest of the world.

References

Dupont, R.R., Ganesan, K., and Theodore, L. (2017). *Pollution Prevention: Sustainability, Industrial Ecology, and Green Engineering*, 2nd Edition. Boca Raton, FL: CRC Press/Taylor & Francis Group.

Matystick, M. (2011). Personal communication to L. Theodore. Bronx, NY.

Shaefer, S. and Theodore, L. (2007). *Probability and Statistics Applications for Environmental Science*. Boca Raton, FL: CRC Press/Taylor & Francis Group.

Theodore, L. (2006). *Nanotechnology: Basic Calculations for Engineers and Scientists*. Hoboken, NJ: John Wiley & Sons.

Theodore, L. (2009). *Air Pollution Control Equipment Calculations*. Hoboken, NJ: John Wiley & Sons.

Theodore, L. and Dupont, R.R. (2012). *Health Risk and Hazard Risk Assessment: Principles and Calculations*. Boca Raton, FL: CRC Press/Taylor & Francis Group.

Theodore, L. and Dupont, R.R. (2019). *Environmental Water Resource Management Issues. Basic Principles and Applications*. Boca Raton, FL: CRC Press/Taylor & Francis Group.

Theodore, L. and Kunz, (2005). *Nanotechnology: Environmental Implications and Solutions*. Hoboken, NJ: John Wiley & Sons.

Theodore, M.K. and Theodore, L. (2021). *Introduction to Environmental Management*, 2nd Edition. Boca Raton, FL: CRC Press/Taylor Francis Group.

Index

Introduction to Desalination: Principles, Processes, and Calculations, First Edition. Louis Theodore and R. Ryan Dupont.

n

Printed and bound by CPI Group (UK) Ltd, Croydon, CR0 4YY
19/09/2026
14965094-0001